Dielectrics in Electric Fields

POWER ENGINEERING

Series Editor

H. Lee Willis

ABB Inc.
Raleigh, North Carolina

ADDITIONAL VOLUMES IN PREPARATION

Dielectrics in Electric Fields

Gorur G. Raju

University of Windsor
Windsor, Ontario, Canada

CRC Press
Taylor & Francis Group
Boca Raton London New York

CRC Press is an imprint of the
Taylor & Francis Group, an **informa** business

First published 2003 by Lawrence Erlbaum Associates, Inc.

Published 2019 by CRC Press
Taylor & Francis Group
6000 Broken Sound Parkway NW, Suite 300
Boca Raton, FL 33487-2742

© 2003 by Taylor & Francis Group, LLC
CRC Press is an imprint of Taylor & Francis Group, an Informa business

First issued in paperback 2019

No claim to original U.S. Government works

ISBN 13: 978-0-367-44682-6 (pbk)
ISBN 13: 978-0-8247-0864-1 (hbk)

**Visit the Taylor & Francis Web site at
http://www.taylorandfrancis.com**

**and the CRC Press Web site at
http://www.crcpress.com**

Library of Congress Cataloging-in-Publication Data
A catalog record for this book is available from the Library of Congress.

TO MY PARENTS.
MY WIFE, PADMINI,
AND OUR SON, ANAND

*WHO GAVE ME ALL I VALUE. SOME DEBTS
ARE NEVER REPAID IN FULL MEASURE.*

SERIES INTRODUCTION

Power engineering is the oldest and most traditional of the various areas within electrical engineering, yet no other facet of modern technology is currently undergoing a more dramatic revolution in both technology and industry structure. This addition to Marcel Dekker's Power Engineering Series addresses a fundamental element of electrical engineering. Dielectric materials are a key element of electric power engineering, one of the most challenging aspects of improving reliability and economy of materials. For an industry pressed hard to increasingly cram more equipment capacity into ever-tighter spaces, to improve reliability, particularly mean time between failures, modern dielectric materials and engineering methods provide a valuable tool to meet these challenges.

Dielectrics in Electric Fields is a well-organized and comprehensive view of both the theory behind and application of dielectric materials in power equipment, industrial equipment, and commercial appliances. At both the introductory and advanced levels, it provides both a solid foundation of theory, fact, nomenclature, and formula, and sound insight into the philosophies of dielectric engineering techniques and their use. Its unifying approach, based on both physics and engineering, makes it useful as a day-to-day reference as well as an excellent tutorial: the book begins with a thorough review of the basics of dielectric and polymer science and builds upon it a comprehensive and very broad presentation of all aspects of modern dielectric theory and engineering, including the lastest analysis and modeling techniques.

As the editor of the Power Engineering Series, I am proud to include *Dielectrics in Electric Fields* among this important group of books. Like all the volumes planned for the series, Professor Raju's book puts modern technology in a context of proven, practical application; useful as a reference book as well as for self-study and advanced

v

classroom use. The series includes books covering the entire field of power engineering, in all its specialties and sub-genres, all aimed at providing practicing power engineers with the knowledge and techniques they need to meet the electric industry's challenges in the 21st century.

H. Lee Willis

PREFACE

Materials that do not normally conduct electricity and have the ability to store electrical charge are known as dielectrics. The behavior of dielectrics in electric fields continues to be an area of study that has fascinated physicists, chemists, material scientists, electrical engineers, and, more recently biologists. Ideas that explain aspects of dielectric behavior in high voltage electrical cables are also applicable to the insulating barrier in metal oxide semiconductors or interlayer insulation of integrated circuits. Microwave drying of milk, dielectric properties of agricultural products such as flour and vegetable oils to determine their moisture content, and the study of curing of cement etc., are some nontraditional applications of dielectric studies that show potential promise. Deeper insight into the interaction between electric fields and molecules has resulted in many new applications. Power engineers are interested in the study of insulating materials to prolong the life of insulation and determine the degree of deterioration in service to plan for future replacements or service maintenance.

Polymer scientists are interested in understanding the role of long chain molecules in varied applications ranging from heat resistant dielectrics to selfrepairing plastics. The intensity of research in this area, after a brief respite, has resumed at a furious pace, the published literature expanding at a rate faster than ever. Advances in instrumentation and theoretical models have also contributed to this renewed interest.

Organic polymers are considered to be stable materials at high temperatures and have the ability to withstand radiation, chemical attacks, and high electrical and mechanical stresses, making them suitable for extreme operating environments as in a nuclear power plant or in outer space. Polymer materials have the ability to store electrical charges. Like a diamond-studded sword, this property is wholly undesirable in applications such as electrical equipment and the petrochemical industry; yet it is a sought-after property in applications such as photocopying and telephones.

This book explains the behavior of dielectrics in electric fields in a fundamentally unifying approach that is based on well-established principles of physics and engineering. Though excellent monographs exist on specialized topics dealing with a relatively narrow area of interest, there is a need for a broader approach to understand dielectrics. It has evolved out of graduate lectures for nearly thirty-five years at the Indian Institute of Science, Bangalore (1966-1980) and the University of Windsor, Windsor, Ontario, Canada (1980-2002). The probing questions of students has helped the author to understand the topics better and to a certain extent dictated the choice of topics.

The book begins with an introductory chapter that explains the ideas that are developed subsequently. The calculation of forces in electric fields in combinations of dielectric media is included because it yields analytical results that are used in the study of the dielectric constant (Ch. 2). The band theory of solids is included because it is required to understand the energy levels of a dielectric, as in the conduction and formation of space charge (Ch. 6-11). The energy distribution function is dealt with because it is a fundamental property that determines the swarm parameters in gaseous breakdown and partial discharges (Ch. 8-9).

Chapter 2 deals with the mechanisms of electrical polarization and their role in determining the value of the dielectric constant under direct voltages. Expressions for the dielectric constant are given in terms of the permanent dipole moment of the molecule and temperature. Several theories of dielectric constant are explained in detail and practical applications are demonstrated. Methods of calculating the dielectric constant of two different media and mixtures of liquids are also demonstrated.

Chapter 3 begins with the definitions of the complex dielectric constant in an alternating electric field. The Debye equations for the complex dielectric constant are explained and the influence of frequency and temperature in determining the relaxation is examined. Functions for representing the complex dielectric constant in the complex plane are given and their interpretation in terms of relaxation is provided. Several examples are taken from the published literature to bring out the salient points.

Chapter 4 continues the discussion of dielectric relaxation from chapter 3. The concept of equivalent circuits is introduced and utilized to derive the set of equations for both Debye relaxation and interfacial polarization. The absorption and dispersion phenomena for electronic polarization are considered, both for damped and undamped situations. These ideas have become very relevant due to developments in fiber optics technology.

Chapter 5 deals with the application of these ideas to understand the experimental results in the frequency domain and with temperature as the main parameter in selected polymers. A brief introduction to polymer science is included to help the reader understand what follows. The terminology used to designate relaxation peaks is explained and methods for interpreting observed results in terms of physics and morphology are presented.

Chapter 6 deals with the measurement of absorption and desorption currents in the time domain in polymers. Though external parameters influence these measurements our concern is to understand the mechanisms of charge generation and drift. Time domain currents may be transformed into the frequency domain complex dielectric constant and the necessary theories are explained. The low frequency, high temperature relaxations observed in several polymers are explained as complementary to the topics in Chapter 5.

The magnitude of electric fields that are employed to study the behavior in dielectrics outlined in Chapters 1–6 is low to moderate. However, the response of a polymer to high electric fields is important from the practical point of view. The deleterious effects of high electric fields and/or high temperatures occur in the form of conduction currents and the complex mechanism of conduction is explained in terms of the band picture of the dielectric. Several examples are selected from the published literature to demonstrate the methods of deciphering the often overlapping mechanisms. Factors that influence the conduction currents in experiments are outlined in Chapter 7.

Chapters 8 deals with the fundamental processes in gaseous electronics mainly in uniform electric fields and again, due to limitation of space, physical principles are selected for discussion in preference to experimental techniques for measuring the cross sections and swarm properties. A set of formulas for representing the relevant properties of several gases, such as the swarm coefficients are provided, from recent published literature.

Chapter 9 is devoted to studies on nonuniform electric field in general and corona phenomenon in particular. These aspects of gaseous breakdown are relevant from practical points of view, for providing better design or to understand the partial discharge phenomena. Both experimental and theoretical aspects are considered utilizing the literature published since 1980, as far as possible. Several computational methods, such as the Boltzmann equation, solutions of continuity equations, and Monte Carlo methods are included. The results obtained from these studies are presented and discussed.

Chapter 10 deals with thermally stimulated processes, mainly in polymers. The theory of thermally stimulated discharge currents and techniques employed to identify the source of charge generation are described to assist in carrying out these experiments

Chapter 11 deals with measurement of the space charges in solids and the different experimental techniques are explained in detail. These nondestructive techniques have largely replaced the earlier techniques of charging a dielectric and slicing it for charge measurements. The Theory necessary to analyze the results of space charge experiments and results obtained is included with each method presented. The author is not aware of any book that systematically describes the experimental techniques and the associated theories in a comprehensive manner.

The book uses the SI units entirely and published literature since 1980 is cited, wherever possible, except while discussing the theoretical aspects. The topics chosen for inclusion has my personal bias, though it includes chapters that interest students and established researchers in a wide range of disciplines, as noted earlier. Partial discharges, breakdown mechanisms, liquid dielectrics, Outdoor insulation and nanodielectrics are not covered mainly due to limitation of space.

I am grateful to a number of graduate students who contributed substantially for a clearer understanding of the topics covered in this volume, by their probing questions. Drs. Raja Rao, G. R. Gurumurthy, S. R. Rajapandiyan, A. D. Mokashi, M. S. Dincer, Jane Liu, M. A. Sussi have contributed in different ways. I am also grateful to Dr. Bhoraskar for reading the entire manuscript and making helpful suggestions. It is a pleasure to acknowledge my association with Drs. R. Gorur, S. Jayaram, Ed Churney, S. Boggs, V. Agarwal, V. Lakdawala, T. Sudarshan and S. Bamji over a number of years. Dr. R. Hackam has been an associate since my graduate student years and it is appropriate to recall the many discussions I have held on various aspects of dielectric phenomena considered in this book. The personal encouragement of Professor Neil Gold, University of Windsor has contributed in no small measure to complete the present book.

Special thanks are due to Dr. N. Srinivas who provided opportunity to complete chapters 8 and 9 during sabbatical leave. Prof. C. N. R. Rao, President of the Jawaharlal Nehru Center for Advanced Scientific Research provided opportunity to spend sabbatical leave during which time I could work on the manuscript. Mr. N. Nagaraja Rao extended generous hospitality on campus making it possible to use the library facilities in Bangalore. Mrs. and Prof. N. Rudraiah of Bangalore University have always extended extraordinary courtesy to me.

This book would not have been completed without the help of Mr. S. Chowdhury who showed me how to make software applications cooperate with each other. Extraordinary help was provided by Alan Johns in keeping the computer system in working condition throughout. Ms. S. Marchand assisted in checking the manuscript and Ms. Ramneek Garewal assisted in the compilation of figures and tables. While acknowledging the help received, I affirm that errors and omissions are entirely my own responsibility.

I have made sincere attempts to secure copyright permission for reproducing every figure and table from the published literature, and acknowledge the prompt response from institutions and individuals. If there are unintentional failures to secure permission from any source, I render apology for the oversight.

Personal thanks are due to Brian Black and B. J. Clark who have patiently suffered my seemingly disconnected communications and, provided great assistance in improving the style and format. Finally the inexhaustible patience of my wife Padmini has been a source of continuous strength all these years.

Gorur G. Raju

CONTENTS

Chapter 6
Absorption and Desorption Currents

Chapter 7
Field Enhanced Conduction

Chapter 8
Fundamental Aspects of Gaseous Breakdown–I

Chapter 9
Fundamental Aspects of Electrical Breakdown–II   **439**

Chapter 10
Thermally Stimulated Processes   **475**

The rich and the poor are two locked caskets of which each contains the key to the other.

Karen Blixen (Danish Writer)

1

INTRODUCTORY CONCEPTS

In this Chapter we recapitulate some basic concepts that are used in several chapters that follow. Theorems on electrostatics are included as an introduction to the study of the influence of electric fields on dielectric materials. The solution of Laplace's equation to find the electric field within and without dielectric combinations yield expressions which help to develop the various dielectric theories discussed in subsequent chapters. The band theory of solids is discussed briefly to assist in understanding the electronic structure of dielectrics and a fundamental knowledge of this topic is essential to understand the conduction and breakdown in dielectrics. The energy distribution of charged particles is one of the most basic aspects that are required for a proper understanding of structure of the condensed phase and electrical discharges in gases. Certain theorems are merely mentioned without a rigorous proof and the student should consult a book on electrostatics to supplement the reading.

1.1 A DIPOLE

A pair of equal and opposite charges situated close enough compared with the distance to an observer is called an electric dipole. The quantity

$$\mu = Qd \tag{1.1}$$

where d is the distance between the two charges is called the electric dipole moment. μ is a vector quantity the direction of which is taken from the negative to the positive charge and has the unit of C m. A unit of dipole moment is 1 Debye = 3.33×10^{-30} C m.

1.2 THE POTENTIAL DUE TO A DIPOLE

Let two point charges of equal magnitude and opposite polarity, +Q and -Q be situated d meters apart. It is required to calculate the electric potential at point P, which is situated at a distance of R from the midpoint of the axis of the dipole. Let R $_+$ and R $_-$ be the distance of the point from the positive and negative charge respectively (fig. 1.1). Let R make an angle θ with the axis of the dipole.

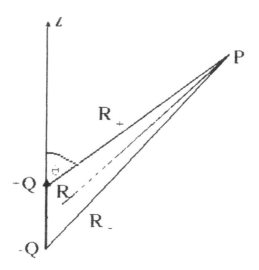

Fig. 1.1 Potential at a far away point P due to a dipole.

The potential at P is equal to

$$\phi = \frac{1}{4\pi\varepsilon_o}(\frac{Q}{R_+} + \frac{-Q}{R_-})$$

$$= \frac{Q}{4\pi\varepsilon_o}\left(\frac{1}{R_+} - \frac{1}{R_-}\right) \tag{1.2}$$

Starting from this equation the potential due to the dipole is

$$\phi \approx \frac{Qd\cos\theta}{4\pi\varepsilon_o R^2} \tag{1.3}$$

Three other forms of equation (1.3) are often useful. They are

$$\phi = \frac{\mu \cdot \mathbf{a_R}}{4\pi\varepsilon_o R^2} \qquad (1.4)$$

$$\phi = \frac{\mu \cdot \mathbf{R}}{4\pi\varepsilon_0 R^3} \qquad (1.5)$$

$$\phi = \frac{\mu z}{4\pi\varepsilon_o R^3} \qquad (1.6)$$

The potential due to a dipole decreases more rapidly than that due to a single charge as the distance is increased. Hence equation (1.3) should not be used when $R \approx d$. To determine its accuracy relative to eq. (1.2) consider a point along the axis of the dipole at a distance of $R=d$ from the positive charge. Since $\theta = 0$ in this case, $\phi = Qd/4\pi\varepsilon_o$ $(1.5d)^2 = Q/9\pi\varepsilon_o d$ according to (1.3). If we use equation (1.2) instead, the potential is $Q/8\pi\varepsilon_o d$, an error of about 12%.

The electric field due to a dipole in spherical coordinates with two variables (r, θ) is given as:

$$\mathbf{E} = \frac{-\partial\phi}{\partial r}\mathbf{a}_r - \frac{1}{r}\frac{\partial\phi}{\partial\theta}\mathbf{a}_\theta \qquad (1.7)$$

$$= E_r\mathbf{a}_r + E_\theta\mathbf{a}_\theta$$

Partial differentiation of equation (1.3) leads to

$$E_r = \frac{Qd\cos\theta}{2\pi\varepsilon_o R^3} \qquad (1.8)$$

$$E_\theta = \frac{Qd\sin\theta}{4\pi\varepsilon_o R^3} \qquad (1.9)$$

Equation (1.7) may be written more concisely as:

$$E = -\nabla \phi \qquad (1.10)$$

Substituting for ϕ from equation (1.5) and changing the variable to r from R we get

$$E = -\frac{\nabla \mu \cdot \mathbf{r}}{4\pi\varepsilon_0 \, r^3} \qquad (1.11)$$

$$= -\frac{\mu \cdot \mathbf{r}}{4\pi\varepsilon_0 \, r^3} \nabla \frac{1}{r^3} - \frac{1}{4\pi\varepsilon_0 \, r^3} \nabla \mu \cdot r \qquad (1.12)$$

We may now make the substitution

$$\nabla \frac{1}{r^3} = -\frac{3}{r^4} \mathbf{a_r} = -\frac{3\mathbf{r}}{r^5}$$

$$\nabla \mu \cdot \mathbf{r} = \mu \, \mathbf{a}_\theta$$

Equation (1.12) now becomes

$$E = \left[\frac{3\mu \cdot \mathbf{r}}{r^5} \mathbf{r} - \frac{\mu}{r^3} \right] \frac{1}{4\pi\varepsilon_0} \qquad (1.13)$$

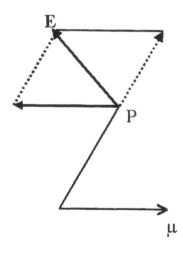

Fig. 1.2 The two components of the electric field due to a dipole with moment μ.

The electric field at P has two components. The first term in equation (1.13) is along the radius vector (figure 1.2) and the second term is along the dipole moment. Note that the second term is anti-parallel to the direction of μ.

In tensor notation equation (1.13) is expressed as

$$\mathbf{E} = \mathbf{T}\mu \tag{1.14}$$

where $\mathbf{T}$ is the tensor $3\mathbf{r}\mathbf{r}r^{-5} - r^{-3}$.

1.3   DIPOLE MOMENT OF A SPHERICAL CHARGE

Consider a spherical volume in which a negative charge is uniformly distributed and at the center of which a point positive charge is situated. The net charge of the system is zero. It is clear that, to counteract the Coulomb force of attraction the negative charge must be in continuous motion. When the charge sphere is located in a homogeneous electric field $\mathbf{E}$, the positive charge will be attracted to the negative plate and vice versa. This introduces a dislocation of the charge centers, inducing a dipole moment in the sphere.

The force due to the external field on the positive charge is

$$\mathbf{F} = Ze\mathbf{E} \tag{1.15}$$

in which Ze is the charge at the nucleus. The Coulomb force of attraction between the positive and negative charge centers is

$$F' = \frac{ze \cdot e_1}{4\pi\varepsilon_o x^2} \tag{1.16}$$

in which e_1 is the charge in a sphere of radius x and x is the displacement of charge centers. Assuming a uniform distribution of electronic charge density within a sphere of atomic radius R the charge e_1 may be expressed as

$$e_1 = \frac{ze \cdot x^3}{R^3} \tag{1.17}$$

Substituting equation (1.17) in (1.16) we get

$$F' = \frac{(ze)^2 x}{4\pi\varepsilon_o R^3} \tag{1.18}$$

If the applied field is not high enough to overcome the Coulomb force of attraction, as will be the case under normal experimental conditions, an equilibrium will be established when $F = F'$ viz.,

$$ze \cdot E = \frac{(ze)^2 x}{4\pi\varepsilon_o R^3} \tag{1.19}$$

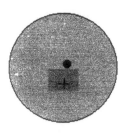

The center of the negative charge coincides with the nucleus.

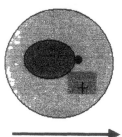

In the presence of an Electric field the center of the electronic charge is shifted towards the positive electrode inducing a dipole moment in the atom.

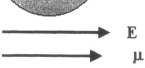

E

μ

Fig. 1.3 Induced dipole moment in an atom. The electric field shifts the negative charge center to the left and the displacement, x, determines the magnitude.

The displacement is expressed as

$$x = \frac{4\pi\varepsilon_o R^3}{ze} E \tag{1.20}$$

The dipole moment induced in the sphere is therefore

$$\mu_{ind} = zex = 4\pi\varepsilon_o R^3 E \tag{1.21}$$

According to equation (1.21) the dipole moment of the spherical charge system is proportional to the radius of the sphere, at constant electric field intensity. If we define a quantity, polarizability, α, as the induced dipole moment per unit electric field intensity, then α is a scalar quantity having the units of Farad meter. It is given by the expression

$$\alpha = \frac{\mu_{ind}}{E} = 4\pi\varepsilon_o R^3 \tag{1.22}$$

1.4 LAPLACE'S EQUATION

In spherical co-ordinates (r,θ,ϕ) Laplace's equation is expressed as

$$\nabla^2 V = \frac{1}{r^2}\frac{\partial}{\partial r}\left(r^2\frac{\partial V}{\partial r}\right) + \frac{1}{r^2\sin\theta}\frac{\partial}{\partial\theta}\left(\sin\theta\frac{\partial V}{\partial\theta}\right) + \frac{1}{r^2\sin^2\theta}\frac{\partial^2 V}{\partial\phi^2} = 0 \tag{1.23}$$

If there is symmetry about ϕ co-ordinate, then equation (1.23) becomes

$$\frac{\partial}{\partial r}\left(r^2\frac{\partial V}{\partial r}\right) + \frac{1}{\sin\theta}\frac{\partial}{\partial\theta}\left(\sin\theta\frac{\partial V}{\partial\theta}\right) = 0 \tag{1.24}$$

The general solution of equation (1.24) is

$$V = \left(Ar + \frac{B}{r^2}\right)\cos\theta \tag{1.25}$$

in which A and B are constants which are determined by the boundary conditions. It is easy to verify the solution by substituting equation (1.25) in (1.24).

The method of finding the solution of Laplace's equation in some typical examples is shown in the following sections.

1.4.1 A DIELECTRIC SPHERE IMMERSED IN A DIFFERENT MEDIUM

A typical problem in the application of Laplace's equation towards dielectric studies is to find the electric field inside an uncharged dielectric sphere of radius R and a dielectric constant ε_2. The sphere is situated in a dielectric medium extending to infinity and having a dielectric constant of ε_1 and an external electric field is applied along Z direction, as shown in figure 1.4. Without the dielectric the potential at a point is, $\phi = - E Z$.

There are two distinct regions: (1) Region 1 which is the space outside the dielectric sphere; (2) Region 2 which is the space within. Let the subscripts 1 and 2 denote the two regions, respectively. Since the electric field is along Z direction the potential in each region is given by equation (1.24) and the general solution has the form of equation (1.25).

Thus the potential within the sphere is denoted by ϕ_2. The solutions are:

Region 1:

$$\phi_1 = \left(A_1 r + \frac{B_1}{r^2} \right) \cos\theta \qquad (1.26)$$

Region 2:

$$\phi_2 = \left(A_2 r + \frac{B_2}{r^2} \right) \cos\theta \qquad (1.27)$$

To determine the four constants A_1 ..... B_2 the following boundary conditions are applied.

(1)   Choosing the center of the sphere as the origin, ϕ_2 is finite at $r = 0$. Hence $B_2=0$ and

$$\phi_2 = A_2 r \cos\theta \qquad (1.28)$$

(2)   In region 1, at $r \to \infty$, ϕ_1 is due to the applied field is only since the influence of the sphere is negligible. i.e.,

$$d\phi_1 = -Edz \qquad (1.29)$$

which leads to

$$\phi_1 = -Ez \tag{1.30}$$

Since $r\cos\theta = z$ equation (1.30) becomes

$$\phi_1 = -Er\cos\theta \tag{1.31}$$

Substituting this in equation (1.26) yields $A_1 = -E$, and

$$\phi_1 = \left(-Er + \frac{B_1}{r^2}\right)\cos\theta \tag{1.32}$$

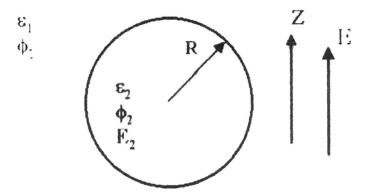

Fig. 1.4 Dielectric sphere embedded in a different material and an external field is applied.

(3)   The normal component of the flux density is continuous across the dielectric boundary, *i.e.*, at $r = R$,

$$\varepsilon_o\varepsilon_2 E_2 = \varepsilon_o\varepsilon_1 E_1 \tag{1.33}$$

resulting in

$$\left(\varepsilon_2\frac{\partial\phi_2}{\partial r}\right)_{r=R} = \left(\varepsilon_1\frac{\partial\phi_1}{\partial r}\right)_{r=R} \tag{1.34}$$

Differentiating equations (1.28) and (1.32) and substituting in (1.34) yields

$$\varepsilon_2 A_2 = -\varepsilon_1 \left(E + \frac{2B_1}{R^3} \right)$$ (1.35)

leading to

$$B_1 = \frac{-R^3(\varepsilon_2 A_2 + \varepsilon_1 E)}{2\varepsilon_1}$$ (1.36)

(4)   The tangential component of the electric field must be the same on each side of the boundary, *i.e.*, at r = R we have $\phi_1 = \phi_2$. Substituting this condition in equation (1.26) and (1.28) and simplifying results in

$$A_2 R = \frac{B_1}{R^2} - ER$$ (1.37)

Further simplification yields

$$B_1 = R^3(A_2 + E)$$ (1.38)

Equating (1.36) and (1.38), A_2 is obtained as

$$A_2 = -\frac{3\varepsilon_1}{2\varepsilon_1 + \varepsilon_2} E$$ (1.39)

Hence

$$B_1 = R^3 \left(\frac{\varepsilon_2 - \varepsilon_1}{2\varepsilon_1 + \varepsilon_2} \right) E$$ (1.40)

Substituting equation (1.39) in (1.28) the potential within the dielectric sphere is

$$\phi_2 = -\frac{3\varepsilon_1}{2\varepsilon_1 + \varepsilon_2} Ez$$ (1.41)

From equation (1.41) we deduce that the potential inside the sphere varies only with z, *i.e.*, the electric field within the sphere is uniform and directed along E. Further,

$$E_2 = -\frac{\partial \phi_2}{\partial z} = \frac{3\varepsilon_1}{2\varepsilon_1 + \varepsilon_2} \mathbf{E} \qquad (1.42)$$

(a) If the inside of the sphere is a cavity, i.e., $\varepsilon_2 = 1$ then

$$\mathbf{E_2} = \frac{3\varepsilon_1}{2\varepsilon_1 + 1} \mathbf{E} \qquad (1.43)$$

resulting in an enhancement of the field.

(b) If the sphere is situated in a vacuum, ie., $\varepsilon_1 = 1$ then

$$\mathbf{E_2} = \frac{3}{2 + \varepsilon_2} \mathbf{E} \qquad (1.44)$$

resulting in a reduction of the field inside.

Substituting for A_1 and B_1 in equation (1.26) the potential in region (1) is expressed as

$$\phi_1 = \left[\frac{(\varepsilon_2 - \varepsilon_1)}{(2\varepsilon_1 + \varepsilon_2)} \frac{R^3}{r^3} - 1 \right] EZ \qquad (1.45)$$

The changes in the potentials ϕ_1 and ϕ_2 are obviously due to the apparent surface charges on the dielectric. If we represent these changes as $\Delta\phi_1$ and $\Delta\phi_2$ in region 1 and 2 respectively by defining

$$\Delta\phi_1 = \phi + \phi_1 \qquad (1.46)$$

$$\Delta\phi_2 = \phi + \phi_2 \qquad (1.47)$$

where ϕ is the potential applied in the absence of the dielectric sphere, then

$$\Delta\phi_1 = \frac{\varepsilon_2 - \varepsilon_1}{2\varepsilon_1 + \varepsilon_2} \frac{R^3}{r^3} EZ \qquad (1.48)$$

$$\Delta\phi_2 = \frac{\varepsilon_2 - \varepsilon_1}{2\varepsilon_1 + \varepsilon_2} EZ \tag{1.49}$$

Fig. 1.5 shows the variation of $\mathbf{E}_2$ for different values of ε_2 with respect to ε_1.

The increase in potential within the sphere, equation (1.49), gives rise to an electric field

$$\frac{\Delta\phi_2}{\Delta z} = \Delta\mathbf{E} = -\frac{\varepsilon_2 - \varepsilon_1}{2\varepsilon_1 + \varepsilon_2} \mathbf{E} \tag{1.50}$$

The total electric field within the sphere is

$$\mathbf{E}_2 = \Delta\mathbf{E} + \mathbf{E} \tag{1.51}$$

$$= \frac{3\varepsilon_1}{2\varepsilon_1 + \varepsilon_2} \mathbf{E} \tag{1.52}$$

Equation (1.52) agrees with equation (1.42) verifying the correctness of the solution.

1.4.2 A RIGID DIPOLE IN A CAVITY WITHIN A DIELECTRIC

We now consider a hollow cavity in a dielectric material, with a rigid dipole at the center and we wish to calculate the electric field within the cavity. The cavity is assumed to be spherical with a radius R. A dipole is defined as rigid if its dipole moment is not changed due to the electric field in which it is situated. The material has a dielectric constant ε (Fig. 1.6).

The boundary conditions are:

(1) $(\phi_1)_{r \to \infty} = 0$ because the influence of the dipole decreases with increasing distance from it according to equation (1.3). Substituting this boundary condition in equation (1.26) gives $A_1 = 0$ and therefore

$$\phi_1 = \frac{B_1 \cos\theta}{r^2} \tag{1.53}$$

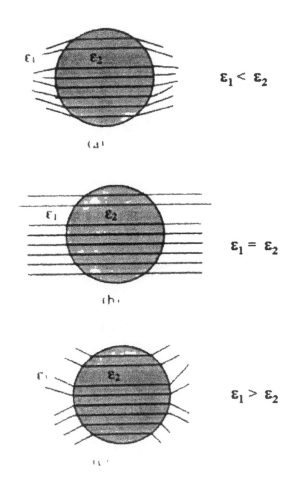

$\varepsilon_1 < \varepsilon_2$

$\varepsilon_1 = \varepsilon_2$

$\varepsilon_1 > \varepsilon_2$

Fig. 1.5 Electric field lines in two dielectric media. The influence of relative dielectric constants of the two media are shown. (a) $\varepsilon_1 < \varepsilon_2$ (b) $\varepsilon_1 = \varepsilon_2$ (c) $\varepsilon_1 > \varepsilon_2$

(2)   At any point on the boundary of the sphere the potential is the same whether we approach the point from infinity or the center of the sphere. This condition gives

$$(\phi_1)_{r=R} = (\phi_2)_{r=R} \tag{1.54}$$

leading to

$$B_1 = A_2 R^3 + B_2 \tag{1.55}$$

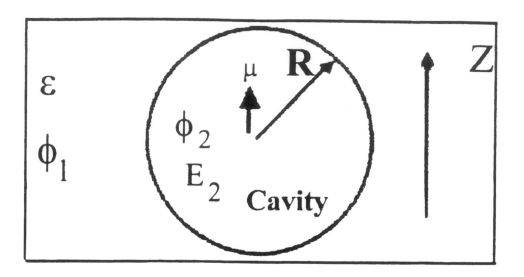

Fig. 1.6 A rigid dipole at the center of a cavity in a dielectric material. There is no applied electric field.

(3)   The normal component of the flux density across the boundary is continuous, expressed as

$$\left(\frac{\partial \phi_2}{\partial r}\right)_{r=R} = \varepsilon \left(\frac{\partial \phi_1}{\partial r}\right)_{r=R}$$ (1.56)

Applying this condition to the pair of equations (1.26) and (1.27) leads to

$$A_2 = \frac{2}{R^3}(B_2 - B_1 \varepsilon)$$ (1.57)

(4)   If the boundary of the sphere is moved far away *i.e.*, R→∞ the potental at any point is given by equation (1.3),

$$(\phi_2)_{R\to\infty} = \frac{\mu \cos\theta}{4\pi\varepsilon_o r^2}$$ (1.58)

and

$$A_2 = 0 \tag{1.59}$$

Substituting equations (1.58) and (1.59) in equation (1.27) results in

$$B_2 = \frac{\mu}{4\pi\varepsilon_o} \tag{1.60}$$

Equation (1.57) now becomes

$$A_2 = \frac{2\mu}{4\pi\varepsilon_o R^3} \frac{(1-\varepsilon)}{(1+2\varepsilon)} \tag{1.61}$$

Substituting equation (1.61) in (1.55) gives

$$B_1 = \frac{\mu}{4\pi\varepsilon_o} \left(\frac{3}{1+2\varepsilon}\right) \tag{1.62}$$

For convenience the other two constants are collected here:

$$B_2 = \frac{\mu}{4\pi\varepsilon_o} \tag{1.60}$$

$$A_1 = 0$$

The potential in the two regions are:

$$\phi_1 = \frac{\mu\cos\theta}{4\pi\varepsilon_o r^2} \left(\frac{3}{2\varepsilon+1}\right) \tag{1.63}$$

$$\phi_2 = \frac{\mu\cos\theta}{4\pi\varepsilon_o} \left[\frac{1}{r^2} + \frac{2r(1-\varepsilon)}{R^3(2\varepsilon+1)}\right] \tag{1.64}$$

Let ϕ_r be the potential at r due to the dipole in vacuum. The change in potential in the presence of the dielectric sphere is due to the presence of apparent charges on the walls of the sphere. These changes are:

$$\Delta\phi_1 = \phi_1 - \phi_r$$

$$= \frac{2(1-\varepsilon)}{2\varepsilon+1} \frac{\mu\cos\theta}{4\pi\varepsilon_o r^2} \tag{1.65}$$

$$\Delta\phi_2 = \frac{2(1-\varepsilon)}{4\pi\varepsilon_o R^3 (2\varepsilon+1)} \mu r\cos\theta \tag{1.66}$$

Since ε is greater than unity equations (1.63) and (1.64) show that there is a decrease in potential in both regions. The apparent surface charge has a dipole moment μ_a given by

$$\mu_a = \frac{2(1-\varepsilon)}{2\varepsilon+1} \mu \tag{1.67}$$

Equation (1.66) shows that the electric field in the cavity has increased by **R,** called the reaction field according to

$$\mathbf{R} = \frac{2(\varepsilon-1)}{4\pi\varepsilon_o R^3 (2\varepsilon+1)} \mu \tag{1.68}$$

It is interesting to calculate the approximate magnitude of this field at molecular level. Substituting $\mu = 1$ Debye $= 3.3 \times 10^{-30}$ C m, $R = 1 \times 10^{-10}$ m, and $\varepsilon = 3$, the reaction field is of the order of 10^{10} V/m which is very high indeed. The field reduces rapidly with distance, at $R = 1 \times 10^{-9}$ m, it is 10^7 V/m, a reduction by a factor of 1000. This is due to the fact that the reaction field changes according to the third power of R.

The converse problem of a dipole situated in a dielectric sphere which is immersed in vacuum may be solved similarly and the reaction field will then become

$$\mathbf{R} = \frac{2(1-\varepsilon)\mu}{4\pi\varepsilon_0\varepsilon R^3 (2+\varepsilon)} \tag{1.69}$$

If the dipole is situated in a medium that has a dielectric constant of ε_2 and the dielectric constant of region 1 is denoted by ε_1 the reaction field within the sphere is given by

$$R = \frac{2(\varepsilon_1 - \varepsilon_2)}{4\pi\varepsilon_o\varepsilon_2 r^3 (2\varepsilon_1 + \varepsilon_2)} \mu \tag{1.70}$$

It is easy to see that the relative values of ε_1 and ε_2 determine the magnitude of the reaction field. The reaction field is parallel to the dipole moment.

The general result for a dipole within a sphere of dielectric constant ε_2 surrounded by a second dielectric medium ε_1 is

$$\phi_1 = \frac{\mu\cos\theta}{4\pi\varepsilon_o\varepsilon_2 r^2}\left(\frac{3\varepsilon_2}{\varepsilon_2 + 2\varepsilon_1}\right)$$

$$\phi_2 = \frac{\mu\cos\theta}{4\pi\varepsilon_o\varepsilon_2}\left[\frac{1}{r^2} + \frac{2r(\varepsilon_2 - \varepsilon_1)}{R^3(\varepsilon_2 + 2\varepsilon_1)}\right]$$

If $\varepsilon_2 = 1$ then these equations reduce to equations (1.63) and (1.64). If R→∞ then ϕ_2 reduces to a form given by (1.3).

1.4.3 FIELD IN A DIELECTRIC DUE TO A CONDUCTING INCLUSION

When a conducting sphere is embedded in a dielectric and an electric field **E** is applied the field outside the sphere is modified. The boundary conditions are:

(1) At $r\to\infty$ the electric field is due to the external source and $\phi_1 \to -Er\cos\theta$.

Substituting this condition in equation (1.26) gives $A_1 = -E$ and therefore

$$\phi_1 = \left(-Er + \frac{B_1}{r^2}\right)\cos\theta \tag{1.71}$$

(2) Since the sphere is conducting there is no field within. Let us assume that the surface potential is zero. *i.e.*,

$$(\phi_1)_{r=R} = (\phi_2)_{r=R} = 0 \tag{1.72}$$

This condition when applied to equation (1.26) gives

$$\left(-ER + \frac{B_1}{R^2}\right)\cos\theta = 0 \tag{1.73}$$

Equation (1.73) is applicable for all values of θ and therefore

$$B_1 = ER^3 \tag{1.74}$$

The potential in region 1 is obtained as

$$\phi_1 = \left(-Er + \frac{ER^3}{r^2}\right)\cos\theta \tag{1.75}$$

We note that the presence of the inclusion increases the potential by an amount given by the second term of the eq. (1.75). Comparing it with equation (1.3) it is deduced that the increase in potential is equivalent to $4\pi\varepsilon_0$ times the potential due to a dipole of moment of value ER^3.

1.5 THE TUNNELING PHENOMENON

Let an electron of total energy ε eV be moving along x direction and the forces acting on it are such that the potential energy in the region $x < 0$ is zero (fig. 1.7). So its energy is entirely kinetic. It encounters a **potential barrier** of height ε_{pot} which is greater than its energy. According to classical theory the electron cannot overcome the potential barrier and it will be reflected back, remaining on the left side of the barrier. However according to quantum mechanics there is a finite probability for the electron to appear on the right side of the barrier.

To understand the situation better let us divide the region into three parts:

(1) Region I from which the electron approaches the barrier.
(2) Region II of thickness d which is the barrier itself.
(3) Region III to the right of the barrier.

The Schröedinger's equation may be solved for each region separately and the constants in each region is adjusted such that there are no discontinuities as we move from one region to the other.

A traveling wave encountering an obstruction will be partly reflected and partly transmitted. The reflected wave in region I will be in a direction opposite to that of the incident wave and a lower amplitude though the same frequency. The superposition of the two waves will result in a standing wave pattern. The solution in this region is of the type[1]

$$\psi_1(x) = A_1 \exp(\frac{jpx}{h}) + A_2 \exp(-\frac{jpx}{h}) \tag{1.76}$$

where we have made the substitutions:

$$h = \frac{h}{2\pi}; \qquad \varepsilon = \frac{1}{2}mv^2; \qquad p = mv = \sqrt{2m\varepsilon} \tag{1.77}$$

The first term in equation (1.76) is the incident wave, in the x direction; the second term is the reflected wave, in the $-x$ direction.

Within the barrier the wave function decays exponentially from $x = 0$ to $x = d$ according to:

$$\psi_{II} = A_2 \exp\left(-\frac{p_1 x}{h}\right) + B_2 \exp\left(-\frac{p_1 x}{h}\right); \quad 0 \le x \le d \tag{1.78}$$

$$p_1 = \sqrt{2m(\varepsilon_{pot} - \varepsilon)} \tag{1.79}$$

Since $\varepsilon_{pot} > \varepsilon$ the probability density is real within the region $0 \le x \le d$ and the density decreases exponentially with the barrier thickness. The central point is that in the case of a sufficiently thin barrier (< 1 nm) we have a finite, though small, probability of finding the electron on the right side of the barrier. This phenomenon is called the **tunneling effect**.

The relative probability that tunneling will occur is expressed as the **transmission co-efficient** and this is strongly dependent on the energy difference $(\varepsilon_{pot}-\varepsilon)$ and d. After tedious mathematical manipulations we get the **transmission co-efficient** as

$$T = \exp(-2p_1 d) \qquad (1.80)$$

in which p_1 has already been defined in connection with eq. (1.79). A co-efficient of T=0.01 means that 1% of the electrons impinging on the barrier will tunnel through. The remaining 99% will be reflected.

The tunnel effect has practical applications in the tunnel diode, Josephson junction and scanning tunneling microscope.

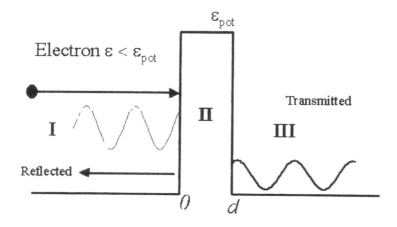

Fig. 1.7 An electron moving from the left has zero potential energy. It encounters a barrier of ε_{pot} Volts and the electron wave is partly reflected and partly transmitted. The transmitted wave penetrates the barrier and appears on the right of the barrier[2]. (with permission of McGraw Hill Ltd., Boston).

1.6 BAND THEORY OF SOLIDS

A brief description of the band theory of solids is provided here. For greater details standard text books may be consulted.

1.6.1 ENERGY BANDS IN SOLIDS

If there are N atoms in a solid sufficiently close we cannot ignore the interaction between them, that is, the wave functions associated with the valence electrons can not be treated as remaining distinct. This means that the N wave functions combine in $2N$ different ways. The wave functions are of the form

$$\psi_1 = \psi_1{}^i + \psi_2{}^i + \psi_3{}^i + \ldots\ldots\ldots + \psi_N{}^i$$

$$\psi_2 = \psi_1{}^i + \psi_2{}^i + \psi_3{}^i + \ldots\ldots\ldots - \psi_N{}^i$$

$$\ldots$$

$$\psi_{2N-1} = \psi_1{}^i - \psi_2{}^i + \psi_3{}^i + \ldots\ldots\ldots + \psi_N{}^i$$

$$\psi_{2N} = -\psi_1{}^i + \psi_2{}^i + \psi_3{}^i + \ldots\ldots\ldots + \psi_N{}^i$$

(1.81)

Each orbital is associated with a particular energy and we have 2N energy levels into which the isolated level of the electron splits. Recalling that $N \approx 10^{28}$ atoms per m^3 the energy levels are so close that they are viewed as an energy band. The energy bands of a solid are separated from each other in the same way that energy levels are separated from each other in the isolated atom.

1.6.2 THE FERMI LEVEL

In a metal the various energy bands overlap resulting in a single band which is partially full. At a temperature of zero Kelvin the highest energy level occupied by electrons is known as the Fermi level and denoted by ε_F. The reference energy level for *Fermi energy* is the bottom of the energy band so that the *Fermi energy* has a positive value. The probability of finding an electron with energy ε is given by the Fermi-Dirac statistics according to which we have

$$P(\varepsilon) = \frac{1}{1 + \exp\left[\dfrac{\varepsilon - \varepsilon_F}{kT}\right]}$$

(1.81)

At $\varepsilon = \varepsilon_F$ the the probability of finding the electron is ½ for all values of kT so that we may also define the Fermi Energy at temperature T as that energy at which the probability of finding the electron is ½ . The occupied energy levels and the probability are shown for four temperatures in figure (1.8). As the temperature increases the probability extends to higher energies. It is interesting to compare the probability given by the Boltzmann classical theory:

$$P(\varepsilon) = A \exp\left(\frac{-\varepsilon}{kT}\right)$$

(1.82)

The fundamental idea that governs these two equations is that, in classical physics we do not have to worry about the number of electrons having the same energy. However in quantum mechanics there cannot be two electrons having the same energy due to Pauli's exclusion principle. For $(\varepsilon - \varepsilon_F) \gg kT$ equation (1.81) may be approximated to

$$P(\varepsilon) = \exp{-\left(\frac{\varepsilon - \varepsilon_F}{kT}\right)} \tag{1.83}$$

which has a similar form to the classical equation (1.82).

The elementary band theory of solids, when applied to semi-conductors and insulators, results in a picture in which the conduction band and the valence bands are separated by a forbidden energy gap which is larger in insulators than in semi-conductors. In a perfect dielectric the forbidden gap cannot harbor any electrons; however presence of impurity centers and structural disorder introduces localized states between the conduction band and the valence band. Both holes and electron traps are possible[3].

Fig. 1.9 summarizes the band theory of solids which explains the differences between conductors, semi-conductors and insulators. A brief description is provided below.

A: In metals the filled valence band and the conduction band are separated by a forbidden band which is much smaller than kT where k is the Boltzmann constant and T the absolute temperature.
B: In semi-conductors the forbidden band is approximately the same as kT.
C: In dielectrics the forbidden band is several electron volts larger than kT. Thermal excitation alone is not enough for valence electrons to jump over the forbidden gap.
D: In p-type semi-conductor acceptors extend the valence band to lower the forbidden energy gap.
E: In n-type semi-conductor donors lower the unfilled conduction band again lowering the forbidden energy gap.
F: In p-n type semi-conductor both acceptors and donors lower the energy gap.

An important point to note with regard to the band theory is that the theory assumes a periodic crystal lattice structure. In amorphous materials this assumption is not justified and the modifications that should be incorporated have a bearing on the theoretical magnitude of current. The fundamental concept of the individual energy levels transforming into bands is still valid because the interaction between neighboring atoms is still present in the amorphous material just as in a crystalline lattice. Owing to irregularities in the lattice the edges of the energy bands lose their sharp character and become rather foggy with a certain number of allowed states appearing in the tail of each band. If the tail of the valence band overlaps the tail of the conduction band then the material behaves as a semi-conductor[+] (fig. 1.10).

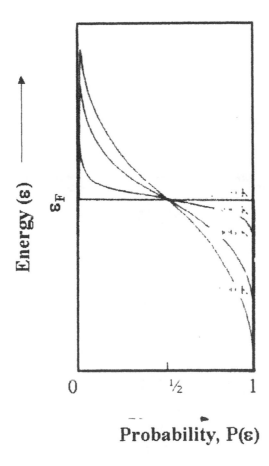

Fig. 1.8 The probability of filling is a function of energy level and temperature. The probability of filling at the Fermi energy is ½ for all temperatures. As T increases P(ε) extends to higher energies.

The amorphous semi-conductor is different from the normal semi-conductor because impurities do not substantially affect the conductivity of the former. The weak dependence of conductivity on impurity is explained on the basis that large fluctuations exist in the local arrangement of atoms. This in turn will provide a large number of localized trap levels, and impurity or not, there is not much difference in conductivity. Considering the electron traps, we distinguish between shallow traps closer to the conduction band, and deep traps closer to the valence band. The electrons in shallow traps have approximately the same energy as those in the conduction band and are likely to be thermally excited in to that band. Electrons in the ground state, however, are more likely to recombine with a free hole rather than be re-excited to the conduction band (fig.

1.9). The recombination time may be relatively long. Thus electrons in the ground state act as though they are deep traps and recombination centers.

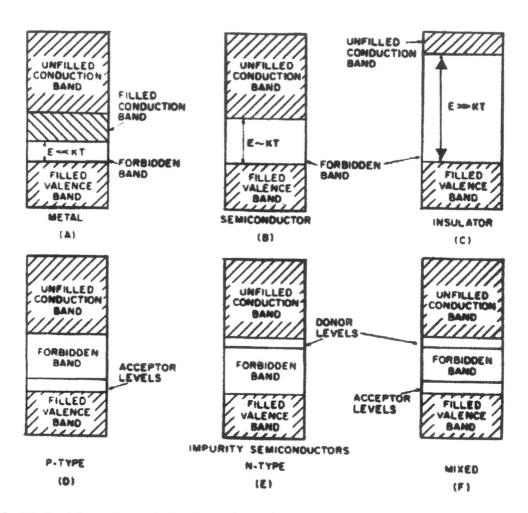

Fig. 1.9 Band theory for conduction in metals, semi-conductors, and Insulators [After A. H. Wilson, Proc. Roy. Soc., **A 133** (1931) 458] (with permission of the Royal Society).

Shallow traps and the ground states are separated by an energy level which corresponds to the Fermi level in the excited state. This level is the steady-state under excitation. To distinguish this level from the Fermi level corresponding to that in the metal the term 'dark Fermi level' has been used (Eckertova, 1990). Electrons in the Fermi level have the same probability of being excited to the conduction band or falling into the ground state. Free electrons in the conduction band and free holes in the valence band can move under the influence of an electric field, though the mobility of the electrons is much higher. Electrons can also transfer between localized states, eventually ending up in the conduction band. This process is known as "conduction by hopping". An electron transferring from a trap to another localized state under the influence of an electric field is known as the Poole-Frenkel effect. Fig. 1.11 [5, 6] shows the various possible levels for both the electrons and the holes.

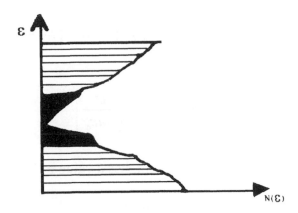

Fig. 1.10 Energy diagram of an amorphous material with the valence band and conduction band having rough edges (Schematic diagram).

1.6.3 ELECTRON EMISSION FROM A METAL

Electrons can be released from a metal by acquiring energy from an external source. The energy may be in the form of heat, by rising the temperature or by electromagnetic radiation. The following mechanisms may be distinguished:

(1) Thermionic Emission (Richardson-Dushman equation):

$$J = (1 - R) B_o T^2 \exp\left(-\frac{\phi}{kT}\right); \quad B_o = \frac{4\pi e m k^2}{h^3} \tag{1.84}$$

where J is the current density, ϕ the work function, T the absolute temperature and R the reflection co-efficient of the electron at the surface. The value of R will depend upon the surface conditions. B_o, called the Richardson-Dushman constant, has a value of 1.20×10^6 A m^{-2} K^{-2}. The term $(1-R)B_o$ can be as low as 1×10^2 A m^{-2} K^{-2}.

(2) Field assisted thermionic emission (Schottky equation):

In the presence of a strong electric field the work function is reduced according to

$$\phi_{eff} = \phi - \left[\frac{e^3 E}{4\pi\varepsilon_o} \right]^{1/2} \tag{1.85}$$

where ε_o is the permittivity of free space $= 8.854 \times 10^{-12}$ F/m and E the electric field. The current density is given by:

$$J = B_o T^2 \exp\left[-\frac{(\phi - \beta_s E^{1/2})}{kT} \right]; \quad \beta_s = \left[\frac{e^3}{4\pi\varepsilon_o} \right]^{1/2} \tag{1.86}$$

where and β_s, called the Schottky co-efficient, has a value of 3.79×10^{-5} [eV/ ($V^{1/2}$ $m^{-1/2}$)]

(3) Field emission (Fowler Nordheim equation)

In strong electric fields tunneling can occur even at room temperature and the current density for field emission is given by the expression:[7]

$$J = \frac{e^3 E^2}{8\pi h\phi} \exp\left\{ -\frac{4}{3}\left(\frac{2m}{h^2} \right)^{\frac{1}{2}} \frac{(\phi - E_F)^{\frac{1}{2}}}{eE} \right\} \tag{1.87}$$

where J is the current density in A/m^2, e the electronic charge in Coulomb, E is the electric field in V/m, h the Planck's constant in eV, ϕ the work function in eV and m the electron rest mass.

For practical applications equation (1.87) may be simplified to

$$I_{FN} = C_3 E^2 \exp(-\frac{C_4}{E}) \tag{1.88}$$

Improvements have been worked out to this equation but it is accurate enough for our purposes.

The effect of temperature is to multiply the current density by a factor:

$$\frac{f(\phi)T}{\sin[f(\phi)T]} \qquad (1.89)$$

where

$$f(\phi) = \frac{2\pi k(2m^*\phi)^{\frac{1}{2}}}{heE} \qquad (1.90)$$

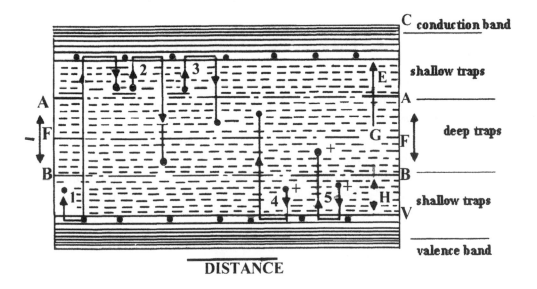

Fig. 1.11 Band-gap model and trapping events. C: conduction band, V: valence band, F-F: dark Fermi level, A-A: electron Fermi level (under photo-excitation), B-B: hole Fermi level (under photo-excitation), E: shallow electron traps, H: shallow hole traps, G: ground states (retrapping centers, deep traps), 1: photoexcitation of molecule; hole is captured in neutral shallow trap; electron is lifted to conduction band and captured in shallow electron trap; 2: shallow trapped electron is thermally activated into the conduction band, recombines into ground state; 3-shallow trapped electron, thermally activated into conduction band, is captured by deep trap; 4-4-shallow trapped hole receives electron from valence band and recombines into ground state; 5: shallow trapped hole receives electron from ground state and is captured in deep trap. The figure shows carrier movement under an applied electric field (Johnson, 1972; with permission of IEEE ©).

Here m^* is the effective rest mass of the electron. The temperature correction for the field emission current is small and given in Table 1.1.

Table 1.1
Temperature correction for the field emission current[8]b

	MULTIPLYING FACTOR			
T(K)	ϕ=1 eV	ϕ=2 eV	ϕ=3 eV	ϕ=4 eV
100	1.013	1.026	1.040	1.067
200	1.053	1.111	1.172	1.312
300	1.126	1.275	1.454	1.943
400	1.239	1.570	2.048	4.050
500	1.411	2.123	3.570	

1.6.4 FIELD INTENSIFICATION FACTOR

The electric field at the cathode is a macroscopic field and hence an average field at all points on the surface. It is idealized assuming that the cathode surface is perfectly smooth which is impossible to realize in practice. The surface will have imperfections and the electric field at the tip of these micro-projections will be more intense depending upon the tip radius; the smaller the radius, that is, the sharper the micro-projection greater will be the electric field at the tip. The effect of a micro-projection may be taken into account by introducing a field intensification factor β. The field emission current may now be expressed as:

$$Ln(\frac{I_{FN}}{E^2}) = LnK_3 + LnA + 2Ln\beta - \frac{C_4}{\beta E}$$
(1.91)

We see that a plot of $Ln(\frac{I_{FN}}{E^2})$ against 1/E yields a straight line from the slope of which

β may be calculated. The field intensification factor depends upon the ratio h/r where h is the height and r the radius of the projection. Experimentally observed values of β can be as high as 1000[9].

Recent measurements on electron emission in a vacuum have been carried out by Juttner et. al.[10, 11] with a time resolution of 100 ns and the vacuum gap exposed to a second gap in which high current arcing ~15A occurs. Higher emission currents are observed with a time constant of 1-3 μs and mechanical shocks are also observed to increase emission.

Field emission from atomic structures that migrate to the surface under the influence of the electric field is believed to increase the emission current, which renders the spark breakdown of the vacuum gap lower. Mechanical shocks are also believed to increase the migration, explaining the observed results.

1.7 ENERGY DISTRIBUTION FUNCTION

We consider now the distribution of energies of particles in a gas. A detailed discussion of energy distribution of electrons is given in chapter 9 and the present discussion is a simplified approach because inelastic collisions, which result in energy loss, are neglected. The energy distribution of molecules in a gas is given by the well known Maxwell distribution or Boltzmann distribution. The velocity distribution function according to Maxwell is given by

$$f(v) = \frac{4}{\pi^{\frac{1}{2}}} \left(\frac{m}{2kT} \right)^{\frac{3}{2}} v^2 \exp \left[-\frac{mv^2}{2kT} \right] \tag{1.92}$$

The arithmetic mean speed which is also called the **mean thermal velocity** v_{th} is given by

$$v_{th} = \left(\frac{8kT}{\pi m} \right)^{\frac{1}{2}} \tag{1.93}$$

The velocity distribution is expressed in terms of the energy by substituting

$$\varepsilon = \frac{1}{2} mv^2$$

However the Maxwell distribution may be considered as a particular case of a general distribution function of the form[12]

$$F(\varepsilon)d\varepsilon = A \frac{\varepsilon^{\frac{1}{2}}}{\bar{\varepsilon}^{\frac{3}{2}}} \exp \left[-B \left(\frac{\varepsilon}{\bar{\varepsilon}} \right)^{2(p+1)} \right] \tag{1.94}$$

in which $\bar{\varepsilon}$ is the mean energy of the electrons and, A and B are expressed by the following functions.

$$A = 2(p+1)\left[\Gamma\frac{5}{4(p+1)}\right]^{3/2} \times \left[\Gamma\frac{3}{4(p+1)}\right]^{-5/2} \qquad (1.95)$$

$$B = \left[\Gamma\frac{5}{4(p+1)} / \Gamma\frac{3}{4(p+1)}\right]^{2(p+1)} \qquad (1.96)$$

in which the symbol Γ stands for **Gamma function**. The Gamma function is defined according to

$$\Gamma(n+1) = n! = 1 \cdot 2 \cdot 3 \cdot \ldots\ldots(n-1) \cdot n$$
$$\Gamma(n+1) = n\Gamma n$$

Some values of the gamma functions are:

$$\Gamma\frac{1}{4} = 3.6256; \quad \Gamma\frac{1}{3} = 2.6789; \quad \Gamma\frac{1}{2} = 1.7724; \quad \Gamma\frac{3}{4} = 1.2254$$

Values of the Gamma function are tabulated in Abramowitz and Stegun[13]. The energy distribution function given by the expression (1.94) is known as the p-set[14, 15] and is quite simple to use. $p = -1/2$ gives the Maxwellian distribution identical to (1.92), the difference being that the latter expression is expressed in terms of the temperature T. The distribution functions for various values of p are shown in figure (1.12) for a mean energy of 4 eV. The use of the distribution function for calculating the swarm properties of the electron avalanche will be demonstrated in chapter 8.

1.8 THE BOLTZMANN FACTOR

We adopt a relatively easy procedure to derive the Boltzmann factor for a gas with equilibrium at a uniform absolute temperature T^2. Consider a large number of similar particles which interact weakly with each other. Gas molecules in a container are a typical example. It is not necessary that all particles be identical, but if they are not identical, then there must be a large number of each species. Electrons in a gas may be a typical example. We shall derive the probability that a molecule has an energy E. Suppose that the molecules can assume energy values in the ascending values E_1, E_2,

E_3, ...E_n. The energy may vary discreetly or continuously as in the case of ideal gas molecules. In either case we assume that there are no restrictions on the number of molecules that can assume a given energy E_n.

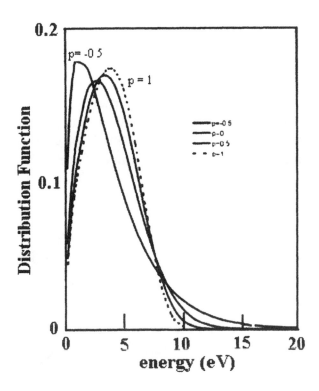

Fig. 1.12 Energy distribution function for various values of p for mean energy of 4 eV. p = - 0.5 gives Maxwellian distribution (author's calculation).

Let us consider two particles with energy E_1 and E_2 before collision, and their energy becomes E_3 and E_4, respectively, after collision. Conservation of energy dictates that

$$E_1 + E_2 \Leftrightarrow E_3 + E_4 \qquad (1.97)$$

where the symbol $\Leftrightarrow$ means that the equality holds irrespective of which side is before and after the event of collision. In equilibrium condition let the probability of an electron having an energy E be P(E) where P(E) is the fraction of electrons with energy E. The probability of collision between two particles of energies E_1 and E_2 is P(E_1) P(E_2). After collision the energies of the two particles are E_3 and E_4 (Fig. 1.13). The probability of collision is P(E_3) P(E_4). The probability of collisions from left to right should be equal to the probability of collisions from right to left.

$$P(E_1)P(E_2) = P(E_3)\,P(E_4) \tag{1.98}$$

The solution of equations (1.97) and (1.98) is

$$P = A\exp(-E/kT) \tag{1.99}$$

where A is a constant. Equation (1.99) is perhaps one of the most fundamental equations of classical physics. Instead of assuming that all particles are of the same kind, we could consider two different kinds of particles, say electrons and gas molecules. By a similar analysis we would have obtained the same constant k for each species. It is therefore called the universal **Boltzmann constant** with a value of $k = 1.381 \times 10^{-23}$ J/K $= 8.617 \times 10^{-5}$ eV/K.

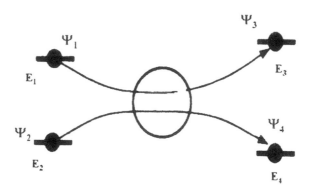

Fig. 1.13 Particles of wave functions Ψ_1 and Ψ_2 interact and end up with energies E3 and E4. Their corresponding wave functions are Ψ_3 and Ψ_4 (Kasap, 1997). (with permission of McGraw Hill Co.,)

1.9 A COMPARISON OF DISTRIBUTION FUNCTIONS

The Boltzmann distribution function given by equation (1.99), n/N_c, does not impose any restriction on the number of molecules or electrons having the same energy, except that the number should be large. However, in quantum electronics, Pauli's exclusion principle forbids two electrons having the same energy and one adopts the Fermi distribution (equation (1.83)) instead. The differences between the two distribution functions are shown in Table 1.2.

Table 1.2
Comparison of Boltzmann and Fermi Distributions
Extracted from Ref. [16] (with permission)

	Boltzmann	Fermi
Distribution Function	$A \exp(-E/kT)$	$\{\mathrm{Exp}[(E-E_F)/kT]+1\}^{-1}$
Basic characteristic.	Applies to distinguishable particles	Applies to indistinguishable particles obeying the exclusion principle.
Example of system.	Distinguishable particles or approximation to quantum distribution at $E \gg kT$	Identical particles of odd half integral spin.
Behavior of the distribution function.	Exponential function of E/kT	For $E \gg kT$, exponential where $E \gg E_F$. If $E_F \gg kT$, decreases abruptly near E_F.
Specific problems applied to in this book.	Distribution of dipoles in ch. 2; Energy distribution of electrons in Ch. 8	Electrons in dielectrics, Ch. 1, Ch.7-12.

1.10 REFERENCES

1   S. Brandt and H.D. Dahmen, "The Picture Book of Quantum Mechanics", second edition, Springer- Verlag, New York, 1995, p. 70.

2 S. O. Kasap, "Principles of Electrical Engineering Materials", McGraw Hill, Boston, 1997, p. 184.

3 B. Gross, "Radiation-induced Charge Storage and Polarization Effects", in "Electrets", Topics in Applied Physics, Ed: G. M. Sessler, Springer-Verlag, Berlin, 1980.

4 Physics of Thin Films: L. Eckertova, Plenum Publishing Co., New York, 1990.

5 W. C. Johnson, IEEE Trans., Nuc Sci., **NS-19 (6)** (1972) 33.

6 H. J. Wintle, IEEE Trans., Elect. Insul., **EI-12,** (1977) 12.

7 Electrical Degradation and Breakdown in polymers, L. A. Dissado and J. C. Fothergill, Peter Perigrinus, London, 1992, p. 226.

8 Electrical Degradation and Breakdown in polymers, L. A. Dissado and J. C. Fothergill, Peter Perigrinus, London, 1992, p. 227.

9 R. W. Strayer, F. M. Charbonnier, E. C. Cooper, L. W. Swanson, Quoted in ref. 3.

10 B. Jüttner, M. Lindmayer and G. Düning, J. Phys. D.: Appl. Phys., **32** (1999) 2537-2543.

11 B. Jüttner, M. Lindmayer and G. Düning, J. Phys. D.: Appl. Phys., **32** (1999) 2544-2551.

12 Morse, P. M., Allis, W. P. and E. S. Lamar, Phys. Rev., 48 (1935) p. 412.

13 Handbook of Mathematical Functions: M. Abramowitz and I. A. Stegun, Dover, New York, 1970.

14 A. E. D. Heylen, Proc. Phys. Soc., 79 (1962) 284.

15 G. R. Govinda Raju and R. Hackam, J. Appl. Phys., 53 (1982) 5557-5564.

16 R. Eisberg and R. Resnick, "Quantum Physics", John Wiley & Sons, New York, 1985.

Seek simplicity, and distrust it.
-Alfred North Whitehead

2

POLARIZATION and STATIC DIELECTRIC CONSTANT

The purposes of this chapter are (i) to develop equations relating the macroscopic properties (dielectric constant, density, etc.) with microscopic quantities such as the atomic radius and the dipole moment, (ii) to discuss the various mechanisms by which a dielectric is polarized when under the influence of a static electric field and (iii) to discuss the relation of the dielectric constant with the refractive index. The earliest equation relating the macroscopic and microscopic quantities leads to the so-called **Clausius-Mosotti** equation and it may be derived by the approach adopted in the previous chapter, i.e., finding an analytical solution of the electric field. This leads to the concept of the internal field which is higher than the applied field for all dielectrics except vacuum. The study of the various mechanisms responsible for polarizations lead to the **Debye equation** and **Onsager theory**. There are important modifications like Kirkwood theory which will be explained with sufficient details for practical applications. Methods of Applications of the formulas have been demonstrated by choosing relatively simple molecules without the necessity of advanced knowledge of chemistry.

A comprehensive list of formulas for the calculation of the dielectric constants is given and the special cases of heterogeneous media of several components and liquid mixtures are also presented.

2.1 POLARIZATION AND DIELECTRIC CONSTANT

Consider a vacuum capacitor consisting of a pair of parallel electrodes having an area of cross section A m^2 and spaced d m apart. When a potential difference V is applied between the two electrodes, the electric field intensity at any point between the electrodes, perpendicular to the plates, neglecting the edge effects, is $E = V/d$. The capacitance of the vacuum capacitor is $C_0 = \varepsilon_0\, A/d$ and the charge stored in the capacitor is

$$Q_0 = A\varepsilon_0 E \tag{2.1}$$

in which ε_0 is the permittivity of free space.

If a homogeneous dielectric is introduced between the plates keeping the potential constant the charge stored is given by

$$Q = \varepsilon_0 \varepsilon\, AE \tag{2.2}$$

where ε is the dielectric constant of the material. Since ε is always greater than unity $Q_1 > Q$ and there is an increase in the stored charge given by

$$Q_1 - Q_0 = AE\varepsilon_0(\varepsilon - 1) \tag{2.3}$$

This increase may be attributed to the appearance of charges on the dielectric surfaces. Negative charges appear on the surface opposite to the positive plate and vice-versa (Fig. 2.1)[1]. This system of charges is apparently neutral and possesses a dipole moment

$$\mu = AE\varepsilon_0(\varepsilon - 1)d \tag{2.4}$$

Since the volume of the dielectric is $v = Ad$ the dipole moment per unit volume is

$$P = \frac{\mu}{Ad} = E\varepsilon_0(\varepsilon - 1) = \chi\varepsilon_0 E \tag{2.5}$$

The quantity P, is the polarization of the dielectric and denotes the dipole moment per unit volume. It is expressed in C/m^2. The constant $\chi = (\varepsilon-1)$ is called the susceptability of the medium.

The flux density D defined by

$$D = \varepsilon_0 \varepsilon E \tag{2.6}$$

becomes, because of equation (2.5),

$$D = \varepsilon_0 \varepsilon + P \tag{2.7}$$

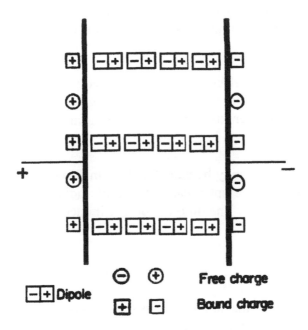

Fig. 2.1 Schematic representation of dielectric polarization [von Hippel, 1954]. (With permission of John Wiley & Sons, New York)

Polarization of a dielectric may be classified according to

1. **Electronic or Optical Polarization**
2. **Orientational Polarization**
3. **Atomic or Ionic Polarization**
4. **Interfacial Polarization.**

We shall consider the first three of these in turn and the last mechanism will be treated in chapter 4.

2.2 ELECTRONIC POLARIZATION

The classical view of the structure of the atom is that the center of the atom consists of positively charged protons and electrically neutral neutrons. The electrons move about the nucleus in closed orbits. At any instant the electron and the nucleus form a dipole with a moment directed from the negative charge to the positive charge. However the axis of the dipole changes with the motion of the electron and the time average of the dipole moment is zero. Further, the motion of the electron must give rise to electromagnetic radiation and electrical noise. The absence of such effects has led to the concept that the total electronic charge is distributed as a spherical cloud the center of which coincides with the nucleus, the charge density decreasing with increasing radius from the center.

When the atom is situated in an electric field the charged particles experience an electric force as a result of which the center of the negative charge cloud is displaced with respect to the nucleus. A dipole moment is induced in the atom and the atom is said to be **electronically polarized**.

The electronic polarizability α_e may be calculated by making an approximation that the charge is spread uniformly in a spherical volume of radius R. The problem is then identical with that in section 1.3. The dipole moment induced in the atom was shown to be

$$\mu_e = (4\pi\varepsilon_0 R^3)E \tag{1.42}$$

For a given atom the quantity inside the brackets is a constant and therefore the dipole moment is proportional to the applied electric field. Of course the dipole moment is zero when the field is removed since the charge centers are restored to the undisturbed position.

The electronic polarizability of an atom is defined as the dipole moment induced per unit electric field strength and is a measure of the ease with which the charge centers may be dislocated. α_e has the dimension of F m^2. Dipole moments are expressed in units called **Debye** whose pioneering studies in this field have contributed so much for our present understanding of the behavior of dielectrics. 1 **Debye unit** = 3.33 x 10^{-30} C m.

α_e can be calculated to a first approximation from atomic constants. For example the radius of a hydrogen atom may be taken as 0.04 nm and α_e has a value of 10^{-41} F m^2. For a field strength of 1 MV/m which is a high field strength, the displacement of the negative charge center, according to eq. (1.42) is 10^{-16} m; when compared with the atomic radius the displacement is some 10^{-5} times smaller. This is due to the fact that the internal electric field within the atom is of the order of 10^{11} V/m which the external field is required to overcome.

Table 2.1

Electronic polarizability of atoms[2]

Element	radius (10^{-10} m)	α_e (10^{-40} F m^2)
He	0.93	0.23
Ne	1.12	0.45
Ar	1.54	1.84
Xe	1.9	4.53
I	1.33	6.0
Cs	3.9	66.7

(with permission of CRC Press).

Table 2.1 shows that the electronic polarizability of rare gases is small because their electronic structure is stable, completely filled with 2, 10, 18 and 36 electrons. As the radius of the atom increases in any group the electronic polarizability increases in accordance with eq. (1.42). Unlike the rare gases, the polarizability of alkali metals is more because the electrons in these elements are rather loosely bound to the nucleus and therefore they are displaced relatively easily under the same electric field. In general the polarizability of atoms increases as we move down any group of elements in the periodic table because then the atomic radius increases.

Fig. 2.2[3] shows the electronic polarizability of atoms. The rare gas atoms have the lowest polarizability and Group I elements; alkali metals have the highest polarizability, due to the single electron in the outermost orbit. The intermediate elements fall within the two limits with regularity except for aluminum and silver.

The ions of atoms of the elements have the same polarizability as the atom that has the same number of electrons as the ion. Na^+ has a polarizability of 0.2×10^{-40} F m^2 which is of the same order of magnitude as α_e for Ne. K^+ is close to Argon and so on. The polarizability of the atoms is calculated assuming that the shape of the electron is spherical. In case the shape is not spherical then α_e becomes a tensor quantity; such refinement is not required in our treatment.

Molecules possess a higher α_e in view of the much larger electronic clouds that are more easily displaced. In considering the polarizability of molecules we should take into account the **bond polarizability** which changes according to the axis of symmetry. Table 2.2[2] gives the polarizabilities of molecules along three principal axes of symmetry in units of 10^{-40} Fm^2. The mean polarizability is defined as $\alpha_m = (\alpha_1 + \alpha_2 + \alpha_3)/3$. Table 2.3 gives the polarizabilities of chemical bonds parallel and normal to the bond axis and also the mean value for all three directions in space, calculated according to $\alpha_m = (\alpha_{||} + 2 \alpha_{\perp})/3$. The constant 2 appears in this equation because there are two mutually perpendicular axes to the bond axis.

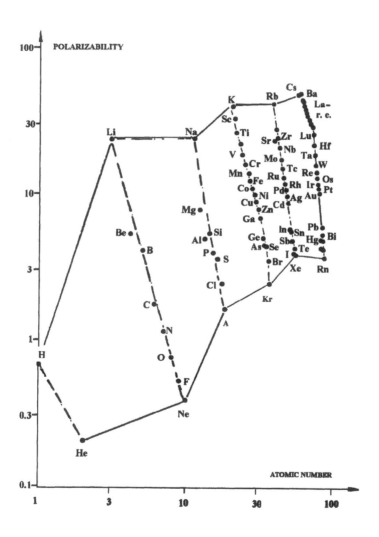

Fig. 2.2 Electronic polarizability (F/m^2) of the elements versus the atomic number. The values on the y axis must be multiplied by the constant $4\pi\varepsilon_o \times 10^{-30}$. (Jonscher, 1983: With permission of the Chelsea Dielectric Press, London).

It is easy to derive a relationship between the dielectric constant and the electronic polarizability. The dipole moment of an atom, by definition of α_e, is given by α_e E and if N is the number of atoms per unit volume then the dipole moment per unit volume is $N\alpha_e$ E. We can therefore formulate the equation

$$P = N\alpha_e \mathbf{E} \tag{2.8}$$

Substituting equation (2.5) on the left side and equation (1.22) on the right yields

$$\varepsilon = 4\pi N R^3 + 1 \tag{2.9}$$

This expression for the dielectric constant in terms of N and R is the starting point of the dielectric theory. We can consider a gas at a given pressure and calculate the dielectric constant using equation (2.9) and compare it with the measured value. For the same gas the atomic radius R remains independent of gas pressure and therefore the quantity (ε-1) must vary linearly with N if the simple theory holds good for all pressures.

Table 2.4 gives measured data for hydrogen at various gas pressures at 99.93° C and compares with those calculated by using equation (2.9)[4]. At low gas pressures the agreement between the measured and calculated dielectric constants is quite good. However at pressures above 100 M pa (equivalent to 1000 atmospheric pressures) the calculated values are lower by more than 5%. The discrepancy is due to the fact that at such high pressures the intermolecular distance becomes comparable to the diameter of the molecule and we can no longer assume that the neighboring molecules do not influence the polarizability.

Table 2.2

Polarizability of molecules [3]

Molecule	α_1	α_2	α_3	α_m
H_2	1.04	0.80	0.80	0.88
O_2	2.57	1.34	1.34	5.25
N_2O	5.39	2.30	2.30	3.33
CCl_4	11.66	11.66	11.66	11.66
HCl	3.47	2.65	2.65	2.90

(with permission from Chelsea dielectric press).

Table 2.3
Polarizability of molecular bonds [3]

Bond	$\alpha_{\parallel}$	$\alpha_{\perp}$	α_m	comments
H-H	1.03	0.80	0.88	
N-H	0.64	0.93	0.83	NH_3
C-H	0.88	0.64	0.72	aliphatic
C-Cl	4.07	2.31	2.90	
C-Br	5.59	3.20	4.0	
C-C	2.09	0.02	0.71	aliphatic
C-C	2.50	0.53	1.19	aromatic
C=C	3.17	1.18	1.84	
C=O	2.22	0.83	1.33	carbonyl

(with permission from Chelsea dielectric press).

The increase in the electric field experienced by a molecule due to the polarization of the surrounding molecules is called the **internal field, E_i**. When the internal field is taken into account the induced dipole moment due to electronic polarizability is modified as

$$\mu_e = \alpha_e E_i \qquad (2.10)$$

The internal field is calculated as shown in the following section.

2.3 THE INTERNAL FIELD

To calculate the internal field we imagine a small spherical cavity at the point where the internal field is required. The result we obtain varies according to the shape of the cavity; Spherical shape is the least difficult to analyze. The radius of the cavity is large enough in comparison with the atomic dimensions and yet small in comparison with the dimensions of the dielectric.

Let us assume that the net charge on the walls of the cavity is zero and there are no short range interactions between the molecules in the cavity. The internal field, E_i at the center of the cavity is the sum of the contributions due to
1. The electric field due to the charges on the electrodes (free charges), E_1.
2. The field due to the bound charges, E_2.
3. The field due to the charges on the inner walls of the spherical cavity, E_3. We may also view that E_3 is due to the ends of dipoles that terminate on the surface of the sphere. We have shown in chapter 1 that the polarization of a dielectric **P** gives rise to a surface

charge density. Note the direction of P_n which is due to the negative charge on the cavity wall.

4. The field due to the atoms within the cavity, E_4.

Table 2.4

Measured and calculated dielectric constant [4]. $R=91 \times 10^{-12}$ m

pressure (MPa)	density Kg/m^3	N $(m^3) \times 10^{26}$	ε equation (2.9)	ε (measured)
1.37	0.439	2.86	1.00266	1.00271
4.71	1.482	9.82	1.00898	1.00933
8.92	2.751	18.62	1.01670	1.01769
14.35	4.305	29.91	1.02628	1.02841
22.43	6.484	46.80	1.03966	1.04446
48.50	12.496	101.21	1.07750	1.09615
93.81	20.374	195.78	1.12840	1.18599
124.52	24.504	259.86	1.15620	1.24687
144.39	26.833	301.32	1.17232	1.28625

We can express the internal field as the sum of its components:

$$E_i = E_1 + E_2 + E_3 + E_4 \tag{2.11}$$

The sum of the field intensity E_1 and E_2 is equal to the external field,

$$E = E_1 + E_2 \tag{2.12}$$

E_3 may be calculated by considering a small element of area dA on the surface of the cavity (Fig. 2.3). Let θ be the angle between the direction of E and the charge density P_n. P_n is the component of P normal to the surface, i.e.,

$$P_n = P\cos\theta \tag{2.13}$$

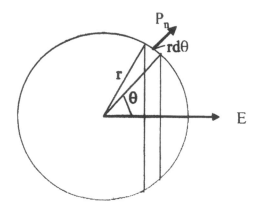

Fig. 2.3 Calculation of the internal field in a dielectric.

The charge on dA is

$$dq = P\cos\theta dA \qquad (2.14)$$

The electric field at the center of the cavity due to charge dq is

$$d\mathbf{E}_3' = \frac{P\cos\theta dA}{4\pi\varepsilon_0 r^2} \qquad (2.15)$$

We are interested in finding the field which is parallel to the applied field. The component of dE_3' along $\mathbf{E}$ is

$$d\mathbf{E}_3 = d\mathbf{E}_3'\cos\theta$$

$$= \frac{P\cos^2\theta dA}{4\pi\varepsilon_0 r^2} \qquad (2.16)$$

All surface elements making an angle θ with the direction of E give rise to the same dE_3. The area dA is equal to

$$dA = 2\pi r\sin\theta \cdot rd\theta$$

$$= 2\pi r^2 \sin\theta d\theta \qquad (2.17)$$

The total area so situated is the area of the annular ring having a radii of r and $r + d\theta$. i.e., Substituting equation (2.16) in (2.17) gives

$$dE_3 = \frac{P\cos^2\theta}{4\pi\varepsilon_0 r^2} \cdot 2\pi r^2 \sin\theta d\theta \qquad (2.18)$$

Because of symmetry the components perpendicular to E cancel out. We therefore get

$$E_3 = \int_0^\pi \frac{P\cos^2\theta}{2\varepsilon_0} \sin\theta d\theta \qquad (2.19)$$

$$= \frac{P}{3\varepsilon_0} \qquad (2.20)$$

The charge on the element considered also gives rise to an electric field in a direction perpendicular to E and this component is

$$dE_3' \sin\theta = \frac{P\cos\theta \sin\theta dA}{4\pi\varepsilon_0 r^2}$$

Substituting expression (2.17) in this expression and integrating we get

$$\int_0^\pi \frac{P\cos\theta \sin^2\theta d\theta}{2} = 0$$

Hence we consider only the parallel component of dE_3' in calculating the electric field according to equation (2.16).

Because of symmetry the short range forces due to the dipole moments inside the cavity become zero, $E_4 = 0$, for cubic crystals and isotropic materials. Substituting equation (2.20) in (2.11) we get

$$E_i = E + \frac{P}{3\varepsilon_0} \qquad (2.21)$$

E_i is known as the **Lorentz field.** Substituting equation (2.5) in (2.21) we get

$$E_i = E(\frac{2+\varepsilon}{3})$$

$$(2.22)$$

Since

$$P = E\varepsilon_0(\varepsilon - 1) = N\alpha_e E_i$$

we get, after some simple algebra

$$\frac{\varepsilon - 1}{\varepsilon + 2} = \frac{N\alpha_e}{3\varepsilon_0} = \frac{N_A\alpha_e}{3\varepsilon_0 V}$$

$$(2.23)$$

where V is the molar volume, given by M/ρ. By definition N is the number of molecules per unit volume. If ρ is the density (kg/m³), M the molecular weight of dielectric (kg/mole), and N_A the Avagadro number, then

$$N_A = \frac{N \times M}{\rho} = N \times V$$

$$(2.24)$$

Substituting equation (2.24) in equation (2.23) we get,

$$\frac{\varepsilon - 1}{\varepsilon + 2} \frac{M}{\rho} = \frac{N_A\alpha_e}{3\varepsilon_0} = R$$

$$(2.25)$$

in which R is called the **molar polarizability.** Equation (2.25) is known as the Clausius-Mosotti equation. The left side of equation (2.25) is often referred to as C-M factor. In this equation all quantities except α_e may be measured and therefore the latter may be calculated.

Maxwell deduced the relation that $\varepsilon = n^2$ where n is the refractive index of the material. Substituting this relation in equation (2.25), and ignoring for the time being the restriction that applies to the Maxwell equation, the discussion of which we shall postpone for the time being, we get

$$\frac{n^2 - 1}{n^2 + 2} \frac{M}{\rho} = \frac{N_A\alpha_e}{3\varepsilon_0}$$

$$(2.26)$$

Combining equations (2.25) and (2.26) we get

$$\frac{\varepsilon - 1}{\varepsilon + 2} = \frac{n^2 - 1}{n^2 + 2} \qquad (2.27)$$

Equation (2.27) is known as the **Lorentz-Lorenz** equation. It can be simplified further depending upon particular parameters of the dielectric under consideration. For example, gases at low pressures have $\varepsilon \approx 1$ and equation (2.23) simplifies to

$$\varepsilon = 1 + \frac{N\alpha_e}{\varepsilon_0} \qquad (2.28)$$

If the medium is a mixture of several gases then

$$\frac{\varepsilon - 1}{\varepsilon + 2} = \sum_{j=1}^{j=k} \frac{N_j \alpha_{ej}}{3\varepsilon_0} \qquad (2.29)$$

where N_j and α_{ej} are the number and electronic polarizability of each constituent gas.

Equation (2.23) is applicable for small densities only, because of the assumptions made in the derivation of Clausius-Mosotti equation. The equation shows that the factor $(\varepsilon - 1) / (\varepsilon + 2)$ increases with N linearly assuming that α_e remains constant. This means that ε should increase with N faster than linearly and there is a critical density at which ε should theoretically become infinity. Such a critical density is not observed experimentally for gases and liquids.

It is interesting to calculate the displacement of the electron cloud in practical dielectrics. As an example, Carbon tetrachloride (CCl_4) has a dielectric constant of 2.24 at 20°C and has a density of 1600 kg/m^3. The molecular weight is 156 x 10^{-3} Kg/mole. The number of molecules per m^3 is given by equation (2.24) as:

$$N = N_A \frac{\rho}{M} = 6.03 \times 10^{23} \times \frac{1600}{156 \times 10^{-3}}$$

$$= 6.2 \times 10^{27} m^{-3}$$

Let us assume an electric field of 1MV/m which is relatively a high field strength. Since $P = N\mu = E\,\varepsilon_o\,(\varepsilon_r- 1)$ the induced dipole moment is equal to

$$\mu = \frac{10^6 \times 8.854 \times 10^{-12} \times 1.24}{6.2 \times 10^{27}} = 1.78 \times 10^{-33} C\,m$$

CCl_4 has 74 electrons. Hence each electron-proton pair, on the average has a dipole moment of $(1.78 \times 10^{-32})\ /74= 2.4 \times 10^{-35}$ Cm. The average displacement of the electron cloud is obtained by dividing the dipole moment by the electronic charge, 1.5×10^{-16} m. This is roughly 10^{-6} times that of the molecular size.

Returning to equation (2.25) a rearrangement gives

$$\varepsilon = \frac{1 + 2\rho R / M}{1 - \rho R / M} \tag{2.30}$$

where the molar polarizability R refers to the compound. Denoting the molar polarizability and atomic weight of individual atoms as r and m respectively we can put $R = \Sigma r$ and $M = \Sigma m$. Table 2.5 shows the application of equation (2.30) to an organic molecule.

As an example we consider the molecule of heptanol: Formula $CH_3\,(CH_2)_5\,(CH_2OH)$, $\rho = 824$ kg/m^3 (20°C), b. p. = 176° C, m. p. = -34.1° C.

$\alpha_T = \Sigma\,\alpha = (7 \times 1.06 + 16 \times 0.484 + 1 \times 0.67) \times 10^{-40} = 15.83 \times 10^{-40}$ F/m^2
$M = \Sigma\,m = 7 \times 12.01 + 16 \times 1.01 + 1 \times 16.00 = 116.23$
$N_A\alpha_T\rho\,/\,3\varepsilon_0 M = 0.2556$, $\varepsilon = 2.03$, Measured $n^2 = (1.45)^2 = 2.10$

Equation (2.25) is accurate to about 1% when applied to non-polar polymers. Fig. 2.4[5] shows the variation of the dielectric constant with density in non-polar polymers and the relation that holds may be expressed as

$$\frac{1}{\rho}\left(\frac{\varepsilon - 1}{\varepsilon + 2}\right) \cong 0.325 \tag{2.31}$$

The Clausius-Mosotti function is linearly dependent on the density.

Table 2.5
Calculation of molar polarizability of a compound[6]

Atom structure	α (x 10^{-40} F m^2)	m
C	1.064	12.01
H	0.484	1.01
O (alcohol)	0.671	16.00
O(carbonyl)	0.973	-
O(ether)	0.723	-
O(ester)	0.722	-
Si	4.17	28.08
F (one/carbon)	0.418	19.00
Cl	2.625	35.45
Br	3.901	79.91
I	6.116	126.90
S	3.476	32.06
N	1.1-1.94	14.01
Structural effects		
Double bond	1.733	
Triple bond	2.398	
3-member ring	0.71	
4-member ring	0.48	

(with permission from North Holland Co.).

2.4 ORIENTATIONAL POLARIZATION

The Clausius-Mosotti equation is derived assuming that the relative displacement of electrons and nucleus is elastic, i.e., the dipole moment is zero after the applied voltage is removed. However molecules of a large number of substances possess a dipole moment even in the absence of an electric field. In the derivation of equation (2.23) the temperature variation was not considered, implying that polarization is independent of temperature. However the dielectric constant of many dielectrics depends on the temperature, even allowing for change of state. The theory for calculating the dielectric constant of materials possessing a permanent dipole moment is given by Debye.

Dielectrics, the molecules of which possess a permanent dipole moment, are known as polar materials as opposed to non-polar substances, the molecules of which do not possess a permanent dipole moment. Di-atomic molecules like H_2, N_2, Cl_2, with homopolar bonds do not possess a permanent dipole moment. The majority of molecules that are formed out of dissimilar elements are polar; the electrons in the valence shell tend to acquire or lose some of the electronic charge in the process of formation of the molecule. Consequently the center of gravity of the electronic charge is displaced with respect to the positive charges and a permanent dipole moment arises.

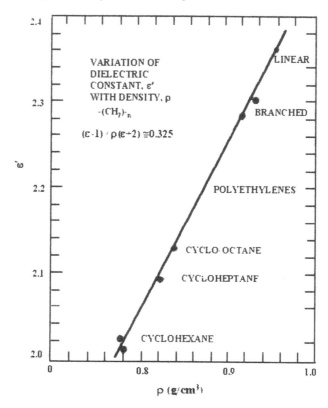

Fig. 2.4 Linear variation of dielectric constant with density in non-polar polymers [Link, 1972]. (With permission from North Holland Publishing Co.)

For example, in the elements of HCl, the outer shell of a chlorine atom has seven electrons and hydrogen has one. The chlorine atom, on account of its high electronegativity, appropriates some of the electronic charge from the hydrogen atom with the result that the chlorine atom becomes negatively charged and the hydrogen atom is depleted by the same amount of electronic charge. This induces a dipole moment in the molecule directed from the chlorine atom to the hydrogen atom. The distance between the atoms of hydrogen and chlorine is 1.28×10^{-10} m and it possesses a dipole moment of 1.08 D.

Since the orientation of molecules in space is completely arbitrary, the substance will not exhibit any polarization in the absence of a external field. Due to the fact that the electric field tends to orient the molecule and thermal agitation is opposed to orientation, not all molecules will be oriented. There will be a preferential orientation, however, in the direction of the electric field. Increased temperature, which opposes the alignment, is therefore expected to decrease the orientational polarization. Experimentally this fact has been confirmed for many polar substances.

If the molecule is symmetrical it will be non-polar. For example a molecule of CO_2 has two atoms of oxygen distributed evenly on either side of a carbon atom and therefore the CO_2 molecule has no permanent dipole moment. Carbon monoxide, however, has a dipole moment. Water molecule has a permanent dipole moment because the O-H bonds make an angle of 105° with each other.

Hydrocarbons are either non-polar or possess a very small dipole moment. But substitution of hydrogen atoms by another element changes the molecule into polar. For example, Benzene (C_6H_6) is non-polar, but monochlorobenzene (C_6H_5Cl), nitrobenzene ($C_6H_5NO_2$), and monoiodobenzene (C_6H_5I) are all polar. Similarly by replacing a hydrogen atom with a halogen, a non-polar hydrocarbon may be transformed into a polar substance. For example methane (CH_4) is non-polar, but chloroform ($CHCl_3$) is polar.

2.5 DEBYE EQUATIONS

We have already mentioned that the polarization of the dielectric is zero in the absence of an electric field, even for polar materials, because the orientation of molecules is random with all directions in space having equal probability. When an external field is applied the number of dipoles confined to a solid angle $d\Omega$ that is formed between θ and $\theta + d\theta$ is given by the Boltzmann distribution law:

$$n(\theta) = A \exp(-\frac{\upsilon}{kT})d\Omega \tag{2.32}$$

in which υ is the potential energy of the dipole and $d\Omega$ is the solid angle subtended at the center corresponding to the angle θ, where k, is the Boltzmann constant, T the absolute temperature and A is a constant that depends on the total number of dipoles. Consider the surface area between the angles θ and $\theta + d\theta$ on a sphere of radius r (fig. 2.5),

$$ds = 2\pi \, r \sin\theta \cdot r d\theta$$

$$= 2\pi r^2 \sin\theta \, d\theta \tag{2.33}$$

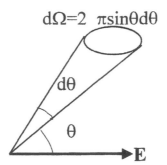

Fig. 2.5 Derivation of the Debye equation. The solid angle is dΩ.

Since the solid angle is defined as ds/r^2 the solid angle between θ and θ +dθ is

$$d\Omega = 2\pi \sin\theta\, d\theta \tag{2.34}$$

We recall that the total solid angle subtended at the center of the sphere is 4π steridians. This may be checked by the formula,

$$d\Omega = \int_0^\pi 2\pi \sin\theta\, d\theta = 4\pi$$

The potential energy of a dipole in an electric field is

$$\upsilon = -\mu E \tag{2.35}$$

Since the dipoles in the solid angle are situated at an angle of θ the potential energy is reduced to

$$\upsilon = -\mu E \cos\theta \tag{2.36}$$

Substituting equations (2.33) and (2.34) in (2.32) we get

$$n(\theta) = A\exp\left(\frac{\mu E \cos\theta}{kT}\right)2\pi \sin\theta\, d\theta \tag{2.37}$$

Since a dipole of permanent moment μ making an angle θ with the direction of the electric field contributes a moment $\mu \cos\theta$, the contribution of all dipoles in $d\Omega$ is equal to

$$\mu(\theta) = n(\theta)\mu\cos\theta \tag{2.38}$$

Therefore the average moment per dipole in the direction of the electric field is given by the ratio of the dipole moment due to all molecules divided by the number of dipoles,

$$\mu_o = \frac{\int_0^\pi n(\theta)\mu\cos\theta\,d\theta}{\int_0^\pi n(\theta)\,d\theta} \tag{2.39}$$

Note that the average dipole moment cannot be calculated by dividing equation (2.38) by (2.37), because we must consider all values of θ from 0 to π before averaging. Substituting equations (2.37) and (2.38) in (2.39) we get

$$\mu_0 = \frac{\int_0^\pi A\exp\left(\dfrac{\mu E\cos\theta}{kT}\right)(2\pi\sin\theta)\mu\cos\theta\,d\theta}{\int_0^\pi A\exp\left(\dfrac{\mu E\cos\theta}{kT}\right)2\pi\sin\theta\,d\theta} \tag{2.40}$$

To avoid long expressions we make the substitution

$$\frac{\mu E}{kT} = x \tag{2.41}$$

and

$$\cos\theta = y \tag{2.42}$$

The substitution simplifies equation (2.40) to

$$\frac{\mu_0}{\mu} = \frac{\int_{-1}^{-1} y \exp(xy)\,dy}{\int_{-1}^{-1} \exp(xy)\,dy} \tag{2.43}$$

Recalling that

$$\int \exp(xy)\,dy = \frac{\exp(xy)}{x}$$

$$\int y \exp(xy)\,dy = \frac{e^{xy}}{x^2}(xy-1)$$

equation (2.43) reduces to

$$\frac{\mu_o}{\mu} = \frac{(e^x + e^{-x})}{(e^x - e^{-x})} - \frac{1}{x} \tag{2.44}$$

Since the first term on the right side of equation (2.44) is equal to $coth\,x$ we get

$$\frac{\mu_0}{\mu} = \coth x - \frac{1}{x} = L(x) \tag{2.45}$$

$L(x)$ is called the **Langevin function** and it was first derived by Langevin in calculating the mean magnetic moment of molecules having permanent magnetism, where similar considerations apply.

The Langevin function is plotted in Fig. 2.6. For small values of x, i.e., for low field intensities, the average moment in the direction of the field is proportional to the electric field. This can be proved by the following considerations:

Substituting the identities for the exponential function in equation (2.44) we have

$$e^x = 1 + x + \frac{x^2}{2!} + \frac{x^3}{3!} + \dots$$

$$e^{-x} = 1 - x + \frac{x^2}{2!} - \frac{x^3}{3!} + \dots \tag{2.46}$$

$$L(x) = \coth x - \frac{1}{x} = \frac{1}{x}(\frac{x^2}{3} - \frac{x^4}{45} + \dots) \tag{2.47}$$

For small values of x higher powers of x may be neglected and the Langevin function may be approximated to

$$L(x) = \frac{1}{3}x \cong \frac{\mu E}{3kT} \tag{2.48}$$

For large values of x however, i.e., for high electric fields or low temperatures, $L(x)$ has a maximum value of 1, though such high electric fields or low temperatures are not practicable as the following example shows.

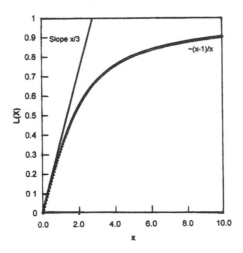

Fig. 2.6 Langevin function with x defined according to eq. (2.41). For small values of x, $L(x)$ is approximately equal to $x/3$.

Let us consider the polarizability of HCl in an electric field of 150 kV/m. The dipole moment of HCl molecule is $1D = 3.3 \times 10^{-30}$ C m so that, at room temperature

$$x = \frac{\mu E}{kT} = \frac{3.3 \times 10^{-30} \times 150 \times 10^3}{1.38 \times 10^{-23} \times 293} = 1.25 \times 10^{-4}$$

Note that x is a dimensionless quantity. The Langevin function for this small value of x, because of approximation (2.48) is

$$L(x) = \frac{x}{3} = 4 \times 10^{-5}$$

The physical significance of this parameter is that, on the average we can say that 0.004% of molecules are oriented in the direction of the applied field. At higher electric fields or lower temperatures $L(x)$ will be larger. Increase of electric field, of course, is equivalent to applying higher torque to the dipoles. Decrease of temperature reduces the agitation velocity of molecules and therefore rotating them on their axis is easier. An example is that it is easier to make soldiers who are standing in attention obey a command than people in a shopping mall. Table 2.6 gives $L(x)$ for select values of x.

The field strength required to increase the $L(x)$ to, say 0.2, may be calculated with the help of equation (2.48). Substituting the appropriate values we obtain $E = 7 \times 10^8$ V/m, which is very high indeed. Clearly such high fields cannot be applied to the material without causing electrical breakdown. Hence for all practical purposes equation (2.48) should suffice.

Table 2.6

Langevin Function for select values of x

x	coth x	$L(x)$
10^{-4}	10^4	0
0.01	100.003	0.0033
0.1	10.033	0.0333
0.15	6.716	0.0499
0.18	5.615	0.0598
0.21	4.83	0.0698

Since $L(x)$ was defined as the ratio of μ_0/μ in equation (2.45) we can express equation (2.48) as

$$\mu_0 = \frac{\mu^2 E}{3kT} \tag{2.49}$$

Therefore the polarizability due to orientational polarization is

$$\alpha_0 = \frac{\mu_0}{E} = \frac{\mu^2}{3kT} \tag{2.50}$$

The meanings associated with μ_0 and μ should not create confusion. The former is the average contribution of a molecule to the polarization of the dielectric; the latter is its inherent dipole moment due to the molecular structure. A real life analogy is that μ represents the entire wealth of a rich person whereas μ_0 represents the donation the person makes to a particular charity. The latter can never exceed the former; in fact the ratio $\mu_0 / \mu \ll 1$ in practice as already explained.

The dipole moment of many molecules lies, in the range of 0.1-3 Debye units and substituting this value in equation (2.48) gives a value $\alpha_0 \approx 10^{-40}$ F m^2, which is the same order of magnitude as the electronic polarizability. The significance of equation (2.49) is that although the permanent dipole moment of a polar molecule is some 10^6 times larger than the induced dipole moment due to electronic polarization of non-polar molecules, the contribution of the permanent dipole moment to the polarization of the material is of the same order of magnitude, though always higher (i.e., $\alpha_o > \alpha_e$). This is due to the fact that the measurement of dielectric constant involves weak fields.

The polarization of the dielectric is given by

$$\mathbf{P} = N\alpha_0 \mathbf{E} = \frac{N\mu^2 \mathbf{E}}{3kT} \tag{2.51}$$

If the dielectric is a mixture of several components then the dielectric constant is given by

$$\mathbf{P} = \frac{\mathbf{E}}{3kT} \sum_{j=1}^{j=n} N_j \mu_j^2 \tag{2.52}$$

where N_j and μ_j are the number and dipole moment of constituent materials respectively. In the above equation **E** should be replaced by E_i, equation (2.22) if we wish to include the influence of the neighboring molecules.

2.6 EXPERIMENTAL VERIFICATION OF DEBYE EQUATION

In deriving the **Debye equation** (2.51) the dipole moment due to the electronic polarizability was not taken into consideration. The electric field induces polarization P_e and this should be added to the polarization due to orientation. The total polarization of a polar dielectric is therefore

$$P = NE(\alpha_e + \frac{\mu^2}{3kT})$$ (2.53)

Equation (2.25) now becomes

$$R = \frac{\varepsilon - 1}{\varepsilon + 2} \frac{M}{\rho} = \frac{N_A}{3\varepsilon_0}\left(\alpha_e + \frac{\mu^2}{3kT}\right)$$ (2.54)

For mixtures of polar materials the Debye equation becomes

$$\frac{\varepsilon - 1}{\varepsilon + 2} = \frac{1}{3\varepsilon_0} \sum_{j=1}^{j=k} N_j(\alpha_{ej} + \frac{\mu_j^2}{3kT})$$ (2.55)

According to equation (2.54) a plot of R as a function of 1/T yields a straight line with an intercept

$$\frac{N_A\alpha_e}{3\varepsilon_0} \quad and \ a \ slope \ \frac{N_A\mu^2}{9k\varepsilon_0}$$ (2.56)

Fig. 2.7 shows results of measurements of C-M factor as a function of 1/T for silicone fluids[7]. Silicone fluid, also known as poly(dimethyl siloxane) has the formula $(CH_3)_3$ Si $- [OSi(CH_3)_2]_x$ $OSi(CH_3)_3$. It has a molecular weight $(162.2 + 72.2 \ x)$ g /mole, $x = 1$ to

1000 and an average density of 960 kg/m^3. The advantage of using silicone fluid is that liquids of various viscosities with the same molecular weight can be used to examine the influence of temperature on the dielectric constant. The polarizabilities calculated from the intercept is 1.6×10^{-37}, 2.9×10^{-37}, and 3.7×10^{-37} Fm^2 for 200, 500 and 1000 cSt viscosity. These are the sum of the electronic and atomic polarizabilities. The calculated α_e using data from Table 5 for a monomer ($x = 1$, No. of atoms: C – 8, H – 24, Si – 3, O – 2) is 3.4×10^{-39} F m^2. For an average value of $x = 100$ (for transformer grade x = 40) electronic polarizability alone has a value of the same order mentioned above and the liquid is therefore slightly polar.

The slopes give a Dipole moment of 5.14, 8.3 and 9.4 D respectively. Sutton and Mark[8] give a dipole moment of 8.47 D for 300 cSt fluid which is in reasonable agreement with the value for 200 cSt. Application of Onsagers theory (see section 2.8) to these liquids gives a value for the dipole moment of 8.62, 12.1 and 13.1 D respectively. To obtain agreement of the dipole moment obtained from fig. 2.7 and theory, a correlation factor of g = 2.8, 2.1 and 2.1 are employed, respectively.

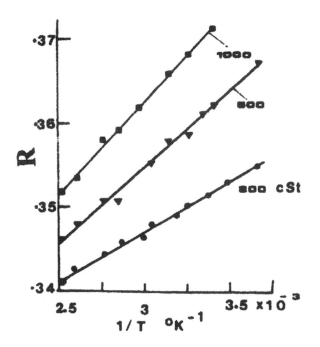

Fig. 2.7 Molar polarizability in silicone fluids versus temperature for various viscosities (Raju, ©1988, IEEE.).

Fig. 2.8 shows the calculated R-T variation for some organic liquids using the data shown in Table 2.7. From plots similar to fig. 2.8 we can separate the electronic polarizability and the permanent dipole moment. For non-polar molecules the slope is zero because, $\mu = 0$.

The Debye equation is valid for high density gases and weak solutions of polar substances in non-polar solvents. However the dipole moments calculated from ε for the gas (vapor) phase differ appreciably from the dipole moments calculated using measurements of ε of the liquid phase. As an example the data for $CHCl_3$ (Chloroform) are given in Table 2.8. The dielectric constant of the vapor at 1 atmospheric pressure at 100°C is 1.0042. In the gaseous phase it has a dipole moment of 1.0 D. In the liquid phase, at 20°C, $N = 7.47 \times 10^{27}$ m^{-3} and equation (2.54) gives a value of 1.6 D.

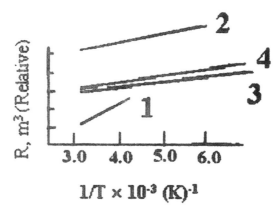

Fig. 2.8 Molar polarizability versus temperature for several polar molecules. The dipole moment may be calculated from the slope. 1-water (H_2O), 2-Methyl alcohol (CH_4O), 3-Ethyl alcohol (C_2H_6OH). 4-Proponal (C_3H_8O).

Table 2.7
Dielectric properties of selected liquids

Name	Formula	M ($\times 10^{-3}$ kg/mole)	ρ Kg/m^3	ε	μ (Gas Phase) D	n	ε_∞
Water	H_2O	18	1000	80.4	1.85	1.33	3.48
Nitrobenzene	$C_6H_5NO_2$	123.1	1210	35.7	4.1	1.55	
Methanol	CH_4O	32.04	791.4	33.62	1.70	1.33	1.33
Chlorobenzene	C_6H_5Cl	112.56	1105.8	5.62	1.7	1.52	
Chloroform	$CHCl_3$	119.38	1483.2	4.81	1.0	1.44	
Ethyl alcohol	C_2H_6OH	46.07	789.3	24.35	1.69	1.36	1.36
1-propanol	C_3H_8O	60.11	803.5	20.44	1.68	1.45	1.45

1 D = 3.3 x 10^{-30} Cm

2.7 SPONTANEOUS POLARIZATION

Neglecting electronic polarization we can write equation (2.54) as

$$\frac{\varepsilon - 1}{\varepsilon + 2} = \frac{N\mu^2}{9\varepsilon_0 kT} \qquad (2.57)$$

For a given material at a given temperature, the right side of equation (2.57) is a constant and we can make a substitution

$$\frac{N\mu^2}{9\varepsilon_0 k} = T_c \qquad (2.58)$$

The constant T_c is called the **critical temperature** and has the dimension of K. Equation (2.57) then simplifies to

$$\frac{\varepsilon - 1}{\varepsilon + 2} = \frac{T_c}{T} \qquad (2.59)$$

For values of $T \le T_c$ equation (2.59) can be satisfied only if ε becomes infinitely large. The physical significance is that the material possesses a very large value of ε at temperatures lower than a certain critical temperature. The material is said to be spontaneously polarized at or below T_c, and above T_c it behaves as a normal dielectric. A similar behavior obtains in the theory of magnetism. Below a certain temperature, called the **Curie Temperature,** the material becomes spontaneously magnetized and such a material is called ferromagnetic. In analogy, dielectrics that exhibit spontaneous polarization are called ferroelectrics.

The slope of the ε -T plots (dε /dT) changes sign at the Curie temperature as data for several ferroelectrics clearly show. Certain polymers such as poly(vinyl chloride) also exhibit a change of slope in dε /dT as T is increased[9] and the interpretation of this data in terms of the Curie temperature is deferred till ch. 5.

Table 2.8

Dielectric constant of Chloroform and Water[2]

T (° C)	ε (CHCl$_3$)	T (° C)	ε (H$_2$O)	ρ for H$_2$O (kg/m^3)
180	2.92	75	-	974.89
140	3.32	60	66.8	983.24
100	3.72	50	69.9	988.07
20	4.81	40	73.3	992.24
-20	5.61	30	76.6	995.67
-40	6.12	20	80.3	998.23
-60	6.76	10	84.0	999.73
		0 (liquid)	88.1	998.97

We may obtain an idea of magnitude for the critical temperature for polar dielectrics with the help of equation (2.58). Water is a polar dielectric and a molecule has a permanent dipole moment of 6.23 x 10^{-30} C m (1.87 D). The number of molecules is given by equation (2.24) as

$$N = \frac{6.03 \times 10^{23} \times 10^3}{18 \times 10^{-3}} = 3.35 \times 10^{28} m^{-3}$$

Substituting these values in equation (2.58) we get T$_c$ $\approx$ 930° C. Apparently, water at room temperature should be ferroelectric according to Debye theory, but this is not so. Hence an improvement to the theory has been suggested by Onsager which is dealt with in the next section.

Ferroelectric materials with high permittivity are used extensively in microwave devices, high speed microelectronics, radar and communication systems. Ceramic barium titanate (BaTiO$_3$) exhibits high permittivity, values ranging from 2000-10,000, depending upon the method of preparation, grain size, etc[10]. Bismuth titanate ceramics (Bi$_2$Ti$_2$O$_7$) exhibit even larger permittivity near the Curie temperature. Fig. 2.9[11] shows the measured permittivity at various temperatures and frequencies. If we ignore the frequency dependence for the time being, as this will be discussed in chapters 3 and 4, the permittivity remains relatively constant up to ~200°C and increases fast beyond 300-400°C. Permittivity as high as 50,000 is observed at the Curie temperature of ~700°C. Table 2.9 lists some well known ferro-electric crystals.

2.8 ONSAGER'S THEORY

As mentioned before, the Debye theory is satisfactory for gases, vapors of polar liquids and dilute solutions of polar substances in non-polar solvents. For pure polar liquids the value of μ calculated from equation (2.54) does not agree with the dipole moments calculated from measurements on the vapor phase where the Debye equation is known to apply, as Table 2.7 shows. Further, the prediction of the Curie point below where the liquid is supposed to be spontaneously polarized, does not hold true except in some special cases.

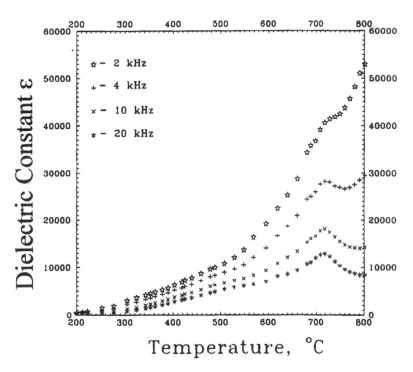

Fig. 2.9 Dielectric constant ε of Bismuth titanate ceramic which is ferro-electric as a function of temperature. The dependence of ε on frequency is discussed in chapter 3. (Yardonov et. al., 1998: With permission of J. Phys. D.: Appl. Phys.).

Onsager[12] attributed these difficulties to the inaccuracy caused by neglecting the reaction field which was introduced in chapter 1. The field which acts upon a molecule in a polarized dielectric may be decomposed into a cavity field and a reaction field which is proportional to the total electric moment and depends on the instantaneous orientation of the molecule. The mean orientation of a molecule is determined by the orienting force couple exerted by the cavity field upon the electric moment of the molecule. The approach of Debye, though a major step in the development of dielectric theory, is equivalent to the assumption that the effective orienting field equals the average cavity field plus the reaction field. This is inaccurate because the reaction field does not exert a torque on the molecule. The Onsager field is therefore lower than that considered by Debye by an amount equal to the reaction field.

Onsager considered a spherical cavity of the dimension of a molecule with a permanent dipole moment μ at its center. It is assumed that the molecule occupies a sphere of radius r, i.e., $\frac{3}{4} \pi r^3 N = 1$ and its polarizability is isotropic. The field acting on a molecule is made up of three components:

(1) The externally applied field $\mathbf{E}$ along Z direction.

(2) A field due to the polarization of the dielectric.

(3) A reaction field R due to the dipole moment μ of the molecule itself.

The three components combined together give rise to the **Lorentz field** E_i which was shown to be

$$\mathbf{E}_i = \mathbf{E}(\frac{2+\varepsilon}{3})$$
(2.22)

The reaction field of a dipole in the center of a dielectric sphere is given by (1.70)

$$\mathbf{R} = \frac{2(\varepsilon_1 - \varepsilon_2)}{4\pi\varepsilon_0\varepsilon_2 r^3 (2\varepsilon_1 + \varepsilon_2)}\mu$$
(1.70)

in which ε_1 and ε_2 are the dielectric constants of the medium outside and inside the sphere. For a cavity, $\varepsilon_2 = 1$ and omitting the suffix for the other dielectric constant we obtain

$$\mathbf{R} = \frac{1}{4\pi\varepsilon_0} \frac{2(\varepsilon-1)}{(2\varepsilon+1)} \frac{\mu}{r^3}$$
(1.68)

All the quantities on the right side of this equation are constants and we can make the substitution

Table 2.9
Selected Ferro-electric Crystals

NAME	FORMULA	T_c (K)	$P \times 10^{-4}$ ($\mu C/m^2$) at	T (K)
Potassium dihydrogen phosphate	KH_2PO_4	123	4.75	96
Potassium dideuterium phosphate	KD_2PO_4	213	4.83	180
Rubidium dihydrogen phosphate	RbH_2PO_4	147	5.6	90
Rubidium dideuterium phosphate	RbD_2DO_4	218	-	-
Barium titanate	$BaTiO_3$	393	26.0	300
Lead titanate	$PbTiO_2$	763	>50	300
Cadmium titanate	$CdTiO_2$	55	-	-
Pottassium niobate	$KnbO_3$	708	30.0	523
Rochelle salt	$NaKC_4H_4O_6. \ 4D_2O$	297[A] 255	0.25	278
Deuterated Rochelle salt	$NaKC_4H_2D_2O_6. \ 4D_2O$	308[A] 251	0.35	279

[A] There are upper and lower T_c

Source: F. Jona and G. Shirane, Ferroelectric Crystals, Pergamon Press, NY, 1962, p. 389 (with permission).

$$\mathbf{R} = f\mu \qquad (2.60)$$

in which

$$f = \frac{1}{4\pi\varepsilon_0 r^3} \frac{2(\varepsilon - 1)}{(2\varepsilon + 1)} \qquad (2.61)$$

Equations (1.69) and (1.70) were derived assuming that the dipole was rigid, *i.e.,* its dipole moment was constant. This is only an approximation because the reaction field increases the dipole moment, the increase being αR. This in turn will increase the reaction field to R_m and equation (2.60) will be modified as

$$\mathbf{R}_m = f(\mu + \alpha \mathbf{R}_m) \qquad (2.62)$$

in which α is the polarizability of the molecule. Therefore

$$\mathbf{R_m} = \frac{f}{1 - f\alpha}\mu \tag{2.63}$$

Substituting the value of f from equation (2.61) we can rewrite this equation as

$$\mathbf{R}_m = \frac{2(\varepsilon - 1)}{4\pi\varepsilon_0 r^3(2\varepsilon + 1) - 2\alpha(\varepsilon - 1)}\mu \tag{2.64}$$

We can now use Equation (2.26) to substitute for α,

$$\alpha = \frac{3\varepsilon_0}{N}\frac{n^2 - 1}{n^2 + 2} \tag{2.26}$$

We further note that

$$\frac{4}{3}\pi N r^3 = 1 \tag{2.65}$$

Substituting equations (2.26) and (2.65) in (2.64) we get

$$\mathbf{R}_m = \frac{2N(\varepsilon - 1)(n^2 + 2)}{9\varepsilon_0(n^2 + 2\varepsilon)}\mu \tag{2.66}$$

Similarly we can deduce the modified dipole moment from equation

$$\mu_m = \mu + \alpha\,\mathbf{R}_m \tag{2.67}$$

Combining equations (2.67) and (2.63) we get

$$\mu_m = \frac{\mu}{1 - f\alpha} \tag{2.68}$$

Substituting for f from equation (2.61) and for r from equation (2.65) we get

$$\mu_m = \frac{2\varepsilon + 1}{3(2\varepsilon + n^2)}(n^2 + 2)\mu \qquad (2.69)$$

Kirkwood[13] gives an alternate expression for the modified dipole moment μ_m on the assumption that the total moment of the molecule consists of a point dipole at the center of a sphere of radius a of dielectric constant unity, as opposed to n^2 implied in equation (2.69),

$$\frac{\mu_m}{\mu} = \frac{1}{[1 - \dfrac{2(\varepsilon - 1)}{(2\varepsilon + 1)}\dfrac{\alpha N}{3\varepsilon_0}]}$$

Substituting equation (2.26) in this equation yields

$$\frac{\mu_m}{\mu} = \frac{1}{1 - \left[\dfrac{2(\varepsilon - 1)}{(2\varepsilon + 1)}\right]\left[\dfrac{n^2 - 1}{n^2 + 2}\right]} \qquad (2.70)$$

At the beginning of this section it was mentioned that a part of the internal field is due to the reaction field R. When the dipole is directed by an external field, the average value of the reaction field in the direction of E is R_m <cos θ> where the symbols enclosing cos θ signifies average value, assuming that the reaction field follows the direction of μ instantaneously. Since R_m has the same direction as μ at any instant, R_m cos θ does not contribute to the directing torque. We have to, therefore, apply a correction to the internal field as

$$E = E_i - R_m < \cos\theta > \qquad (2.71)$$

The average value of cos θ is given by the expression (2.39) as

$$< \cos\theta > = \frac{\mu_0}{\mu} = \frac{\int_0^\pi n(\theta)\cos\theta\, d\theta}{\int_0^\pi n(\theta)\, d\theta} \qquad (2.39)$$

$$= \frac{\mu E}{3kT} \qquad (2.72)$$

Substituting equation (2.71) in (2.72) we get

$$< \cos\theta >= \frac{\mu}{3kT}(E_i - R_m < \cos\theta >) \qquad (2.73)$$

Rearrangement gives

$$< \cos\theta >= \frac{\mu}{\mu R_m + 3kT} E_i \qquad (2.74)$$

Equation (2.54) now becomes

$$\frac{\varepsilon - 1}{\varepsilon + 2} = \frac{N}{3\varepsilon_o} \left(\alpha + \frac{\mu^2}{3kT + \mu R_m} \right) \qquad (2.75)$$

Substituting for R_m from equation (2.66) we get

$$\frac{(\varepsilon - 1)}{(\varepsilon + 2)} = \frac{N}{3\varepsilon_0} \left(\alpha_e + \frac{\mu^2}{3kT + \dfrac{2N\mu^2}{9\varepsilon_0} \dfrac{(\varepsilon - 1)(n^2 + 2)}{(n^2 + 2\varepsilon)}} \right) \qquad (2.76)$$

Because of equation (2.26) the above equation may be rewritten as

$$\frac{(\varepsilon - 1)}{(\varepsilon + 2)} - \frac{(n^2 - 1)}{(n^2 + 2)} = \frac{N}{3\varepsilon_0} \left(\frac{\mu^2}{3kT + \dfrac{2N\mu^2}{9\varepsilon_0} \dfrac{(\varepsilon - 1)(n^2 + 2)}{(n^2 + 2\varepsilon)}} \right) \qquad (2.77)$$

After lengthy algebra to obtain the relation between ε and μ Onsager derives the expression

$$\frac{N\mu^2}{9\varepsilon_o kT} = \frac{(\varepsilon - n^2)(2\varepsilon + n^2)}{(n^2 + 2)^2} \qquad (2.78)$$

leading to

$$\frac{(\varepsilon-1)}{(\varepsilon+2)} - \frac{(n^2-1)}{(n^2+2)} = \frac{N\mu^2}{3\varepsilon_0 kT}\frac{\varepsilon(n^2+2)}{(2\varepsilon+n^2)(\varepsilon+2)} \qquad (2.79)$$

We note here that the left side of equation (2.79) is not zero because, the relationship, $\varepsilon = n^2$ does not hold true for polar substances at steady fields.

We may approximate equation (2.79) with regard to specific conditions as follows:

(1) For non-polar materials $\mu = 0$. We then obtain

$$\frac{(\varepsilon-1)}{(\varepsilon+2)} = \frac{(n^2-1)}{(n^2+2)} \qquad (2.27)$$

which is the Lorenz-Lorentz equation.

(2) For polar gases at low pressures the following approximations apply:
    $\varepsilon \approx 1$, $(\varepsilon - 1) \ll 1$, $(n^2 -1) \ll 1$, $(n^2 +1) \approx 3$ and

equation (2.79) simplifies to

$$(\varepsilon-1) = \frac{N\mu^2}{3\varepsilon_0 kT} \qquad (2.57)$$

which is the Debye equation for polar gases.

If we view the Onsager's equation (2.79) as a correction to the Debye equation then it is of interest to calculate the magnitude of the modified reaction field R_m and the modified dipole moment μ_m from equations (2.66) and (2.69). Table 2.8 gives the appropriate data and the calculated values. R_m has a magnitude of the order of 10^9 V/ m. To obtain the significance of this field we compare the electric field that exists due to the dipole along its axis and along the perpendicular to the axis. Eq. (1.26) gives the potential due to a dipole at a point with co-ordinates (ϕ, θ) as

$$\phi(r,\theta) = \frac{1}{4\pi\varepsilon_o}\frac{\mu\cos\theta}{r^2} \qquad (1.26)$$

The field due to the dipole has two components given by

$$E_r = -\frac{d\phi}{dr}a_r \tag{2.80}$$

$$E_\theta = -\frac{1}{r}\frac{d\phi}{d\theta}a_\theta \tag{2.81}$$

The field on the axis of a dipole is therefore

$$E_{(r,0)} = \frac{\mu}{2\pi\varepsilon_o r^3} \tag{2.82}$$

The field perpendicular to the axis is

$$E_{(r,0)} = \frac{\mu}{4\pi\varepsilon_o r^3} \tag{2.83}$$

For a dipole of moment 1D the two fields at a distance of $r = 50$ nm are $E_{(r,0)} = 500$ MV/m and $E_{(r,\pi/2)} = 250$ MV/m. Comparing these values with R_m we find that the latter is 10 - 20 times larger, making a significant difference to the calculation of the dipole moment. For the effective increase in the dipole moment, the ratio μ_m/μ is between 25-50%.

The Onsager's equation (2.78) may be used to determine whether a polar monomer contributes all of its dipole moment to the polarization of the polymer. This will be true if the dipole moment of the polymer is higher than that of the monomer. The monomer of poly(vinyl acetate) (PVAc) is polar, the measured dielectric constant at 55°C is 9.5[14], $\varepsilon_\infty = n^2 = 2$, mol. wt. = 86 g/mole. In the case of a polymer equation, (2.78) may be rewritten as:

$$\mu^2 = \frac{9\varepsilon_0 kT(\varepsilon_s - \varepsilon_\infty)(2\varepsilon_s + \varepsilon_\infty)M_w}{N_A\rho\varepsilon_s(\varepsilon_\infty + 2)^2} \tag{2.84}$$

in which N_A is the avagadro number, M_w is the molecular wt. of the monomer, and ρ the density of the polymer. Collecting the values, $\varepsilon_s = 9.5$, $\varepsilon_\infty = 2$, $T = 328$ K, $k = 1.38 \times 10^{-23}$ J/K, $\rho = 1100$ Kg/m^3, $N_A = 6.03 \times 10^{23}$, eq. (2.77) gives $\mu = 2.1$ D. Since the monomer has a dipole moment of 1.79 D the monomer contributes 100% to the polarization. Another example of the application of Onsager's equation is given in Ch. 5.

2.9 THEORY OF KIRKWOOD

Kirkwood[15] has developed a theory for the polarization of a non-polar dielectric in a homogenous electric field from a statistical point of view. We have already discussed the relation of the dielectric constant to the structural properties of the constituent molecules; this relation depends upon the internal field which is different from the externally applied field.

The internal field was calculated assuming that the contribution to the field of the molecules within the cavity was zero, i.e., $E_4 = 0$. Lorentz showed that this was true for cubic crystals and remarked that E_4 also becomes zero in a fluid for a random distribution of other molecules around the central one. These conclusions are equivalent to assuming that the electric moment of the molecule retains its average value, that is that the dipole remains rigid, through the various phases of thermal fluctuation.

Kirkwood suggests that the fluctuations in the induced dipole moments of a molecule require a correction to the Lorentz field, equation (2.22) and the formulas derived from it change appropriately. The approach is statistical and the dielectric is treated from the outset as a system of molecules rather than a continuum, thus eliminating the 'artificial' approach of a cavity.

In this approach the effects arising from inhomogeneity of the local field in the region occupied by the molecule is neglected. Further the dipole moment of a molecule is the product of its polarizability (α_T) in vacuum and the average field acting upon it. This field is partly due to the charges on the electrodes and partly due to other molecules surrounding it.

Kirkwood's theory considers an assembly of N spherically symmetrical particles placed in an electric field. At any instant a set of N vector equations gives the dipole moments of all the particles. These N vectorial equations are expressed as 3N scalar equations provided the co-ordinates of the points are known at that instant. In this model the polarizability is assumed to be at the center of the spherical molecule.

The electric field at a radial distance r due to a point dipole of moment μ, according to equation (1.34) is

$$\mathbf{E} = \mathbf{T}\mu \qquad\qquad\qquad\qquad (1.14)$$

This field adds to the applied field and the dipole moment of the ith molecule in an assembly changes to

$$\mu_i = \alpha E + \alpha \sum_{\substack{j=1 \\ j \neq i}}^{N} T_{ij}\mu_j \tag{2.85}$$

Since α has a magnitude comparable to the molecular volume and T_{ij} varies as $1/r^3$ (equation 1.14) we can expand the summation term as a power series with the expectation that it converges fast.

The internal field is obtained as a series:

$$(E_i)_{kirkwood} = [1 + (1 + \frac{\alpha}{4a^3})H + (\frac{1}{16} + \frac{3\alpha}{2a^3})H^2 + \dots]E \tag{2.86}$$

A comparison of equation (2.86) with equation (2.23) shows the Kirkwood theory is a correction that is applied to the internal field. A simplified version assumes that a geometrical correlation factor g may be used as a correction to the Onsager's equation.

If we consider only the first term on the right side of equation (2.86) then we get the expression

$$\frac{\varepsilon - 1}{\varepsilon + 2} = \frac{N\alpha}{3\varepsilon_0} \tag{2.23}$$

which is the Clausius-Mossotti equation[16]. The first correction to the Clausius-Mosotti equation is suggested by Brown[17] on the concept that the C-M parameter may be expressed as a polynomial of ρ rather than a linear function as implied by equation (2.25),

$$y = \frac{A_0}{\rho} + A_1 + A_2\rho + \dots. \tag{2.87}$$

where

$$y = \frac{\varepsilon + 2}{\varepsilon - 1}$$

$$A_0 = \frac{3\varepsilon_0}{\alpha_e} \frac{M}{N_A} \tag{2.88}$$

and, A_1 and A_2 are constants that depend on the parameter $\alpha/\varepsilon_0 a^3$ where a is the molecular diameter.

An alternate form for ε is given by Bottcher as:

$$\varepsilon = 1 + 3\left(\frac{N\alpha}{3\varepsilon_0}\right) + 3\left(1 + \frac{\alpha}{4\pi a^3}\right)\left(\frac{N\alpha}{3\varepsilon_0}\right)^2 + 3\left(\frac{1}{16} + \frac{3\alpha}{2a^3}\right)\left(\frac{N\alpha}{3\varepsilon_0}\right)^3 + ... \tag{2.89}$$

Returning to the Brown version of Kirkwood theory, equation (2.87), we first note that the left side is expressed as the inverse of the same part in the Clausius-Mossotti equation. The ratio

$$\frac{(\varepsilon - 1)}{(\varepsilon + 2)\rho}$$

should be constant for a given substance according to equation (2.25). If the measured and calculated dielectric constants differ to an unacceptable extent the Brown expression given by (2.87) should be tested.

To test this equation Brown[18] assumed that molecules were spherically symmetrical in their mechanical and electrical properties and interact electrostatically as point dipoles. These assumptions are reasonably appropriate even if the molecules have a shape that departs from spherical symmetry appreciably, if the polarizability is considered as averages over all orientations. For molecules which are close together the neglect of quadruple and higher moments introduces additional errors.

Brown further noted that for highly asymmetric molecules averaging the properties of two molecules individually before they are allowed to interact, rather than of the properties of the interacting pair, is not quite accurate. The same reasoning applies to interacting clusters of higher magnitude. Moreover, one has to take into account the fact that electrostatic interactions are far from being of a dipole-dipole nature, that it is not even permissible to regard the molecule as equivalent to a dipole and a quadruple. For long chain molecules, as in polymers, these considerations are particularly important.

Notwithstanding these difficulties which are common to several theories, Brown presented a numerical analysis of equation (2.87) and compared it with more specialized formulas. He concluded that Bottcher's formula, equation (2.89), has no special advantage over the general formula, equation (2.87). To detect deviations of the behavior of molecules from the behavior predicted for spherical molecules we need to have a detailed knowledge of the radial distribution function of molecules[19] and more precise data on dielectric constants.

A summary of the relevant formulas are given below.

(1) Clausius-Mosotti equation

$$\frac{\varepsilon - 1}{\varepsilon + 2} = \frac{N\alpha_e}{3\varepsilon_0} \tag{2.23}$$

(2) The empirical formula of Eykman[20]

$$\frac{\varepsilon - 1}{\varepsilon^{1/2} + 0.4} = \frac{1}{1.4} \frac{N\alpha_e}{3\varepsilon_0} \tag{2.90}$$

(3) Bottcher Formula:

$$\varepsilon = 1 + 3(\frac{N\alpha}{3\varepsilon_0}) + 3(1 + \frac{\alpha}{4\pi a^3})(\frac{N\alpha}{3\varepsilon_0})^2 + 3(\frac{1}{16} + \frac{3\alpha}{2a^3})(\frac{N\alpha}{3\varepsilon_0})^3 + ... \tag{2.89}$$

(4) Kirkwood's Formula: This version is given by Brown [17]

$$\frac{\varepsilon + 2}{\varepsilon - 1} = \frac{3\varepsilon_0}{N\alpha} - \frac{\alpha}{2\pi\varepsilon_0 a^3} + \frac{15N\alpha}{48\varepsilon_0} \tag{2.91}$$

(5) Brown formula :

$$y = \frac{A_0}{\rho} + A_1 + A_2\rho \tag{2.87}$$

Brown applied his equation to carbon disulphide (CS_2) and CO_2 at various densities and made a comparison with other formulas[17, 18]. Here we show how to apply these formulas for one value of the gas density.

The data required for the application of equation (2.87) to CS_2 are:
Mol. Wt. = 76.13, M.P. = - 110.8° C, B. P. = 46.3° C, α_e = 9.622 × 10^{-40} F m^2, ρ = 1247 kg/m^3, ε (measured) = 2.624. The calculated values of ε according to different equations are shown in Table 2.10.

A second example of application of the equations to gases is with regard to ethylene (C_2H_4)[21]. As stated earlier the Clausius-Mosotti parameter

$$\frac{\varepsilon - 1}{(\varepsilon + 2)\rho} = \frac{N_A \alpha_e}{3M\varepsilon_0} \tag{2.92}$$

should be a constant. Also, for gases with non-polar molecules the C-M parameter should be independent of temperature. To verify these conclusions the dielectric constant of ethylene was measured up to 500 atmospheres at temperatures of 25°C and 50°C. The measured dielectric constants are shown in fig. 2-10 along with density data for all gas pressures. The measured dielectric constant at 25°C at a gas pressure of 21.696 atmospheres is 1.0332. The density is 28.78 kg/m^3 and α_e = 4.593 × 10^{-40} Fm^2. Substituting these values the measured C-M parameter is equal to

$$\frac{(1.0332 - 1)}{(1.0332 + 2)28.78} = 3.8 \times 10^{-4} m^3 kg^{-1}$$

The calculated value from the right side of equation (2.92) is 3.71 × 10^{-4} kg/m^3, which is within 2.4% of the measured value. However, the dielectric constant increases to 1.260 at a pressure of 66.9 atmospheres as expected by the theory (fig. 2.10), but the C-M parameter does not remain constant as shown in fig. 2.11.

The suggested reasons are (1) α_e may not remain constant and it may vary with density, (2) the Lorentz molecular model may not be applicable at high densities, requiring a modification of the equation.

Table 2.10
Calculated values of ε in CS_2

Equation	A_0	A_1	A_2	ε	% difference
(2.23)	3550.57	-	-	2.668	1.68
(2.87)	3045.08	479.25	-96.362	2.628	0.152
(2.90)	3201.66			2.581	-1.639
(2.89)	4401.91			2.643	0.724
(2.91)	3782.88			2.637	0.495

W. F. Brown, J. Chem. Phys. 18, 1193, (1950)

Reverting to the formula of Kirkwood, the dielectric constant of a polar substance is expressed as[22]

$$\varepsilon = \varepsilon_\infty + \frac{1}{9\varepsilon_0}(n^2+2)^2\left(\frac{\varepsilon}{2\varepsilon+n^2}\right)\left(\frac{N_A\mu^2}{kTV}\right)g \tag{2.93}$$

where μ is the dipole moment, n the refractive index, V, the molar volume (M/ρ in units of m^3/kg) and g is called the correlation factor. This form of Kirkwood equation is easier to apply.

The correlation factor is a measure of local ordering in the dielectric[23]. It has the value of unity if we imagine a molecule to be held fixed, and the average moment of a spherical region of finite radius within an infinite specimen is also the moment of the molecule. That is to say that when there is local ordering the net dipole moment in the spherical region surrounding the fixed molecule under consideration is zero. This means that fixing the position of a molecule does not influence the position of the surrounding molecules except the influence of the long range electrostatic forces.

On the other hand, if fixing the direction of one dipole tends to align the surrounding dipoles in a parallel direction the correlation factor is greater than one. Likewise, if fixing the direction of one dipole tends to align the surrounding dipoles in an anti-parallel direction the correlation factor is less than one. If the structure of the material is known it is possible to calculate g; otherwise it becomes an empirical parameter which can be calculated by measuring ε and knowing μ.

The calculated values of g will give information about the structure of the material. For water Oster and Kirkwood[24] give a value of 2.65 assuming that $n = 1.33$. However the calculated value of g depends strongly on the assumed value of n. For example, if we assume that $n = 4$ for water (as there is experimental evidence for this in the infrared region) we get $g = 1.09$. This value, close to unity, is interpreted as meaning that the structure of water molecule does not influence the orientational polarization. We also note that substitution of $g = 1$ in equation (2.93) results in an equation identical to that due to Onsager.

2.10 DIELECTRIC CONSTANT OF TWO MEDIA

We shall change our focus somewhat to the formulas that allow us to explain the dielectric constant of more than a single medium, on the basis of concepts that we have established so far. In engineering practice a mixture of several media is a common situation and different methods to calculate the dielectric constant are frequently used by practising engineers. As an example in power transformers celulose materials are extensively used and then impregnated with transformer oil. Polymeric insulation in SF_6 insulated equipment is another example.

There are several different ways in which two different materials may occur in combination: Series, parallel, layered, impregnated, interspersed, etc. Some of these types are shown in fig. 2.12. A useful summary is provided by Morgan[25].

2.10.1 RALEIGH'S FORMULA

One of the earliest analysis of the dielectric constant of two component mixtures is due to Raleigh[26], who assumed that cylindrical particles of dielectric constant ε_p were mixed in a medium of the mixture is given as:

$$\frac{\varepsilon - \varepsilon_m}{\varepsilon + \varepsilon_m} = \frac{V_p(\varepsilon_p - \varepsilon_m)}{\varepsilon_p + \varepsilon_m} \qquad (2.94)$$

where ε is the permittivity of the mixture and V_p is the volume fraction of the particles. The volume fraction of the medium before the mixture is prepared is V_m so that $V_p + V_m = 1$. The above equation may also be rewritten as:

$$\varepsilon = \frac{\varepsilon_m[x(\varepsilon_p - \varepsilon_m) + \varepsilon_p + \varepsilon_m]}{\varepsilon_p + \varepsilon_m - x(\varepsilon_p - \varepsilon_a)} \qquad (2.95)$$

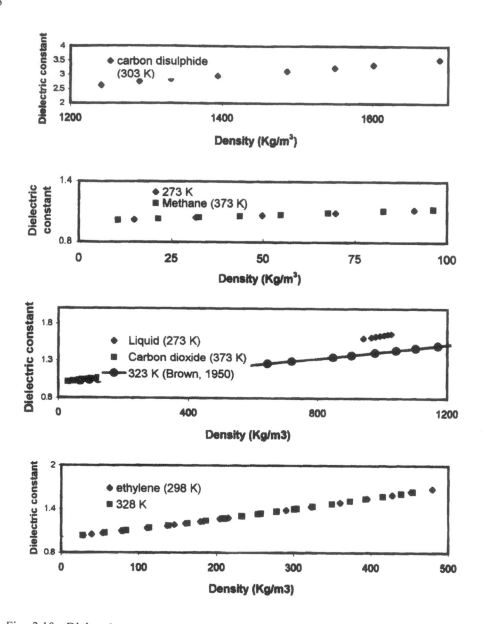

Fig. 2.10 Dielectric constant of non-polar materials versus density. Data are taken from: Methane & CO_2 (Cole, 1965: Gordon and Breach), Ethylene (David et. al. 1951, Gordon and Breach).

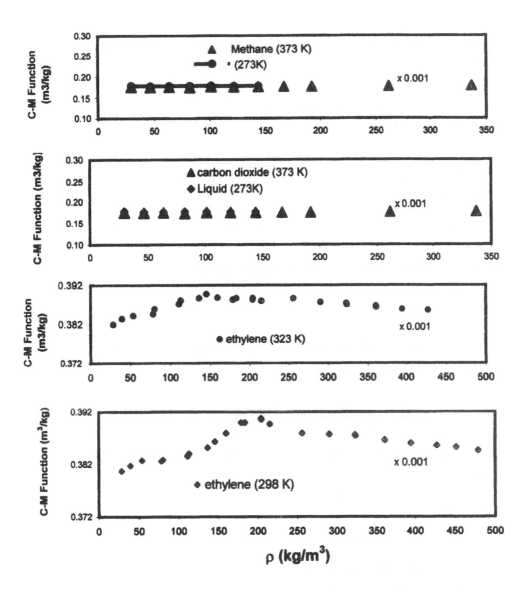

Fig. 2.11 C-M function versus density for non-polar substances. See legend for fig. 2.10 for source of data.

2.10.2 WIENER'S FORMULA

An alternative formulation is due to Wiener[27] who analyzed laminar mixtures and proposed the formula:

$$\frac{1}{\varepsilon + u} = \frac{V_p}{\varepsilon_p + u} + \frac{V_m}{\varepsilon_m + u}; \quad 0 \le u \le \infty \tag{2.96}$$

where u is a factor.

$$\frac{1}{\varepsilon} = \frac{V_p}{\varepsilon_p} + \frac{V_m}{\varepsilon_m} \tag{2.97}$$

For $u = \infty$ the equation for dielectrics in parallel is obtained.

$$\varepsilon = \varepsilon_p V_p + \varepsilon_m V_m \tag{2.98}$$

Bruggeman has suggested that for isotropic materials having randomly arranged grains the best approximation is $u = \sqrt{(\varepsilon_p \, \varepsilon_m)}$

2.10.3 FORMULA OF LICHTENECKER AND ROTHER

For powder and granular materials Lichtenecker and Rother suggested the formula

$$\varepsilon^k = V_p \varepsilon_p^k + V_m \varepsilon_m^k; \quad -1 \le k \le 1 \tag{2.99}$$

where k is a numerical variable.

When k = -1 the formula for dielectrics in series, equation (2.97) is recovered. When k = 1 the formula for parallel dielectrics is obtained, equation (2.98). When k = 0 the exponential terms are expanded in an infinite series and neglecting the higher order terms one obtains:

$$\ln \varepsilon = V_p \ln \varepsilon_p + V_m \ln \varepsilon_m \tag{2.100}$$

With k = 0.5 equation (2.99) reduces to the formula for refractive index upon substituting the Maxwell relation $\varepsilon = n^2$:

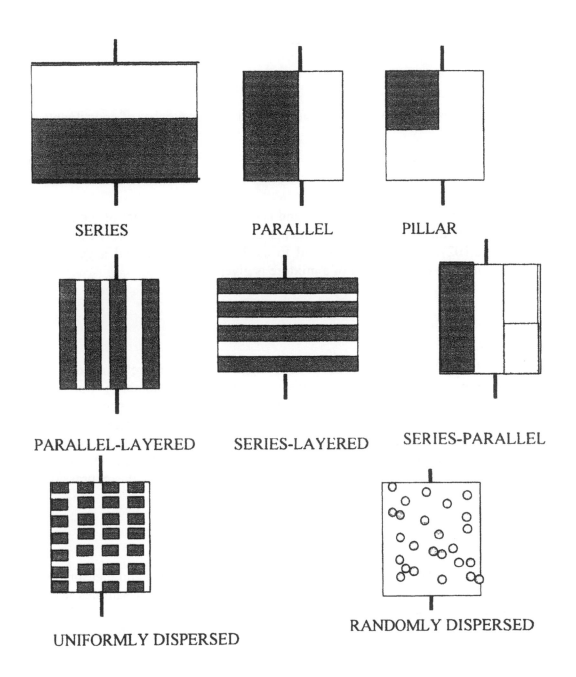

Fig. 2.12 Different arrangements for combination of two media.

For k = 1/3 the so called **Landau, Lifshits and Looyengas'** equation is obtained.

$$n = V_p n_p + V_m n_m \tag{2.101}$$

Lichtenecker's formula (2.99) may be generalized for any number of constituents as

$$\varepsilon^k = \sum_{j=1}^{n} f_j \, \varepsilon_j^k \tag{2.102}$$

where f_j is the relative volume fraction and ε_j the dielectric constant of the jth constituent respectively, n the number of constituents and k is a constant between -1 and $+1$. Equation (2.102) is sometimes referred to as the series model and a geometric mean model for the dielectric constant of a mixture is given by[28]

$$\varepsilon = \prod_{j=1}^{n} \varepsilon_j^{f_j} \tag{2.103}$$

Equations (2.102) and (2.103) have been used quite extensively in soil physics to correlate the measured electrical permittivity of a soil to its volumetric water content. A theoretical basis for the series model is provided by Zakri et. al.[28] who consider that the shapes of the inclusions (particles) is not uniform but follow a beta distribution. This approach is specific to geological mixtures or other mixtures where the particles have a distribution in shapes. In soils there are three constituents, solid, water and air, and using the subscripts s, w and a respectively, the series model gives the expression for the dielectric constant as

$$\varepsilon^k = (1-p)\varepsilon_s^k + (p - f_w)\varepsilon_a^k + f_w \, \varepsilon_w^k \tag{2.104}$$

where p is the porosity of the soil. Fig. 2.13 shows the calculated and measured dielectric constants along with the details of the characteristics of the soil samples according to Lichteneker's formula. To appreciate the differences obtained with other formulas fig. 2.14 presents the values calculated for one sample and it can be seen that Lichtenecker's formula gives the best agreement.

2.10.4 Goldschmidt's Equation

For fibrous materials Goldschmidt[29] has proposed the equation:

$$\varepsilon = \frac{\varepsilon_m [\varepsilon_m + (\varepsilon_p - \varepsilon_m)(V_p + gV_m)]}{\varepsilon_m + gV_m(\varepsilon_p - \varepsilon_m)}; \quad 0 \le g \le 1 \tag{2.105}$$

where g is a constant that assumes different values for each assembly. Since $V_p + V_m = 1$ we can show that this equation reduces to the parallel case, equation (2.98) when $f = 0$.

The equation also reduces to the series equation (2.97) when $f = 1$. For cylindrical particles perpendicular to the electric field Goldschmidt found that $f = 0.5$ was satisfactory.

2.11 THE DISSIPATION FACTOR

Anticipating the discussion of alternating fields presented in chapter 3, the dissipation factor is defined as

$$\tan \delta = \frac{\varepsilon''}{\varepsilon'} \tag{2.106}$$

The dissipation factor of a mixture of two dielectrics having dissipation factors $tan\ \delta_p$ and $tan\ \delta_m$ is given by:

$$\tan \delta = \frac{V_p \varepsilon'_m \tan \delta_p + V_m \varepsilon'_p \tan \delta_m (1 + \tan^2 \delta_p)}{V_p \varepsilon'_m + V_m \varepsilon'_p (1 + \tan^2 \delta_p)} \tag{2.107}$$

where the prime denotes the real part of the dielectric constant. The dissipation factor of dry dielectrics is usually small so that we can make the approximation $tan^2\ \delta << 1$ simplifying this equation into

$$\tan \delta = \frac{V_p \varepsilon'_m \tan \delta_p + V_m \varepsilon'_p \tan \delta_m}{V_p \varepsilon'_m + V_m \varepsilon'_p} \tag{2.108}$$

The dissipation factor measurements are usually carried out to determine the quality of the dielectric and adsorbed moisture is usually the second component. In such a situation the second term in the numerator is usually small when compared with the first term and equation (2.108) may be approximated to:

$$\tan\delta = \frac{V_p \varepsilon'_m \tan\delta_p}{V_p \varepsilon'_m + V_m \varepsilon'_p}$$

(2.109)

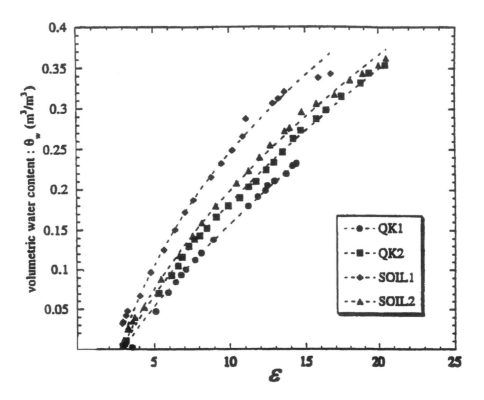

Fig. 2.13 Measured (symbols) and calculated (broken lines) dielectric constant for various soil samples versus water content θ_w at 22°C. Lichtenecker's formula (2.99) is applied for calculation. The characteristics of the samples (porosity ρ and the dielectric constant of the solid fraction are shown below. (Zakri et. al. 1998: With permission of J. Phys. D.: Appl. Phys., UK).

Sample reference	ε	Soil density	Bulk density	porosity
QK1	5.3	2.64	1.72	0.35
QK2	5.95	2.63	1.59	0.40
Soil 1	4.94	2.73	1.46	0.47
Soil 2	4.92	2.79	1.60	0.43

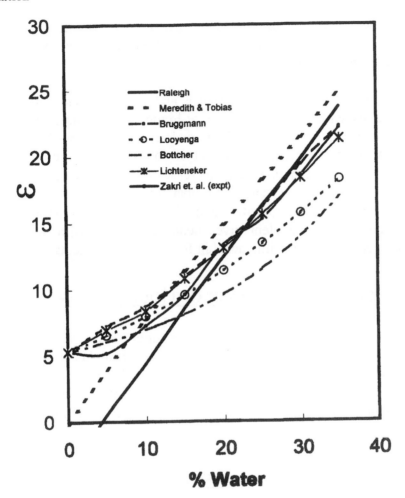

Fig. 2.14 Application of various formulas to soil sample QK1 and comparison with measured values (Zakri et. al. 1998: With permission of J. Phys. D.: Appl. Phys., UK). The calculations have been carried out for ε_{water} = 79.4 and ε_{soil} = 5.3. Agreement with Lichtenecker's formula is much better than with others.

For parallel combination of dielectrics the dissipation factor is given by

$$\tan\delta = \frac{V_p \varepsilon_p' \tan\delta_p + V_m \varepsilon_m' \tan\delta_m}{V_p \varepsilon_p' + V_m \varepsilon_m'} \tag{2.110}$$

Morgan[30] has measured the dissipation factor in layers of dry kraft paper stacked together by compressive force of up to 350 kPa. The dielectric constant of cellulose at 20°C is 6.0-6.5 with a density of 1450 kg/m^3.

Fig. 2.15 shows ε' and ε'' at various frequencies and temperatures. The dielectric constant decreases slightly with frequency except at higher temperatures and above 10 kHz there is a trend towards higher values. This anomalous dispersion has been observed in earlier measurements on Kraft paper. The effect of compressive force is to increase the dielectric constant as expected due to a decrease in thickness of the stack. ε'' increases slowly upto 1 kHz and more rapidly at higher frequencies. This is attributed to the Maxwell-Wagner effect. Calculation of ε' and ε'' by assuming series and parallel arrangement shows that series equivalent gives better agreement with the measured values after increase in density is taken into account for both arrangements.

2.12 DIELECTRIC CONSTANT OF LIQUID MIXTURES

A study of dielectric constant of liquid mixtures is important both from the points of view of dielectric theory and practical applications. Several formulas are available for the calculation of the dielectric constant of binary mixtures of liquids and a collection is given below. We denote the dielectric constants of the individual components by ε_1 and ε_2 and that of the mixture by ε_m (Raju 1988, Van Beck 1967). The volume fractions are denoted by V_1 and V.

2.12.1 RALEIGH'S FORMULA

$$\varepsilon_m = \varepsilon_1 \frac{\dfrac{2\varepsilon_1 + \varepsilon_2}{\varepsilon_2 - \varepsilon_1} + 2V_2 - \dfrac{1.575(\varepsilon_2 - \varepsilon_1)}{4\varepsilon_1 + 3\varepsilon_2} V_2^{3.33}}{\dfrac{2\varepsilon_1 + \varepsilon_2}{\varepsilon_2 - \varepsilon_1} - V_2 - \dfrac{1.575(\varepsilon_2 - \varepsilon_1)}{4\varepsilon_1 + 3\varepsilon_2} V_2^{3.33}}$$

(2.111)

where the subscripts 1, 2, m refer to the two components and the mixture respectively.

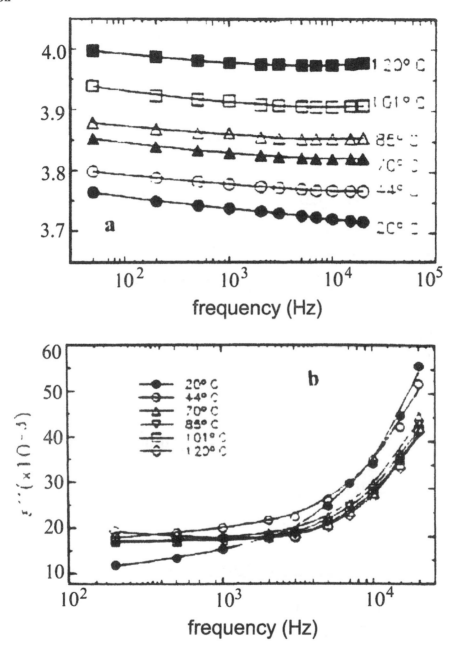

Fig. 2.15 Variation of complex dielectric constant in Kraft paper with frequency and temperature at a compressive stress of 336 kPa. (a) ε′ (b) ε″ (Morgan, ©1998, IEEE).

2.12.2 FORMULA OF MEREDITH AND TOBIAS

$$\varepsilon_m = \varepsilon_1 \frac{A}{B};$$

(2.112)

$$A = \frac{2\varepsilon_1 + \varepsilon_2}{\varepsilon_2 - \varepsilon_1} + 2V_2 - 1.277 \frac{2\varepsilon_1 + \varepsilon_2}{4\varepsilon_1 + 3\varepsilon_2} V_2^{7/3} \frac{1.575(\varepsilon_2 - \varepsilon_1)}{4\varepsilon_1 + 3\varepsilon_2} V_2^{10/3}$$

$$B = \frac{2\varepsilon_1 + \varepsilon_2}{\varepsilon_2 - \varepsilon_1} - V_2 - 1.277 \frac{2\varepsilon_1 + \varepsilon_2}{4\varepsilon_1 + 3\varepsilon_2} V_2^{7/3} \frac{1.575(\varepsilon_2 - \varepsilon_1)}{4\varepsilon_1 + 3\varepsilon_2} V_2^{10/3}$$

2.12.3 BRUGGEMAN'S FORMULA

$$\frac{\varepsilon_2 - \varepsilon_m}{\varepsilon_2 - \varepsilon_1} \left(\frac{\varepsilon_1}{\varepsilon_m} \right)^{1/3} = 1 - V_2$$

(2.113)

where V_2 is the fractional volume of the second component.

2.12.4 LOOYENGA'S FORMULA

$$\varepsilon_m = \left[\varepsilon_1^{1/3} + V_2 \left(\varepsilon_2^{1/3} - \varepsilon_1^{1/3} \right) \right]^3$$

(2.114)

2.12.5 BÖTTCHER'S FORMULA

$$\varepsilon_m = \varepsilon_1 + \frac{3V_2 \varepsilon_m (\varepsilon_2 - \varepsilon_1)}{2\varepsilon_m + \varepsilon_2}$$

(2.115)

The last two formulas are symmetrical. Fig. 2.16 helps to visualize these formulas for a mixture of water and methyl alcohol with $\varepsilon_1 = 78$, $\varepsilon_2 = 32.50$[31]. Looyenga's formula has been proved to be applicable for this mixture[32]. The empirical formulas of Lichtneker, equation (2.99) and Bruggeman, equation (2.113) yield almost identical results while the formulas of Looyenga and Bottcher give values that are higher.

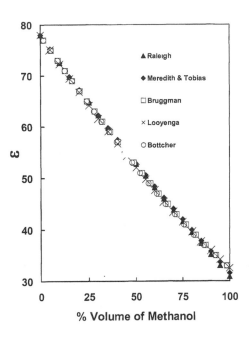

Fig. 2.16 Dielectric constant of mixtures of water and methanol according to various formulas. Looyenga's equation agrees with the measured values of Skai et. al.. (1995: With permission of American Physical Society).

All the formulas listed above are independent of temperature implying that they are applicable at the same temperature. If one wishes to calculate the dielectric constant of a mixture at any other temperature measurements on the individual components have to be carried out at that temperature.

The question has been examined by Govinda Raju (1988) in light of fundamental theories described in the earlier part of this chapter. According to the Debye theory the dielectric constant of a mixture of liquids is given by :

$$\frac{\varepsilon_m - 1}{\varepsilon_m + 2} = \frac{1}{3\varepsilon_0} \sum_{1,2} N\left(\alpha_e + \alpha_a + \frac{\mu^2}{3kT} \right) \tag{2.116}$$

The quantities within the summation sign correspond to each component of the mixture. The Onsager equation for a mixture of liquids is given by

$$(\varepsilon_m - n_m^2)(2\varepsilon + n_m^2) = \sum_{1,2} \frac{N_i(n_i^2 + 2)^2 \mu_i^2}{9\varepsilon_0 kT} \tag{2.117}$$

where n_m is the refractive index of the mixture given by equation (2.114) by substituing ε = n^2. It can also be calculated using equation (2.98) as an approximation. The calculation requires an iterative procedure. Good agreement is obtained between the measured and calculated values of ε.

2.13 EFFECT OF HIGH ELECTRIC FIELDS

Dielectric measurements are usually carried out at low electric field strengths, particularly if frequency is used as a variable. Tan δ measurements using **Schering Bridge** at power frequency are the exceptions. Dielectrics in power equipment are stressed to very high values and at threshold of partial discharges or breakdown, the electric field experienced by insulation can reach several hundred MV/m. Analysis of the change in ε due to high electric field takes into account the higher order terms in the Langevin function and the result shows that the dielectric constant decreases due to high fields. The decrease Δε may be expressed as[33] :

$$\Delta \varepsilon = -\frac{E^2 N_A \mu^4 C}{45 V \varepsilon_0 (kT)^3} \tag{2.118}$$

where C is a constant and V the molar volume. The negative sign indicates that there is a decrease in the dielectric constant. C is given by

$$C = \frac{(\varepsilon + 2)^4}{81} \tag{2.119}$$

An alternative treatment due to Böttcher gives

$$C = \frac{(n^2 + 2)^3 \varepsilon^4}{(2\varepsilon + n^4)(2\varepsilon + n^2)(2n^2 + 1)} \tag{2.120}$$

where n is the refractive index. On the basis of Kirkwood theory a relatively simple equation has been derived:

$$\varepsilon \approx n^2 + \frac{N_A \mu (n^2 + 2)}{3\varepsilon_0 EV} \tag{2.121}$$

To get an idea of the influence of the electric field on the dielectric constant we substitute the values appropriate to water:

$N_A = 6.03 \times 10^{23}$, $n^2 = 1.8$, $\mu = 1.87$ D $= 6.23 \times 10^{-30}$ Cm, V $= 18 \times 10^{-6}$ m^3, ε_0 8.854 $\times 10^{-12}$ F/m, E $= 1 \times 10^9$ V/m gives 31.7 which is less than 50% of the measured static dielectric constant of 81. The decrease of the dielectric constant under high electric field strength has been generally neglected almost entirely in the literature on dielectric breakdown, high field conduction and partial discharges which can hardly be justified.

2.14 ATOMIC POLARIZABILITY

Atomic polarizability arises due to the displacement of the nucleii of atoms forming the molecule, relative to each other, in contrast with the electronic polarizability that arises as a consequence of electronic displacement, in an electric field. Whether a molecule possesses a permanent dipole or not, if it has polar bonds, as in the most cases of dissimilar atoms forming a molecule, the applied field induces a displacement; this displacement is superimposed on the electronic displacement discussed in section 1.3. Though atomic polarization exists in all molecules, it is more pronounced in molecules that have relatively weak bonds.

The charge displacement in an electric field involves changes in bond length and bond angle, in addition to bending or twisting of polar groups with respect to each other. The displacement is restricted by the degree of vibrational freedom of the molecule[34]. Atomic polarizability does not have any simple relationship with electronic polarizability or orientational polarizability. For our purposes, it is sufficiently accurate to determine atomic polarization by the difference between the total polarization and the sum of electronic and orientational polarization. The atomic polarization is usually 1-15% of the electronic polarization, though in certain compounds it can reach a value of about 30% of the electronic polarization.

In the periodic table of elements a noble gas (He, Ne, Ar, Kr, Xe) separates an alkali metal (Li, Na, K etc) from a halogen (F, Cl, Br, I). If the alkali metal loses an electron or the halogen gains one, their electronic structure will be similar to the noble gas in which the outer shell is completely filled. The atom now becomes relatively highly stable. When an electron is added or removed the atom becomes a negative or positive ion. The force of attraction between oppositely charged ions results in an ionic bond. For example, in a molecule of sodium chloride the chlorine atom becomes a negative ion and the sodium atom a positive ion. As a result there will be a relative displacement of the charge centers and the molecule possesses a permanent dipole moment. When there is no electric field the molecules are oriented at random and there is no net dipole moment.

Consider two molecules oriented parallel and anti-parallel to the external field. The molecule parallel to the field will be elongated with an increase in its dipole moment while a molecule anti parallel will have its dipole moment reduced. On average there will be a net contribution to the polarization which may be called as ionic or atomic polarization, depending upon whether the molecule is ionic or not. The polarizability is denoted by α_a and this contribution should be added to the electronic and orientational polarizabiity of the molecule. Thus we have for total polarizability

$$\alpha_T = \alpha_o + \alpha_a + \alpha_e$$

leading to

$$P = NE_i \left(\alpha_e + \alpha_a + \frac{\mu^2}{3kT} \right) \tag{2.122}$$

The Clausius Mosotti function is obtained by replacing α_e by α_T in (2.23).

The average polarizabilities of halogen ions and alkali metal ions are given in Table 2.11, with the electronic polarizability of rare gas atoms included for comparison. Ionic polarizability is generallly higher than the electronic polarizability.

Table 2.11

Average polarizability of halogen ions and alkali metal ions

HALOGENS $(\times 10^{-40}\ F\ m^2)$		NOBLE GAS ATOMS $(\times 10^{-40}\ F\ m^2)$		ALKALI METALS $(\times 10^{-40}\ F\ m^2)$	
		He	0.22	Li^+	0.032
F^-	1.19	Ne	0.45	Na^+	0.26
Cl^-	4.0	Ar	1.84	K^+	1.0
Br^-	4.95	Kr	2.69	Rb^+	1.87
I^-	7.7	Xe	4.37	Cs^+	2.75

CRC handbook (1985-86).

Entries in the same row have the same electronic configuration. However the nuclear charge increases from left to right resulting in a decrease of the polarizability. This is attributed to the fact that the greater Coulomb force between the nucleus and the electrons offers greater resistance to the electronic displacement. We conclude this section by summarizing the average polarizability of selected molecules (Table 2.12).

Table 2.12

Average Polarizability of selected molecules

Molecule	α_e ($\times 10^{-40}$ F m^2)	Molecule	α_e ($\times 10^{-40}$ F m^2)
CO	2.16	N_2	1.89
Cl_2	5.12	NO	33.3
CCl_2F_2	8.67	O_2	1.75
CF_4	4.26	CO_2	3.23
CH_4	2.87	D_2O	1.40
CH_4O	3.65	H_2O	1.61
C_6H_6	11.45	N_2O	3.36
D_2	0.88	NH_3	2.51
F_2	1.52	SF_6	14.65
H_2	0.89	SeF_6	7.91
HCl	2.92	SiF_4	6.05

(CRC Handbook, 1986; with permission of CRC Press).

2.15 REFERENCES

[1] A. Von Hippel, Dielectrics and waves, John Wiley & Sons, New York, 1954.

[2] CRC Handbook, 1985-1986.

[3] Dielectric Relaxation in Solids: A. K. Jonscher, Chelsea Dielectrics Press, London, 1983.

[4] A. Michels, P. Sanders and A. Schipper, Physica, 2 (1935) 753, quoted in Theory of Dielectric Polarization: C. Bottcher, Elsevier, Publishing Co., Amsterdam, 1952.

[5] G. L. Link, "Dielectric properties of polymers", Polymer Science, Vol. 2, Ed: A. D. Jenkins, North Holland Publishing Co., Amsterdam-London, 1972, p. 1281.

[6] Fig. 18.2 of Link (1972).

[7] G. R. Govinda Raju, Conference on Electrical Insulation and Dielectric Phenomena, pp. 357-363, 1988.

[8] C. Sutton and J. B. Mark, J. Chem. Phys., 54 (1971) 5011-5014.

[9] W. Reddish, J. Poly. Sci, part C, pp. 123-137, 1966.

[10] M. P. McNeal, S. Jang and R. E. Newnham, J. Appl. Phys., 83 (1998) 3288.

[11] S. P. Yardonov, I. Ivanov, Ch. P. Carapanov, , J. Phys. D., Appl. Phys., 31 (1998) 800-806.

[12] L. Onsager, J. Am. Chem. Soc., 58 (1936) 1486.

[13] J. G. Kirkwood, J. Chem. Phys., 7 (1939) 911.

[14] M. S. Dionisio, J. J. Moura-Ramos, and G. Williams, Polymer, 34 (1993) 4106-4110.

[15] J. G. Kirkwood, J. Chem. Phys., 4 (1936) 592.

[16] Progress in Dielectrics: Ed: J. B. Birks and J. H. Schulman, John Wiley and Sons, New York, vol.2, "Theory of Polarization and Absorption in Dielectrics", G. Wyllie, p.16.

[17] W. F. Brown, J. Chem. Phys., 18 (1950) 1193.

[18] W. F. Brown, J. Chem. Phys., 18 (1950) 1200.

[19] J. G. Kirkwood, J. Chem. Phys., 7 (1939) 919.

[20] "Documents on Modern Physics-J. G. Kirkwood Collected Works", Ed: R. H. Cole, Gordon and Breach, New York, 1965.

[21] H. G. David, H. D. Hamann and J. G. Pearse, J. Chem. Phys., 19 (1951) 1491.

[22] see ref. 14; This form is from D. W. Davidson and R. H. Cole, J. Chem. Phys.,19 (1951) 1484-1487.

[23] "Dielectric Properties and Molecular Behavior", N. Hill, W. Vaughan, A. H. Price, M. Davies, Van Nostrand Reinhold Co., London 1969, p. 27.

[24] Oster and Kirkwood: J. Chem. Phys. 7 (1939) 911.

[25] V. T. Morgan, IEEE Trans. on Diel. Elec. Insul., 5 (1998) 125-131.

[26] J. W. Rayleigh, Phil. Mag., ser. 5, 34 (1892), 481-502, quoted by V. T. Morgan, IEEE Trans. on Diel. Elec. Insul., 5 (1998) 125-131.

[26] J. W. Rayleigh, Phil. Mag., ser. 5, **34** (1892), 481-502, quoted by V. T. Morgan, IEEE Trans. on Diel. Elec. Insul., **5** (1998) 125-131.

[27] O. Wiener, quoted by V. T. Morgan, IEEE Trans. on Diel. Elec. Insul., **5** (1998) 125-131.

[28] T. Zakri, J. Laurent and M. Vauclin, J. Phys. D: Appl. Phys., **31** (1998)1589-1594.

[29] L. K. H. Van Beck, "Dielectric Behavior of Heterogeneous Systems", Progress in Dielectrics, Vol.7, Chapter 3, C. R. C. Press, pp. 63-117, 1967.
E. Tuncer, S. M. Gubanski, and B. Nettelbald, J. Appl. Phys., **89** (2001) 8092.

[30] V. T. Morgan, IEEE Trans. on Diel. And Elec. Insul., **5** (1998) 125-131.

[31] J. T. Kindt and C. A. Schmuttenmaer, J. Phys. Chem., **100** (1966) 10373-10379.

[32] M. S. Skai and B. M. Ladanyi, J. Chem. Phys., Vol. **102** (1995) 6542-6551.

[33] J. B. Hasted, "Aqueous Dielectrics", Chapman and Hall, London, 1991; "Water: a comprehensive treatise", vol. 1, Ed: Felix Franks, article by J. B. Hasted, "Liquid Water: Dielectric Properties", Plenum Press, 1974.

[34] J. W. Smith, "Electric Dipole Moments", Butterworth Scientific Publications, 1995.

Thou, nature, art my goddess; to thy laws
My services are bound . . .
                              - Carl Friedrich Gauss

3

DIELECTRIC LOSS AND RELAXATION–I

The dielectric constant and loss are important properties of interest to electrical engineers because these two parameters, among others, decide the suitability of a material for a given application. The relationship between the dielectric constant and the polarizability under dc fields have been discussed in sufficient detail in the previous chapter. In this chapter we examine the behavior of a polar material in an alternating field, and the discussion begins with the definition of complex permittivity and dielectric loss which are of particular importance in polar materials.

Dielectric relaxation is studied to reduce energy losses in materials used in practically important areas of insulation and mechanical strength. An analysis of build up of polarization leads to the important Debye equations. The Debye relaxation phenomenon is compared with other relaxation functions due to Cole-Cole, Davidson-Cole and Havriliak-Negami relaxation theories. The behavior of a dielectric in alternating fields is examined by the approach of equivalent circuits which visualizes the lossy dielectric as equivalent to an ideal dielectric in series or in parallel with a resistance. Finally the behavior of a non-polar dielectric exhibiting electronic polarizability only is considered at optical frequencies for the case of no damping and then the theory improved by considering the damping of electron motion by the medium. Chapters 3 and 4 treat the topics in a continuing approach, the division being arbitrary for the purpose of limiting the number of equations and figures in each chapter.

3.1 COMPLEX PERMITTIVITY

Consider a capacitor that consists of two plane parallel electrodes in a vacuum having an applied alternating voltage represented by the equation

$$v = V_m \cos \omega t \tag{3.1}$$

where v is the instantaneous voltage, V_m the maximum value of v and $\omega = 2\pi f$ is the angular frequency in radian per second. The current through the capacitor, i_l is given by

$$i_1 = I_m (\cos \omega t + \frac{\pi}{2}) \tag{3.2}$$

where

$$I_m = \frac{V_m}{z} = \omega c_o V_m \tag{3.3}$$

In this equation C_0 is the vacuum capacitance, some times referred to as geometric capacitance.

In an ideal dielectric the current leads the voltage by 90° and there is no component of the current in phase with the voltage. If a material of dielectric constant ε is now placed between the plates the capacitance increases to $C_0\varepsilon$ and the current is given by

$$i_2 = I_m \cos[\omega t + (\frac{\pi}{2} - \delta)] \tag{3.4}$$

where

$$I_m = \omega c_0 \varepsilon V_m \tag{3.5}$$

It is noted that the usual symbol for the dielectric constant is ε_r but we omit the subscript for the sake of clarity, noting that ε is dimensionless. The current phasor will not now be in phase with the voltage but by an angle (90°-δ) where δ is called the **loss angle.** The dielectric constant is a complex quantity represented by

$$\varepsilon^* = \varepsilon' - j\varepsilon'' \tag{3.6}$$

The current can be resolved into two components; the component in phase with the applied voltage is $I_x = v\omega\varepsilon''c_o$ and the component leading the applied voltage by 90° is $I_y = v\omega\varepsilon'c_o$ (fig. 3.1). This component is the charging current of the ideal capacitor.

The component in phase with the applied voltage gives rise to dielectric loss. δ is the loss angle and is given by

$$\delta = \tan^{-1} \frac{\varepsilon''}{\varepsilon'} \tag{3.7}$$

ε'' is usually referred to as the loss factor and $\tan \delta$ the dissipation factor. To complete the definitions we note that

$$I_y = v\omega C_o \varepsilon'' = \frac{v\omega \, \varepsilon_o \varepsilon'' A}{d} = A\omega \, \varepsilon_o \varepsilon'' E$$

The current density is given by

$$J_1 = \frac{I_x}{A} = \omega \varepsilon_o \varepsilon'' E$$

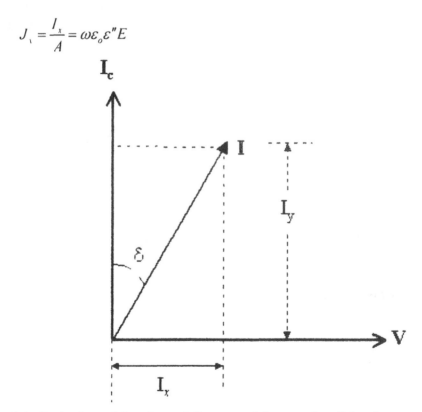

Fig. 3.1 Real (ε') and imaginary (ε'') parts of the complex dielectric constant (ε^*) in an alternating electric field. The reference phasor is along I_c and $\varepsilon^* = \varepsilon' - j\varepsilon''$. The angle δ is shown enlarged for clarity.

The alternating current conductivity is given by

$$\sigma_{ac} = \sigma' + j\sigma'' = \omega\varepsilon_o\left[\varepsilon'' + j(\varepsilon' - \varepsilon_\infty)\right]$$

(3.8)

The total conductivity is given by

$$\sigma_T = \sigma_{ac} + \sigma_{dc} = \omega\varepsilon_o\varepsilon'' + \sigma_{dc}$$

3.2 POLARIZATION BUILD UP

When a direct voltage applied to a dielectric for a sufficiently long duration is suddenly removed the decay of polarization to zero value is not instantaneous but takes a finite time. This is the time required for the dipoles to revert to a random distribution, in equilibrium with the temperature of the medium, from a field oriented alignment. Similarly the build up of polarization following the sudden application of a direct voltage takes a finite time interval before the polarization attains its maximum value. This phenomenon is described by the general term **dielectric relaxation.**

When a dc voltage is applied to a polar dielectric let us assume that the polarization builds up from zero to a final value (fig. 3.2) according to an exponential law

$$P(t) = P_\infty(1 - e^{-\frac{t}{\tau}})$$

(3.9)

Where $P(t)$ is the polarization at time t and τ is called the relaxation time. τ is a function of temperature and it is independent of the time.

The rate of build up of polarization may be obtained, by differentiating equation (3.9) as

$$\frac{d\,P(t)}{dt} = -P_\infty(-\frac{1}{\tau})e^{\frac{-t}{\tau}} = \frac{P_\infty e^{\frac{-t}{\tau}}}{\tau}$$

(3.10)

Substituting equation (3.9) in (3.10) and assuming that the total polarization is due to the dipoles, we get[1]

$$\frac{dP(t)}{dt} = \frac{P_\infty - P(t)}{\tau} \cong \frac{P_\mu - P(t)}{\tau} \tag{3.11}$$

Neglecting atomic polarization the total polarization $P_T(t)$ can be expressed as the sum of the orientational polarization at that instant, $P_\mu(t)$, and electronic polarization, P_e which is assumed to attain its final value instantaneously because the time required for it to attain saturation value is in the optical frequency range. Further, we assume that the instantaneous polarization of the material in an alternating voltage is equal to that under dc voltage that has the same magnitude as the peak of the alternating voltage at that instant.

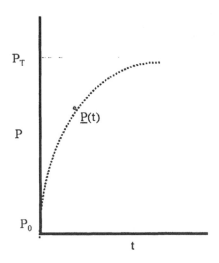

Fig. 3.2 Polarization build up in a polar dielectric.

We can express the total polarization, $P_T(t)$, as

$$P_T(t) = P_\mu(t) + P_e \tag{3.12}$$

The final value attained by the total polarization is

$$P_T = \varepsilon_0(\varepsilon_s - 1)E \tag{3.13}$$

We have already shown in the previous chapter that the following relationships hold under steady voltages:

$$P_e = \varepsilon_0(\varepsilon_\infty - 1)E \tag{3.14}$$

where ε_s and ε_∞ are the dielectric constants under direct voltage and at infinity frequency respectively.

We further note that Maxwell's relation

$$\varepsilon_\infty = n^2 \tag{3.15}$$

holds true at optical frequencies. Substituting equations (3.13) and (3.14) in (3.12) we get

$$P_\mu = \varepsilon_0(\varepsilon_s - 1)E - \varepsilon_0(\varepsilon_\infty - 1)E \tag{3.16}$$

which simplifies to

$$P_\mu = \varepsilon_0(\varepsilon_s - \varepsilon_\infty)E \tag{3.17}$$

Representing the alternating electric field as

$$E = E_{max}e^{j\omega t} \tag{3.18}$$

and substituting equation (3.18) in (3.11) we get

$$\frac{dP(t)}{dt} = \frac{1}{\tau}[\varepsilon_0(\varepsilon_s - \varepsilon_\infty)E_m e^{j\omega t} - P(t)] \tag{3.19}$$

The general solution of the first order differential equation is

$$P(t) = Ce^{-\frac{t}{\tau}} + \varepsilon_0 \frac{(\varepsilon_s - \varepsilon_\infty)E_m e^{j\omega t}}{1 + j\omega\tau} \tag{3.20}$$

where C is a constant. At time t, sufficiently large when compared with τ, the first term on the right side of equation (3.20) becomes so small that it can be neglected and we get the solution for *P(t)* as

$$P(t) = \varepsilon_0 \frac{(\varepsilon_s - \varepsilon_\infty)E_m e^{j\omega t}}{1 + j\omega\tau} \qquad (3.21)$$

Substituting equation (3.21) in (3.12) we get

$$P(t) = \varepsilon_0(\varepsilon_\infty - 1)E_m e^{j\omega t} + \frac{\varepsilon_0(\varepsilon_s - \varepsilon_\infty)}{1 + j\omega\tau} E_m e^{j\omega t} \qquad (3.22)$$

Simplification yields

$$P(t) = [\varepsilon_\infty - 1 + \frac{(\varepsilon_s - \varepsilon_\infty)}{1 + j\omega\tau}]\varepsilon_0 E_m e^{j\omega t} \qquad (3.23)$$

Equation (3.23) shows that $P(t)$ is a sinusoidal function with the same frequency as the applied voltage. The instantaneous value of flux density $\mathbf{D}$ is given by

$$D(t) = \varepsilon_0 \varepsilon^* E_m e^{j\omega t} \qquad (3.24)$$

But the flux density is also equal to

$$D(t) = \varepsilon_0 E_m e^{j\omega t} + P(t) \qquad (3.25)$$

Equating expressions (3.24) and (3.25) we get

$$\varepsilon_0 \varepsilon^* E_m e^{j\omega t} = \varepsilon_0 E_m e^{j\omega t} + P(t) \qquad (3.26)$$

Substituting equation (3.23) in (3.26), and simplifying we get

$$(\varepsilon' - j\varepsilon'') = 1 + [\varepsilon_\infty - 1 + \frac{\varepsilon_s - \varepsilon_\infty}{1 + j\omega\tau}] \qquad (3.27)$$

Equating the real and imaginary parts we readily obtain

$$\varepsilon' = \varepsilon_\infty + \frac{\varepsilon_s - \varepsilon_\infty}{1 + \omega^2\tau^2} \qquad (3.28)$$

$$\varepsilon'' = \frac{(\varepsilon_s - \varepsilon_\infty)\omega\tau}{1 + \omega^2\tau^2} \tag{3.29}$$

It is easy to show that

$$\tan\delta = \frac{\varepsilon''}{\varepsilon'} = \frac{(\varepsilon_s - \varepsilon_\infty)\omega\tau}{\varepsilon_s + \varepsilon_\infty\omega^2\tau^2} \tag{3.30}$$

Equations (3.28) and (3.29) are known as Debye equations[2] and they describe the behavior of polar dielectrics at various frequencies. The temperature enters the discussion by way of the parameter τ as will be described in the following section. The plot of ε'' - ω is known as the relaxation curve and it is characterized by a peak at $\varepsilon''/\varepsilon''_{max}$ = 0.5. It is easy to show $\omega\tau = 3.46$ for this ratio and one can use this as a guide to determine whether Debye relaxation is a possible mechanism. The spectrum of the Debye relaxation curve is very broad as far as the whole gamut of physical phenomena are concerned,[3] though among the various relaxation formulas Debye relaxation is the narrowest. The descriptions that follow in several sections will bring out this aspect clearly.

3.3 DEBYE EQUATIONS

An alternative and more concise way of expressing Debye equations is

$$\varepsilon' = \varepsilon_\infty + \frac{\varepsilon_s - \varepsilon_\infty}{1 + j\omega\tau} \tag{3.31}$$

Equations (3.28)-(3.30) are shown in fig. 3.3. An examination of these equations shows the following characteristics:

(1) For small values of $\omega\tau$, the real part $\varepsilon' \approx \varepsilon_s$ because of the squared term in the denominator of equation (3.28) and ε'' is also small for the same reason. Of course, at $\omega\tau = 0$, we get $\varepsilon'' = 0$ as expected because this is dc voltage.

(2) For very large values of $\omega\tau$, $\varepsilon' = \varepsilon_\infty$ and ε'' is small.

(3) For intermediate values of frequencies ε'' is a maximum at some particular value of $\omega\tau$.

The maximum value of ε'' is obtained at a frequency given by

$$\frac{\partial \varepsilon''}{\partial(\omega\tau)} = (\varepsilon_s - \varepsilon_\infty)\frac{(\omega^2\tau^2 - 1)}{(1 + \omega^2\tau^2)^2} = 0$$

resulting in

$$\omega_p\tau = 1 \tag{3.32}$$

where ω_p is the frequency at ε''_{max}.

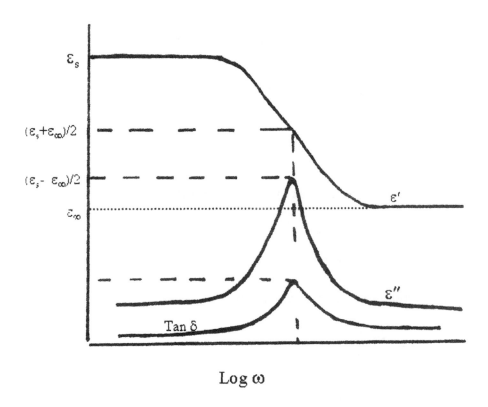

Fig. 3.3 Schematic representation of Debye equations plotted as a function of $\log\omega$. The peak of ε'' occurs at $\omega\tau = 1$. The peak of $\tan\delta$ does not occur at the same frequency as the peak of ε''.

The values of ε' and ε'' at this value of $\omega\tau$ are

$$\varepsilon' = \frac{\varepsilon_s + \varepsilon_\infty}{2} \tag{3.33}$$

$$\varepsilon''_{max} = \frac{\varepsilon_s - \varepsilon_\infty}{2} \tag{3.34}$$

The dissipation factor tan δ also increases with frequency, reaches a maximum, and for further increase in frequency, it decreases. The frequency at which the loss angle is a maximum can also be found by differentiating tan δ with respect to ω and equating the differential to zero. This leads to

$$\frac{\partial(\tan\partial)}{\partial(\omega t)} = (\varepsilon_s - \varepsilon_\infty)\frac{(\varepsilon_\infty\omega^2\tau^2 - \varepsilon_s)}{(\varepsilon_s + \varepsilon_\infty\omega^2\tau^2)} = 0 \tag{3.35}$$

Solving this equation it is easy to show that

$$\omega\tau = \sqrt{\frac{\varepsilon_s}{\varepsilon_\infty}} \tag{3.36}$$

By substituting this value of $\omega\tau$ in equation (3.30) we obtain

$$(\tan\delta)_{max} = \frac{\varepsilon_s - \varepsilon_\infty}{\sqrt{\varepsilon_s\varepsilon_\infty}} \tag{3.37}$$

The corresponding values of ε' and ε'' are

$$\varepsilon' = \frac{2\varepsilon_s\varepsilon_\infty}{\varepsilon_s + \varepsilon_\infty} \tag{3.38}$$

$$\varepsilon'' = \frac{\varepsilon_s - \varepsilon_\infty}{\varepsilon_s + \varepsilon_\infty}\sqrt{\varepsilon_s\varepsilon_\infty} \tag{3.39}$$

Fig. 3.3 also shows the plot of equation (3.30), that is, the variation of tan δ as a function of frequency for several values of τ.

Dividing equation (3.28) from (3.27) and rearranging terms we obtain the simple relationship

$$\frac{\varepsilon''}{\varepsilon' - \varepsilon_\infty} = \omega\tau \tag{3.40}$$

According to equation (3.40) a plot of $\dfrac{\varepsilon''}{\varepsilon' - \varepsilon_\infty}$ against ω results in a straight line passing through the origin with a slope of τ.

Fig. 3.3 shows that, at the relaxation frequency defined by equation (3.32) ε' decreases sharply over a relatively small band width. This fact may be used to determine whether relaxation occurs in a material at a specified frequency. If we measure ε' as a function of temperature at constant frequency it will decrease rapidly with temperature at relaxation frequency. Normally in the absence of relaxation ε' should increase with decreasing temperature according to equation (2.51).

Variation of ε' as a function of frequency is referred to as dispersion in the literature on dielectrics. Variation of ε'' as a function of frequency is called absorption though the two terms are often used interchangeably, possibly because dispersion and absorption are associated phenomena. Fig. 3.4 shows a series of measured ε' and ε'' in mixtures of water and methanol[4]. The question of determining whether the measured data obey Debye equation (3.31) will be considered later in this chapter.

3.4 BI-STABLE MODEL OF A DIPOLE

In the molecular model of a dipole a particle of charge e may occupy one of two sites, 1 or 2, that are situated apart by a distance b[5]. These sites correspond to the lowest potential energy as shown in fig. 3.5. In the absence of an electric field the two sites are of equal energy with no difference between them and the particle may occupy any one of them. Between the two sites, therefore, there is a particle. An applied electric field causes a difference in the potential energy of the sites. The figure shows the conditions with no electric field with full lines and the shift in the potential energy due to the electric field by the dotted line.

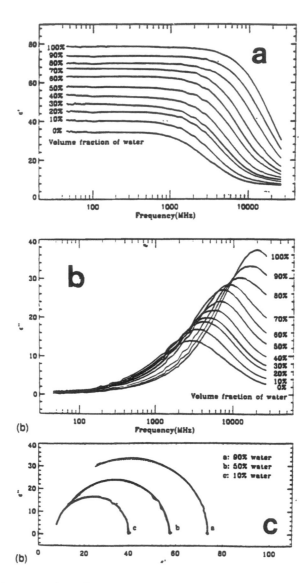

Fig. 3.4 Dielectric properties of water and methanol mixtures at 25°C. (a) Real part, ε' (b) Imaginary part, ε″ (c) Complex plane plot of ε* showing Debye relaxation (Bao et. al., 1996). (with permission of American Physical Society.)

The difference in the potential energy due to the electric field E is

$$\phi_1 - \phi_2 = ebE\cos\theta$$

where θ is the angle between the direction of the electric field and the line joining 1 and 2. This model is equivalent to a dipole changing position by $180°$ when the charge moves from site 1 to 2 or from site 2 to 1. The moment of such a dipole is

$$\mu = \frac{1}{2}eb$$

which may be thought of as having been hinged at the midpoint between sites 1 and 2. This model is referred to as the bistable model of the dipole. We also assume that $\theta = 0$ for all dipoles and that the potential energy of sites 1 and 2 are equal in the absence of an external electric field.

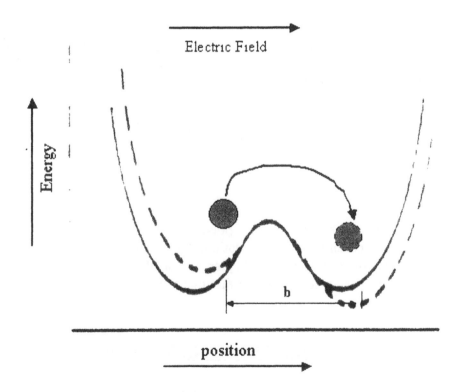

Fig. 3.5 The potential well model for a dipole with two stable positions. In the absence of an electric field (full lines) the dipole spends equal time in each well; this indicates that there is no polarization. In the presence of an electric field (broken lines) the wells are tilted with the 'downside' of the field having a slightly lower energy than the 'up' side; this represents polarization.

We assume that the material contains N number of bistable dipoles per unit volume and the field due to interaction is negligible. A macroscopic consideration shows that the charged particles would not have the energy to jump from one site to the other. However, on a microscopic scale the dipoles are in a heat reservoir exchanging energy with each other and dipoles. A charge in well 1 occasionally acquires enough energy to climb the hill and moves to well 2. Upon arrival it returns energy to the reservoir and remain there for some time. It will then acquire energy to jump to well 1 again.

The number of jumps per second from one well to the other is given in terms of the potential energy difference between the two wells as

$$W_{12} = A \exp^{-\frac{w - \mu E}{kT}}$$

where T is the absolute temperature, k the Boltzmann constant and A is a factor denoting the number of attempts. Its value is typically of the order of 10^{-13} s^{-1} at room temperature though values differing by three or four orders of magnitude are not uncommon. It may or may not depend on the temperature. If it does, it is expressed as B/T as found in some polymers. If the destination well has a lower energy than the starting well then the minus sign in the exponent is valid. The relaxation time is the reciprocal of W_{12} leading to

$$\tau = \tau_0 \exp^{\frac{w}{kT}} \qquad for \, w \gg \mu E$$

The variation of τ with T in liquids and in polymers near the glass transition temperature is assumed to be according to this equation. Other functions of T have also been proposed which we shall consider in chapter 5. The decrease of relaxation time with increasing temperature is attributed to the fact that the frequency of jump increases with increasing temperature.

3.5 COMPLEX PLANE DIAGRAM

Cole and Cole[6] showed that, in a material exhibiting Debye relaxation a plot of ε'' against ε', each point corresponding to a particular frequency yields a semi-circle. This can easily be demonstrated by rearranging equations (3.28) and (3.29) to give

$$(\varepsilon'')^2 + (\varepsilon' - \varepsilon_\infty)^2 = \frac{(\varepsilon_s - \varepsilon_\infty)^2}{1 + \omega^2 \tau^2} \qquad (3.41)$$

The right side of equation (3.41) may be simplified using equation (3.28) resulting in

$$(\varepsilon'')^2 + (\varepsilon' - \varepsilon_\infty)^2 = (\varepsilon_s - \varepsilon_\infty)(\varepsilon' - \varepsilon_\infty) \tag{3.42}$$

Further simplification yields

$$\varepsilon'^2 - \varepsilon'(\varepsilon_s + \varepsilon_\infty) + \varepsilon_s \varepsilon_\infty + \varepsilon''^2 = 0 \tag{3.43}$$

Substituting the algebraic identity

$$\varepsilon_s \varepsilon_\infty = \frac{1}{4}[(\varepsilon_s + \varepsilon_\infty)^2 - (\varepsilon_s + \varepsilon_\infty)^2]$$

equation (3.43) may be rewritten as

$$(\varepsilon' - \frac{\varepsilon_s + \varepsilon_\infty}{2})^2 + (\varepsilon'')^2 = (\frac{\varepsilon_s - \varepsilon_\infty}{2})^2 \tag{3.44}$$

This is the equation of a circle with radius $\frac{\varepsilon_s - \varepsilon_\infty}{2}$ having its center at ($\frac{\varepsilon_s + \varepsilon_\infty}{2}$, 0).

It can easily be shown that (ε_∞, 0) and (ε_s, 0) are points on the circle. To put it another way, the circle intersects the horizontal axis (ε') at ε_∞ and ε_s as shown in fig. 3.6. Such plots of ε'' versus ε' are known as complex plane plots of ε^*.

At $\omega_p \tau = 1$ the imaginary component ε'' has a maximum value of

$$\varepsilon'' = \frac{\varepsilon_s - \varepsilon_\infty}{2}$$

The corresponding value of ε' is

$$\varepsilon' = \frac{\varepsilon_s + \varepsilon_\infty}{2}$$

Of course these results are expected because the starting point for equation (3.44) is the original Debye equations.

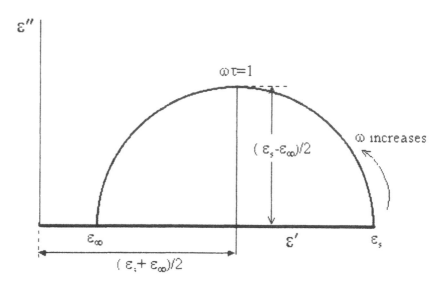

Fig. 3.6 Cole-Cole diagram displaying a semi-circle for Debye equations for ε^*.

In a given material the measured values of ε'' are plotted as a function of ε' at various frequencies, usually from $\omega = 0$ to $\omega = 10^{10}$ rad/s. If the points fall on a semi-circle we can conclude that the material exhibits Debye relaxation. A Cole-Cole diagram can then be used to obtain the complex dielectric constant at intermediate frequencies obviating the necessity for making measurements. In practice very few materials completely agree with Debye equations, the discrepancy being attributed to what is generally referred to as **distribution of relaxation times.**

The simple theory of Debye assumes that the molecules are spherical in shape and therefore the axis of rotation of the molecule in an external field has no influence in deciding the value of ε^*. This is more an exception than a rule because not only the molecules can have different shapes, they have, particularly in long chain polymers, a linear configuration. Further, in the solid phase the dipoles are more likely to be interactive and not independent in their response to the alternating field[7]. The relaxation times in such materials have different values depending upon the axis of rotation and, as a result, the dispersion commonly occurs over a wider frequency range.

3.6 COLE-COLE RELAXATION

Polar dielectrics that have more than one relaxation time do not satisfy Debye equations. They show a maximum value of ε'' that will be lower than that predicted by equation (3.34). The curve of $\tan\delta$ vs $\log\omega\tau$ also shows a broad maximum, the maximum value being smaller than that given by equation (3.37). Under these conditions the plot ε'' vs. ε' will be distorted and Cole-Cole showed that the plot will still be a semi-circle with its center displaced below the ε' axis. They suggested an empirical equation for the complex dielectric constant as

$$\varepsilon^{\cdot} = \varepsilon_{\infty} + \frac{\varepsilon_s - \varepsilon_{\infty}}{1 + (j\omega\tau_{c-c})^{1-\alpha}} ; 0 \le \alpha \le 1; \tag{3.45}$$

$$\alpha = 0 \ for\ Debye\ relaxation$$

where τ_{c-c} is the mean relaxation time and α is a constant for a given material, having a value $0 \le \alpha \le 1$. A plot of equation (3.45) is shown in figs. (3.7) and (3.8) for various values of α. Debye equations are also plotted for the purpose of comparison. Near relaxation frequencies Cole-Cole relaxation shows that ε' decreases more slowly with ω than the Debye relaxation. With increasing α the loss factor ε'' is broader than the Debye relaxation and the peak value, ε_{max} is smaller.

A dielectric that has a single relaxation time, $\alpha = 0$ in this case, equation (3.45) becomes identical with equation (3.29). The larger the value of α, the larger the distribution of relaxation times.

To determine the geometrical interpretation of equation (3.45) we substitute $1-\alpha = n$ and rewrite it as

$$\varepsilon' - j\varepsilon'' = \varepsilon_{\infty} + \frac{\varepsilon_s - \varepsilon_{\infty}}{1 + (\omega\tau_{c-c})^n (\cos\dfrac{n\pi}{2} + j\sin\dfrac{n\pi}{2})} \tag{3.46}$$

Equating real and imaginary parts we get

$$\varepsilon' = \varepsilon_\infty + (\varepsilon_s - \varepsilon_\infty)\frac{1 + (\omega\tau_{c-c})^n \cos(\frac{n\pi}{2})}{1 + 2(\omega\tau_{c-c})^n \cos(\frac{n\pi}{2}) + (\omega\tau_{c-c})^{2n}} \tag{3.47}$$

$$\varepsilon'' = (\varepsilon_s - \varepsilon_\infty)\frac{(\omega\tau_{c-c})^n \sin(n\pi/2)}{1 + 2(\omega\tau_{c-c})^n \cos(n\pi/2) + (\omega\tau_{c-c})^{2n}} \tag{3.48}$$

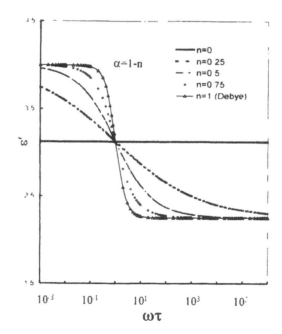

Fig. 3.7 Real part of ε* in a polar dielectric according to Cole-Cole relaxation. α =0 gives Debye relaxation.

Using the identity

$$j^\alpha \equiv e^{j\alpha\pi/2} \equiv \cos\alpha\pi/2 + \sin\alpha\pi/2$$

equations (3.47) and (3.48) may be expressed alternatively as[8]

$$\frac{\varepsilon' - \varepsilon_\infty}{\varepsilon_s - \varepsilon_\infty} = \frac{1}{2}\left(1 - \frac{\sinh ns}{\cosh ns + \sin(\alpha\pi/2)}\right) \tag{3.46a}$$

$$\frac{\varepsilon''}{\varepsilon_s - \varepsilon_\infty} = \frac{1}{2}\left(\frac{\cos(\alpha\pi/2)}{\cosh ns + \sin(\alpha\pi/2)}\right) \tag{3.47a}$$

where

$$s = Ln\omega\tau$$

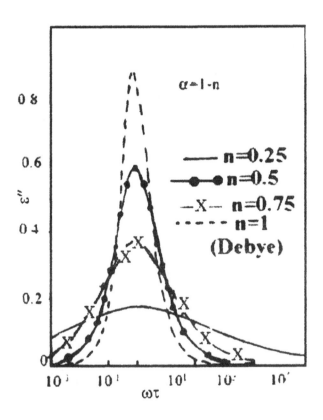

Fig. 3.8 Imaginary part of ε* in a polar dielectric according to Cole-Cole relaxation. α =0 gives Debye relaxation.

Eliminating $\omega\tau_{c\text{-}c}$ from equations (3.47) and (3.48) Cole-Cole showed that

$$(\varepsilon' - \frac{\varepsilon_s + \varepsilon_\infty}{2})^2 + [\varepsilon'' + \frac{\varepsilon_s - \varepsilon_\infty}{2}\cot(n\pi/2)]^2 = [\frac{\varepsilon_s - \varepsilon_\infty}{2}\cosec(n\pi/2)]^2 \tag{3.49}$$

Equation (3.49) represents the equation of a circle with the center at

$$[\frac{\varepsilon_s + \varepsilon_\infty}{2}, \frac{\varepsilon_\infty - \varepsilon_s}{2}\cot(n\pi/2)]$$

and having a radius of

$$[\frac{\varepsilon_s - \varepsilon_\infty}{2}\cosec(n\pi/2)]$$

We note that the y coordinate of the center is negative, that is, the center lies below the ε' axis (fig. 3.9).

Figs. 3.7 and 3.8 show the variation of ε' and ε'' as a function of $\omega\tau$ for several values of α respectively. These are the plots of equations (3.47) and (3.48). At $\omega\tau = 1$ the following relations hold:

$$\varepsilon' = \frac{\varepsilon_s + \varepsilon_\infty}{2}; \qquad \varepsilon'' = \frac{\varepsilon_s - \varepsilon_\infty}{2}\tan\frac{n\pi}{2}$$

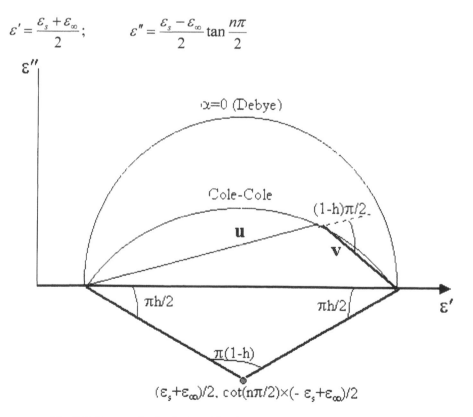

Fig. 3.9 Geometrical relationships in Cole-Cole equation (3.45).

As stated above, the case of $n = 0$ corresponds to an infinitely large number of distributed relaxation times and the behavior of the material is identical to that under dc fields except that the dielectric constant is reduced to $(\varepsilon_s - \varepsilon_\infty)/2$. The complex part of the dielectric constant is also equal to zero at this value of n, consistent with dc fields. As the value of n increases ε' changes with increasing $\omega\tau$, the curves crossing over at $\omega\tau = 1$. At $n=1$ the change in ε' with increasing $\omega\tau$ is identical to the Debye relaxaton, the material then possessing a single relaxation time.

The variation of ε'' with $\omega\tau$ is also dependent on the value of n. As the value of n increases the curves become narrower and the peak value increases. This behavior is consistent with that shown in fig. 3.8.

Let the lines joining any point on the Cole-Cole diagram to the points corresponding to ε_∞ and ε_s be denoted by u and v respectively (Fig. 3.9). Then, at any frequency the following relations hold:

$$u = \varepsilon^* - \varepsilon_\infty; \qquad v = \frac{\varepsilon^* - \varepsilon_\infty}{(\omega\tau)^{1-n}}; \qquad \frac{v}{u} = (\omega\tau)^{1-n}$$

By plotting log ω against (log v-log u) the value of n may be determined. With increasing value of n, the number of degrees of freedom for rotation of the molecules decreases. Further decreasing the temperature of the material leads to an increase in the value of the parameter n.

The Cole-Cole diagrams for poly(vinyl chloride) at various temperatures are shown in fig. 3.10[9]. The Cole-Cole arc is symmetrical about a line through the center parallel to the ε'' axis.

3.7 DIELECTRIC PROPERTIES OF WATER

Debye relaxation is generally limited to weak solutions of polar liquids in non-polar solvents. Water in liquid state comes closest to exhibiting Debye relaxation and its dielectric properties are interesting because it has a simple molecular structure. One is fascinated by the fact that it occurs naturally and without it life is not sustained. Hasted (1973) quotes over thirty determinations of static dielectric constant of water, already referred to in chapter 2.

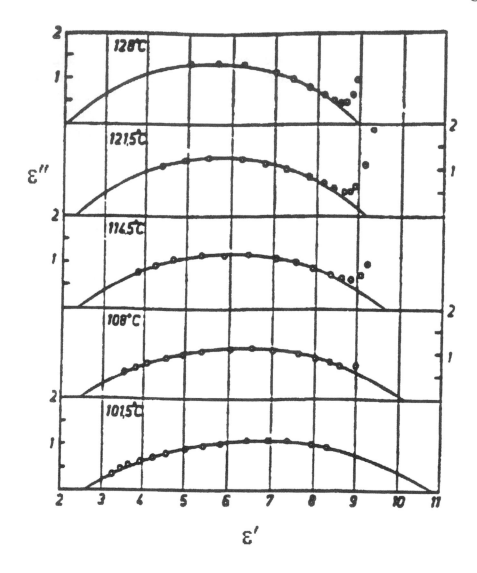

Fig. 3.10 Cole-Cole diagram from measurements on poly (vinyl chloride) at various temperatures (Ishida, 1960). (With permission of Dr. Dietrich Steinkopff Verlag, Darmstadt, Germany).

The dielectric constant is not appreciably dependent on the frequency up to 100 MHz. The measurements are carried out in the microwave frequency range to determine the relaxation frequency, and a particular disadvantage of the microwave frequency is that individual observers are forced, due to cost, to limit their studies to a narrow frequency range. Table 3.1 summarizes the data due to Bottreau et. al. (1975).

Table 3.1

Selective Dielectric Properties of Water at 293 K (Bottreau, et. al., 1975).

$$\varepsilon_\infty = n^2 = 1.78, \ \varepsilon_s = 80.4$$

(With permission of Journal of Chemical Physics, USA)

f (GHz)	Measured complex permittivity		Measured reduced permittivity		Calculated reduced permittivity	
	ε'	ε''	$E_m{}'$	$E_m{}''$	$E_c{}'$	$E_c{}''$
2530.00	3.65	1.35	0.0238	0.0172	0.0230	0.0234
890.00	4.30	2.28	0.0321	0.0290	0.0335	0.0272
300.00	5.48	4.40	0.0471	0.0560	0.0384	0.0582
35.25	20.30	29.30	0.2356	0.3727	0.2230	0.3764
34.88	19.20	30.30	0.2216	0.3854	0.2262	0.3787
24.19	29.64	35.18	0.3544	0.4475	0.3600	0.4478
23.81	30.50	35.00	0.3653	0.4452	0.3667	0.4500
23.77	31.00	35.70	0.3717	0.4541	0.3674	0.4502
23.68	31.00	35.00	0.3717	0.4452	0.3690	0.4507
23.62	30.88	35.75	0.3701	0.4547	0.3701	0.4510
15.413	46.00	36.60	0.5625	0.4655	0.5645	0.4683
9.455	63.00	31.90	0.7787	0.4057	0.7618	0.4003
9.390	61.50	31.60	0.7596	0.4019	0.7641	0.3989
9.375	62.00	32.00	0.7660	0.4070	0.7646	0.3986
9.368	62.80	31.50	0.7761	0.4007	0.7649	0.3984
9.346	61.41	31.83	0.7585	0.4049	0.7656	0.3980
9.346	62.26	32.56	0.7693	0.4141	0.7656	0.3980
9.141	63.00	31.50	0.7787	0.4007	0.7728	0.3934
4.630	74.00	18.80	0.9186	0.2391	0.9215	0.2472
3.624	77.60	16.30	0.9644	0.2073	0.9486	0.2016
3.254	77.80	13.90	0.9669	0.1768	0.9576	0.1835
1.744	79.20	7.90	0.9847	0.1005	0.9868	0.1030
1.200	80.4	7.00	1.000	0.0890	0.9936	0.0717
0.577	80.3	2.75	0.9987	0.0350	0.9985	0.0348
Extrapolated values from a single relaxation of Debye type						
13760	1.98	0.75	0.0026	0.0095	0.0021	0.0096
6880	2.37	1.15	0.0077	0.0146	0.0071	0.0167
3440	3.25	1.45	0.0188	0.0184	0.0179	0.0227

e following definitions apply for the quantities in shown Table 3.1.

$$\varepsilon^* = \varepsilon' - j\varepsilon''; \quad E_m^* = E_m' - jE_m'' = \frac{\varepsilon^* - \varepsilon_\infty}{\varepsilon_s - \varepsilon_\infty}; \qquad E^* = \sum_{i=1}^{3} C_i(1 + j\frac{f}{f_i}) = E_c' - jE_c''$$

and is reproduced from ref.[10]. Fig. 3.11 shows the complex plane plots of $\varepsilon' - j\varepsilon''$ for water (Hasted, 1973) and compared with analysis according to Debye equations and Cole-Cole equations. The relaxation time obtained as a function of temperature from Cole-Cole analysis is shown in Table 3.2 along with ε_∞ used in the analysis.

Table 3.2

Relaxation time in water (Hasted, 1973)

T° C	ε_∞	$\tau\,(10^{-11})$ s	α
0	4.46 ± 0.17	1.79	0.014
10	4.10 ± 0.15	1.26	0.014
20	4.23 ± 0.16	0.93	0.013
30	4.20 ± 0.16	0.72	0.012
40	4.16 ± 0.15	0.58	0.009
50	4.13 ± 0.15	0.48	0.013
60	4.21 ± 0.16	0.39	0.011
75	4.49 ± 0.17	0.32	-

(permission of Chapman and Hall)

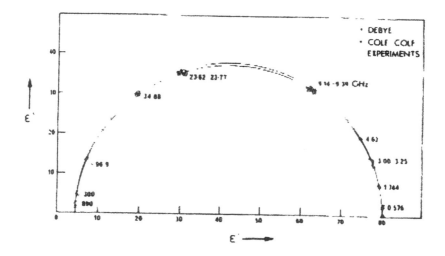

Fig. 3.11 Complex plane plot of ε^* in water at 25°C in the microwave frequency range. Points in closed circles are experimental data. ×, calculations from Cole-Cole plot, +, calculations from Debye equation with optimized parameters [Hasted 1973]. (with permission of Chapman & Hall, London).

Earlier literature on ε^* in water did not extend to as high frequencies as shown in Table 3.1 and it was thought that ε_∞ is much greater than the square of refractive index, $n^2 = 1.8$ (Hasted, 1973), and this was attributed to, possibly absorption and a second dipolar dispersion of ε'' at higher frequencies. However more recent measurements up to 2530 GHz and extrapolation to 13760 GHz shows that the equation $\varepsilon_\infty = n^2$ is valid, as demonstrated in Table 3.1. The relaxation time increases with decreasing temperature in qualitative agreement with the Debye concept. The Cole-Cole parameter α is relatively small and independent of temperature. Recall that as $\alpha \to 0$ the Cole-Cole distribution converges to Debye relaxation.

At this point it is appropriate to introduce the concept of spectral decomposition of the complex plane plot of ε^*. If we suppose that there exist several relaxation processes, each with a characteristic relaxation time and dominant over a specific frequency range, then the Debye equation (3.31) may be expressed as

$$\varepsilon' = \varepsilon_\infty + \sum_{i=1}^{i=n} \frac{\Delta \varepsilon_i}{1 + \omega^2 \tau_i^2} \tag{3.50}$$

$$\varepsilon'' = \sum_{i=1}^{i=n} \frac{\Delta \varepsilon_i}{1 + \omega^2 \tau_i^2} \omega \tau_i \tag{3.51}$$

where $\Delta \varepsilon_i$ and τ_i are the individual amplitude of dispersion ($\varepsilon_{low\ frequency} - \varepsilon_{high\ frequency}$) and the relaxation time, respectively. The assumption here is that each relaxation process follows the Debye equation independent of other processes.

This kind of representation has been used to find the relaxation times in D_2O ice[11]. Polycrystalline ice from water has been shown to have a single relaxation time of Debye type at 270 K[12] and the observed distribution of relaxation times at lower temperatures 165-196 K is attributed to physical and chemical impurities[13]. However the D_2O ice shows a more interesting behavior. Focusing our attention to the point under discussion, namely several relaxation times, fig. 3.12 shows the measured values of ε' and ε'' in the complex plane as well as the three relaxation processes. The $\Delta \varepsilon$ and τ are 88.1, 57.5 and 1.4 (see inset) and, 20 ms, 60 ms and 100 µs respectively.

A method of spectral analysis which is similar in principle to what was described above, but different in procedure, has been adopted by Bottreau et. al. (1975) who use a function of the type

$$\varepsilon^* = \sum_{i=1}^{n} C_i (1 + j\frac{\omega}{\omega_i})^{-1} \tag{3.52}$$

where C_i is the spectral contribution to i^{th} region and ω_i is its relaxation frequency. The condition $\Sigma C_i = 1$ should be satisfied.

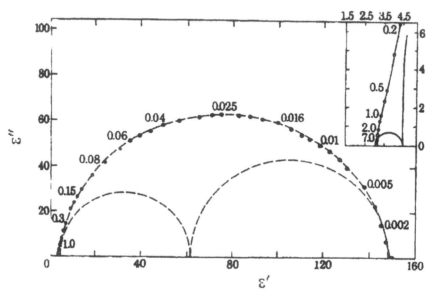

Fig. 3.12 The analysis of Cole-Cole plots into three Debye-type relaxation regions indicated by semi-circles at 191.8 K. The numbers beside the filled data points are frequencies in kHz. Closed circles: low frequency bridge measurements; Open circles: high frequency measurements [11]. (with permission of the Royal Society, England).

This scheme was applied to H_2O data shown in Table 3.1. The results obtained are shown in Table 3.3. Three Debye regions are identified with relaxation times as shown. Application of Cole-Cole relaxation (3.45) equation yields a value of $\omega_p = 107 \times 10^9$ rad s^{-1} with $\alpha = 0.013$ which agrees with the major region of relaxation in Table 3.3.

3.8 DAVIDSON – COLE EQUATION

Davidson – Cole[14] have suggested the empirical equation

$$\varepsilon^* = \varepsilon_\infty + \frac{\varepsilon_s - \varepsilon_\infty}{(1 + j\omega\tau_{d-c})^\beta} \tag{3.53}$$

where $0 \leq \beta \leq 1$ is a constant characteristic of the material.

Separating the real and imaginary parts of equation (3.53), the real and complex parts are expressed as

$$\varepsilon' - \varepsilon_\infty = (\varepsilon_s - \varepsilon_\infty)(\cos\phi)^\beta \cos\phi\beta \tag{3.54}$$

$$\varepsilon'' = (\varepsilon_s - \varepsilon_\infty)(\cos\phi)^\beta \sin\phi\beta \tag{3.55}$$

where $\tan \phi = \omega\tau_0$.

Table 3.3
Spectral contributions and relaxation frequencies of the three Debye constituents of water at 20° C [Bottreau et. al. 1975].

Region	C_i	f_i (GHz)
I	0.0507	5.57 ± 0.50
II	0.9136	17.85 ± 0.30
III	0.0357	3440.3 ± 8.0

(with permission of J. Chem. Phys.)

These equations are plotted in Figs. 3.13 and 3.14 and the Debye curves ($\beta = 1$) are also shown for comparison. The low frequency part of ε' remains unchanged as the value of β increases from 0 to 1. However the high frequency part of ε' becomes lower as β is increased, $\beta = 1$ (Debye) yielding the lowest values.

Similar observations hold gold for ε'' which increases with β in the low frequency part and decreasing with β in the high frequency part. The main point to note is that the curve of ε'' against $\omega\tau$ loses symmetry on either side of the line that is parallel to the ε'' axis and that passes through its peak value.

Expressing equations (3.54) and (3.55) in polar co-ordinates

$$r = [(\varepsilon' - \varepsilon_\infty)^2 + \varepsilon''^2]$$

$$\theta = \tan^{-1} \frac{\varepsilon''}{\varepsilon' - \varepsilon_\infty}$$

Davidson-Cole show, from equations (3.54) and (3.55) that

$$r = (\varepsilon_s - \varepsilon_\infty)[(\cos(\theta / \beta)]^\beta \qquad (3.56)$$

$$\tan \theta = \tan \beta \phi \qquad (3.57)$$

$$\omega \tau_{d-c} = \tan(\frac{\theta}{\beta}) \qquad (3.58)$$

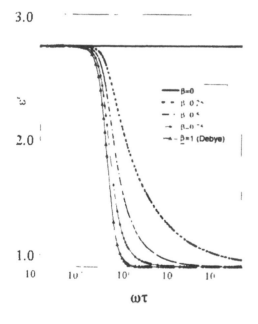

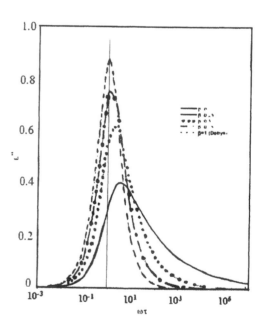

Fig. 3.13 Schematic variation of ε' as a function of $\omega\tau$ for various values of β. The low frequency value of ε' has been arbitrarily chosen.

Fig. 3.14 Schematic variation of ε'' as a function of $\omega\tau$ for various values of β. The value of τ has been arbitrarily chosen.

The locus of equation (3.53) in the complex plane is an arc with intercepts on the ε' axis at ε_s and ε_∞ at the low frequency and high frequency ends respectively (fig. 3-15). As $\omega \to 0$ the limiting curve is a semicircle with center on the ε' axis and as $\omega \to \infty$ the limiting straight line makes an angle of $\beta\pi/2$ with the ε' axis. To explain it another way, at low frequencies the points lie on a circular arc and at high frequencies they lie on a straight line.

If Davidson-Cole equation holds then the values of ε_s, ε_∞ and β may be determined directly, noting that a plot of the right hand quantity of eq. (3.54) against ω must yield a straight line. The frequency ω_p corresponding to tan (θ/β) =1 may be determined and τ may also be determined from the relation $\omega_p \tau$ = 1. We quote two examples to demonstrate Davidson-Cole relaxation in simple systems. Fig. 3-16 shows the measured loss factor in glycerol (b. p. 143-144°C at 300 Pa), over a wide range of temperature and frequency[15]. The asymmetry about the peak can clearly be seen and in the high frequency range, to the right of the peak at each temperature, a power law, $\omega^{-\beta}$ (β<1) holds true.

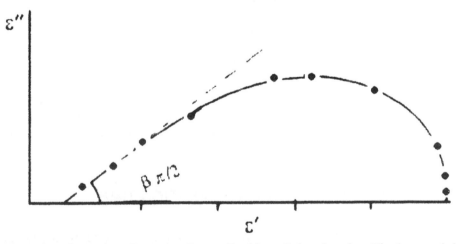

Fig. 3.15 Complex plane plot of $\varepsilon*$ according to Davidson-Cole relaxation. The loss peak is asymmetric and the low frequency branch is proportional to ω. The slope of the high frequency part depends on β.

The second example of Davidson-Cole relaxation is in mixtures of water and ethanol [Bao et. al., 1996] at various fractional contents of each liquid, as shown in fig. 3-17. The Davidson-Cole relaxation is found to hold true though the Debye relaxation may also be applicable if great accuracy is not required. The methods of determining the type of relaxation is dealt with later, but, as noted earlier, the Davidson-Cole relaxation is broader than the Debye relaxation depending upon the value of β.

We need to deal with an additional aspect of the complex plane plot of $\varepsilon*$ which is due to the fact that conductivity of the dielectric introduces anomalous increase of ε'' at both the high frequency (see the inset in fig. 3-12) and low frequency ends of the plots[16] (fig. 3.18). Equation (3.8) shows the contribution of ac conductivity to ε'' and this contribution should be subtracted before deciding upon the relaxation mechanisms.

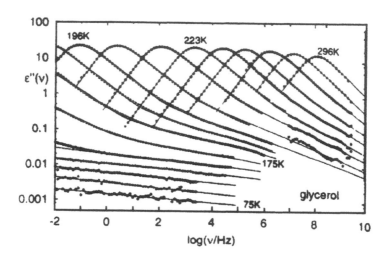

Fig. 3.16 ε'' as a function of ω in glycerol at various temperatures (75, 95, 115, 135, 175, 185, 190, 196, 203, 213, 223, 241, 256, 273 and 296 K) [15]. (with permission of J. Chem. Phys., USA).

3.9 MACROSCOPIC RELAXATION TIME

The relaxation time is a function of temperature according to a chemical rate process defined by

$$\tau = \tau_0 \exp\frac{b}{kT} \tag{3.59}$$

in which τ_0 and b are constants. This is referred to as an Arrhenius equation in the literature.

There is no theoretical basis for dependence of τ on T and in some liquids such as those studied by Davidson and Cole (1951) the relaxation time is expressed as

$$\tau = \tau_0 \exp \frac{b}{k(T - T_c)} \tag{3.60}$$

where T_c is a characteristic temperature for a particular liquid.

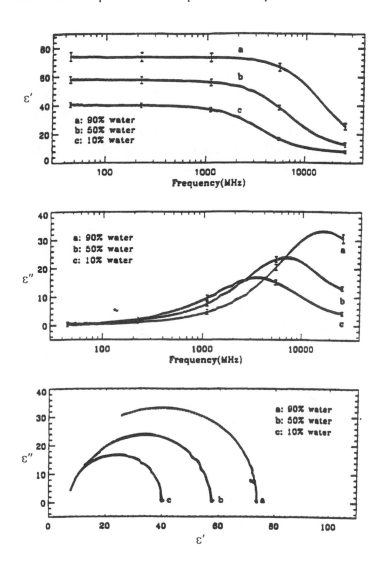

Fig. 3.17 Dielectric properties of water-ethanol mixtures at 25°C. (a) Real part ε' (b) Imaginary part, ε'' (c) Complex plane plot of $\varepsilon*$ exhibiting Davidson-Cole relaxation [Bao et. al. 1996] (with permission of American Inst. of Physics).

In some liquids the viscosity and measured low field conductivity also follow a similar law, the former given by

$$\eta = \eta_o \exp\left[\frac{b_\eta}{k(T - T_c)}\right] \tag{3.61}$$

At $T = T_c$ the relaxation time is infinity according to equation (3.60) which must be interpreted as meaning that the relaxation process becomes infinitely slow as we approach the characteristic temperature. Fig. 3.19 (Johari and Whalley, 1981) shows the plots of τ against the parameter 1000/T in ice. The slope of the line gives an activation energy of 0.58 eV, a further discussion of which is beyond the scope of the book. We will make use of equation (3.60) in understanding the behavior of amorphous polymers near the glass transition temperature T_G in chapter 5.

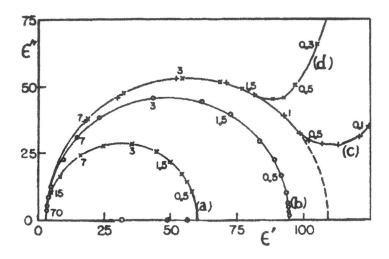

Fig. 3.18 Complex plane plot of ε^* in ice at 262.2 K. (a) sample with interface parallel to the electrodes; Curve (b), true locus; curves (c) and (d), samples with electrode polarization arising from dc conductance. Numbers beside points are frequencies in kHz (Auty and Cole, 1952). (with permission from Am. Inst. Phys.).

The observed correspondence of τ with viscosity is qualitatively in agreement with the molecular relaxation theory of Debye[17] who obtained the equation

$$\tau_m = \frac{3\eta \upsilon}{kT} \tag{3.62}$$

where τ_m is called the molecular relaxation time (see next section), η the viscosity, υ the molecular volume (= $4\pi a^3/3$ where a is the molecular radius), assumed spherical. The molecular volume required to obtain agreement with the relaxtion times is too small for glycerol and propylene glycol ($\sim 10^{-31} m^3$) and of reasonable size for n-propanol (60-900 $\times 10^{-30} m^3$).

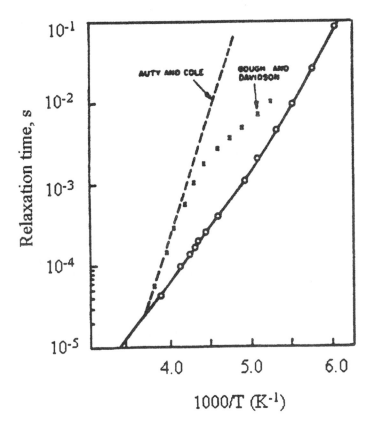

Fig. 3.19 Arrhenius plot of the relaxation time of ice I plotted as a function of 1/T (Johari and Whalley, 1981). (with permission of Am. Inst. Phys.).

The spread in these values arises due to the fact that the molecular relaxation time, τ_D, and the relaxation time, τ, obtained from the dielectric studies may be related in several ways. The inference is that for the first two liquids the units involved are much smaller where as for the third named liquid the unit involved may be the entire molecule. These details are included here to demonstrate the method employed to obtain insight into the relaxation mechanism from measurements of dielectric properties.

3.10 MOLECULAR RELAXATION TIME

The relaxation time τ obtained from dielectric studies that we have discussed so far is a macroscopic quantity and it is quite different from the original relaxation time τ_m used by Debye. τ_m is a microscopic quantity and usually called the internal relaxation time or the molecular relaxation time. τ_m may be expressed in terms of the viscosity of the liquid and the temperature as

$$\tau_m = \frac{4\pi\eta a^3}{kT} \tag{3.63}$$

where a is the molecular radius. The molecular relaxation time is assumed to be due to the inner friction of the medium that hinders the rotation of polar molecules. Hence τ_m is a function of viscosity. As an example of applicability of equation (3.63) we consider water that has a viscosity of 0.01 Poise at room temperature and an effective molecular radius of 2.2×10^{-10} m leading to a relaxation time of 2.5×10^{-11} s. At relaxation, the condition $\omega\tau = 1$ is satisfied and therefore $\omega = 4\times10^{10}$ s^{-1} which is in reasonable agreement with relaxation time obtained from dielectric studies. Equation (3.63) is however expected to be valid only approximately because the internal friction hindering the rotation of the molecule, which is a molecular parameter, is equated to the viscosity, which is a macroscopic parameter.

It is well known that the viscosity of a liquid varies with the temperature according to an empirical law:

$$\eta \propto \exp\frac{c}{kT} \tag{3.64}$$

in which c is a constant for a given liquid. Therefore eq. (3.63) may now be expressed as

$$\tau_m \propto \frac{1}{T}\exp\frac{c}{kT} \tag{3.65}$$

The relaxation time increases with decreasing temperature, as found in many substances. (see Table 3.2).

3.11 STRAIGHT LINE RELATIONSHIPS

There are many convenient methods for measuring the relaxation time experimentally and the formulas for calculating τ have been summarized by Hill et. al[18]. Let $\omega\tau = 1$ and $n^2 = \varepsilon_\infty$. Equation (3.28) and (3.29) may be expressed in several alternative ways:

1. $\dfrac{\varepsilon' - n^2}{\varepsilon_s - n^2} = \dfrac{1}{1 + x^2}$

$\dfrac{\varepsilon''}{\varepsilon_s - n^2} = \dfrac{x}{1 + x^2}$

Dividing the second equation from the first

$$x = \frac{\varepsilon''}{\varepsilon' - n^2} \tag{3.66}$$

2. It is easy to show that

$$x = \frac{\varepsilon_s - \varepsilon'}{\varepsilon''}$$

Equation (3.66) shows that a graph of ε''/x against ε' will be a straight line. The graph of $\varepsilon''x$ as a function of ε' will also be linear. n^2 and ε_s are obtained from the respective intercepts and τ from the slope.

The equations may also be written as :

3. $\quad \dfrac{x}{\varepsilon''} = \dfrac{x^2}{\varepsilon_s - n^2} + \dfrac{1}{\varepsilon_s - n^2}$

4. $\quad \dfrac{1}{\varepsilon''x} = \dfrac{1}{(\varepsilon_s - n^2)x^2} + \dfrac{1}{(\varepsilon_s - n^2)}$

5. $$\frac{1}{\varepsilon' - n^2} = \frac{x^2}{\varepsilon_s - n^2} + \frac{1}{\varepsilon_s - n^2}$$

6. $$\frac{1}{\varepsilon_s - \varepsilon'} = \frac{1}{(\varepsilon_s - n^2)x^2} + \frac{1}{(\varepsilon_s - n^2)}$$

Equations (3)-(6) above have the advantage that they each involve only one of the experimentally measured values of ε' and ε'', but the disadvantage is that the frequency term, x, enters through a squared term distorting the frequency scale. The advantage of these relationships lie in the fact that they may be used to check the standard deviation between computed values and measured values using easily available software.

The relation between the molecular relaxation time, τ_m, and the macroscopic relaxation time, τ, is given by (Hill et. al.)

$$\tau = \frac{\varepsilon_s + 2}{n^2 + 2}\tau_m \qquad\qquad (3.67)$$

The macroscopic relaxation time is higher in all cases than the microscopic relaxation time.

3.12 FROHLICH'S ANALYSIS

It is advantageous at this point to consider the dynamical treatment of Frohlich[19] who visualized a relaxation time that is dependent upon the temperature according to equation (3.63). To understand Frohlich's model let us take the electric field along the + axis and suppose that the dipole can orient in only two directions; one parallel to the electric field, the other anti-parallel. Let us also assume that the dipole is rigidly attached to a molecule. The energy of the dipole is +w when it is parallel to the field, -w when it is anti-parallel, and zero when it is perpendicular. As the field alternates the dipole rotates from a parallel to the anti-parallel position or vice-versa. Only two positions are allowed. A rotation through 180° is considered as a jump.

Frohlich generalized the model which is based on the concept that the activation energies of dipoles vary between two constant values, w_1 and w_2. Frohlich assumed that each process obeyed the Arrhenius relationship and the relaxation times corresponding to w_1 and w_2 are given by

$$\tau_1 = \tau_0 e^{\frac{w_1}{kT}}$$

$$\tau_2 = \tau_0 e^{\frac{w_2}{kT}}$$

The dipoles are distributed uniformly in the energy interval dw and make a contribution to the dielectric constant according to J_ω and H_ω. The analysis leads to fairly lengthy expressions in terms of the difference in the activation energy $w_2 - w_1$, and the final equations are

$$J_\omega = \frac{\varepsilon' - \varepsilon_\infty}{\varepsilon_s - \varepsilon_\infty} = 1 - \frac{1}{2s} Ln\left[\frac{1 + x^2 e^s}{1 + x^2 e^{-s}}\right]$$

$$H_\omega = \frac{\varepsilon''}{\varepsilon_s - \varepsilon_\infty} = \frac{1}{s}(e^{\frac{s}{2}} \tan^{-1} x - e^{\frac{-s}{2}} \tan^{-1} x) \tag{3.68}$$

where

$$s = \frac{w_2 - w_1}{kT}; \qquad x = \frac{\omega_p}{\omega}$$

We have used here the form of expressions given by Williams[20] because they have the advantage of using ω_p which is the radian frequency at which ε'' is a maximum. Since s is a function of temperature the shape of the J_ω and H_ω curves will vary with temperature, tending towards a single relaxation time at higher temperatures. ω_p is related to τ_1 and τ_2,

$$\omega_p = \frac{1}{\tau_1} e^{\frac{-(w_2 - w_1)}{2kT}} = \frac{1}{\sqrt{\tau_1 \tau_2}}$$

Fig. 3.20 shows the factor $\varepsilon''/\varepsilon''_{max}$ as a function of ω/ω_p calculated according to

$$\frac{\varepsilon''}{\varepsilon''_m} = \frac{(\tan^{-1}\frac{\omega}{\omega_p})\sqrt{\frac{\tau_2}{\tau_1}} - (\tan^{-1}\frac{\omega}{\omega_p})\sqrt{\frac{\tau_2}{\tau_1}}}{\tan^{-1}\sqrt{\frac{\tau_2}{\tau_1}} - \tan^{-1}\sqrt{\frac{\tau_1}{\tau_2}}} \tag{3.69}$$

The width of the loss curve increases with increasing $(w_2 - w_1)$, and in the limit the relaxation time will vary from zero to infinity and ε'' will have a constant value over the entire frequency range.

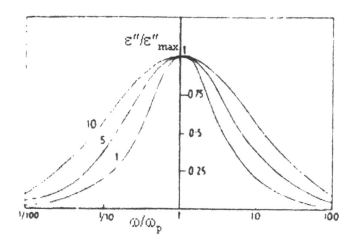

Fig. 3.20 Dependence of dielectric loss $\varepsilon''(\omega)$ on ω according to equation (3.69) for three values of the parameter $\sqrt{(\tau_2/\tau_1)} = 1, 5, 10$. They correspond to a range of heights of the potential barrier. ω_p is the frequency of ε''_{max} [1986]. (with permission of Clarendon Press, Oxford.)

The double potential well of Frohlich leads to a relaxation function that shows the peak of the ε'' - ω plots are independent of the temperature. However, measurements of ε'' in a wide range of materials show that the peak increases with increase in T. For example measurement of dielectric loss in polyimide having adsorbed water shows such a behavior[21].

The temperature dependence of ε''_{max} in the context of Frohlich's theory is often explained by assuming asymmetry in the energy level of the two positions of the dipole. At lower temperatures the lower well is occupied and the higher well remains empty. The number of dipoles jumping from the lower to higher well is zero. However, with increasing temperature the number of dipoles in the higher well increases until both wells occupied equally at kT>> (w_2-w_1). Therefore there will be a temperature range, pending upon the energy difference, in which ε'' increases strongly. According to this del the loss factor is

$$\varepsilon'' \propto \frac{1}{3kT} \cosh^{-2}\left[\frac{w_1 - w_2}{2kT}\right] = f(T)$$

here (w_2-w_1) is the difference in the asymmetry of the two positions. Plots of $f(T)$ rsus T for various values of V are shown in the range of V = 0 to 100 meV (fig. 3.21). comparing the observed variation of ε'' vs T it is possible to estimate the average ergy of asymmetry between the two wells.

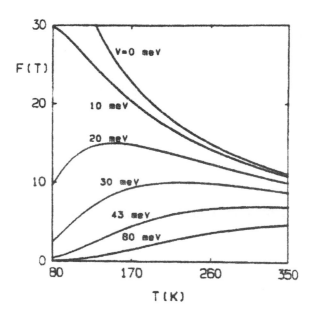

Fig. 3.21 Temperature dependance of the relaxation strength calculated for various values of the asymmetry potential w_1-w_2. The upper curve corresponds to the symmetric case w_1-w_2 = 0 (Melcher et. al., 1989). (with permission of Trans. IEEE on Diel. and El. Insul.)

3.13 FUOSS-KIRKWOOD EQUATION

e Fuoss-Kirkwood[22] dispersion equation is

$$\varepsilon'' = \varepsilon_\infty + (\varepsilon_s - \varepsilon_\infty)\delta \frac{(\omega\tau)^{\delta}}{1+(\omega\tau)^{2\delta}} \qquad (3.70)$$

Let us denote the frequency at which the loss factor is a maximum for Debye relaxation by ω_p. Then the loss factor in terms of its maximum value, $\varepsilon''_{max,}$ is

$$\varepsilon'' = \frac{2\varepsilon''_{max}}{\dfrac{\omega}{\omega_p} + \dfrac{\omega_p}{\omega}}$$

which leads to

$$\varepsilon'' = \varepsilon''_{max} \sec h(Ln\frac{\omega}{\omega_p})$$

In the Fuoss-Kirkwood derivation this expression becomes

$$\varepsilon'' = \varepsilon''_{max} \sec h(\delta\, Ln\frac{\omega}{\omega_p})$$

The empirical factor δ, between 0 and 1, is related to the Cole-Cole parameter α according to

$$\delta\sqrt{2} = \frac{1-\alpha}{\cos\dfrac{(1-\alpha)\pi}{4}}$$

This expression shows that Fuoss-Kirkwood relation holds true for most materials in which the Cole-Cole relaxation is observed; only the value of the parameter will be different.

When $\varepsilon'' = 1/2\ \varepsilon''_{max}$ expression (3.70) leads to $(\omega\tau)^{\delta} = 2 \pm \sqrt{3}$. To apply Fuoss-Kirkwood equation we plot arccosh($\varepsilon''/\varepsilon'$) against ($ln\omega$) to get a straight line. This line intersects the frequency axis at $\omega = \omega_p$ with a slope of δ which is related to the static permittivity in accordance with[23]

$$\delta = \frac{2\varepsilon''_{max}}{\varepsilon_s - \varepsilon_\infty}$$

Fig. 3.22 shows such a plot for PMMA at 353K and the evaluated parameter (ε_s - ε_∞) has value of $\cong 2$.

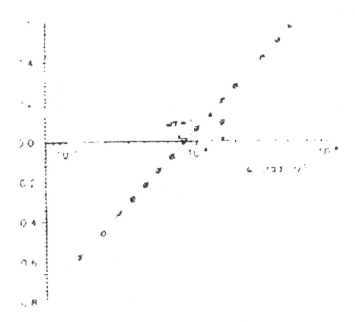

Fig. 3.22 The dependence of arccosh($\varepsilon''_{max}/\varepsilon''$) on log($\omega$) for PMMA at 353 K. τ is the relaxation time. Slope gives the value of β. From the slope (ε_S-ε_∞) is evaluated as $\cong 2$ [Mazur, 1997]. (with permission of Institute of Physics).

Kirkwood and Fuoss[24] also derived the relaxation time distribution function for a freely jointed polyvinyl chloride (PVC) chain, including the molecular weight distribution. Recall that the number of monomer units in a polymer molecule is not a constant and shows a distribution. The theory takes into account the hydrodynamic diffusion of chain segments under an electric field (Williams, 1963). Polyvinyl chloride and poly(acetaldehyde) show common characteristics of the dipole forming a rigid part of the chain backbone. The Kirkwood-Fuoss relation is

$$J_\omega - jH_\omega = \int_0^c \frac{[1 + u + u(u + 2)e^u Ei(-u)][1 - jxu]du}{1 + x^2 u^2} \qquad (3.71)$$

here $u = n / <n>$, n is the degree of polymerization, $<n>$ its average value, $Ei(u)$ the exponential integral, $x = <n>\omega\tau^*$, τ^* is the relaxation time of the monomer unit. H_ω reaches its peak value at $x = 0.1 \times 2\pi$ and is symmetrical about this value. The main feature of Kirkwood-Fuoss distribution is that it is free of any empirical parameter because J_ω and H_ω are expressed in terms of $\log x$.

The shape and height of no-parameter distribution of Kirkwood-Fuoss is independent of temperature and this fact explains the reason for their theory not holding true for PVC. The distribution in PVC is markedly temperature dependent (Kirk and Fuoss, 1941).

Many polymers exhibit a temperature dependent distribution and at higher temperatures some are noticeably temperature independent. At these higher temperatures the Cole-Cole parameter has an approximate value of 0.63 (Kirkwood and Fuoss, 1941). It seems likely that the Kirkwood-Fuoss relation holds when the shape of the distribution is independent of the temperature.

3.14 HAVRILIAK AND NEGAMI DISPERSION

We are now in a position to extend our treatment to more complicated molecular structures, in particular polymer materials. The dispersion in small organic or inorganic molecules is studied by measuring the complex dielectric constant of the material at constant temperature over as wide a range of frequency as possible. The temperature is then varied and the measurements repeated till the desired range of temperature is covered. From each set of isothermal data the complex plane plots are obtained and analyzed to check whether a semi-circular arc in accordance with Cole-Cole equation is obtained or whether a skewed arc in accordance with the Davidson-Cole equation is obtained.

The complex plane plots of polymers obtained by isothermal measurements do not lend themselves to the simple treatment that is used in case of simple molecules. The main reasons for this difficulty are: (1) The dispersion in polymers is generally very broad so that data from a fixed temperature are not sufficient for analysis of the dispersion. Data from several temperatures have to be pooled to describe dispersions meaningfully. (2) The shapes of the plots in the complex plane are rarely as simple as that obtained with molecules of simpler structure rendering the determination of dispersion parameters very uncertain.

In an attempt to study the α-dispersion in many polymers, Havriliak and Negami[25] have measured the dielectric properties of several polymers. α-dispersion in a polymer is the process associated with the glass transition temperatures where many physical properties change in a significant way. In several polymers the complex plane plot is linear at high frequencies and a circular arc at low frequencies. Attempts to fit a circular arc (Cole-Cole) is successful at lower frequencies but not at higher frequencies. Likewise, an attempted fit with a skewed circular arc (Davidson-Cole) is successful at higher frequencies but not at lower frequencies.

The two dispersion equations, reproduced here for convenience, are represented by:

$$\frac{\varepsilon^* - \varepsilon_\infty}{\varepsilon_s - \varepsilon_\infty} = [1 + (j\omega\tau)^{1-\alpha}]^{-1} : \quad \text{circular arc} \quad \text{(Cole - Cole)}$$

$$\frac{\varepsilon^* - \varepsilon_\infty}{\varepsilon_s - \varepsilon_\infty} = (1 + j\omega\tau_0)^{-\beta} : \text{Skewed semicircle} \quad \text{(Davidson - Cole)}$$

Combining the two equations, Havriliak and Negami proposed a function for the complex dielectric constant as

$$\frac{\varepsilon^* - \varepsilon_\infty}{\varepsilon_s - \varepsilon_\infty} = [1 + (j\omega\tau_{H-N})^{1-\alpha}]^{-\beta} \tag{3.72}$$

This function generates the previously discussed relaxations as special cases. When β = 1 the circular arc shown above is generated. When α = 0 the skewed semicircle is obtained. When α = 0 and β = 1 the Debye function is obtained. For convenience we omit the subscript H-N hereafter.

To test the relaxation function given by equation (3.72) we apply successively the DeMoivre's theorem and rationalize the denominator to obtain the expressions

$$\varepsilon' - \varepsilon_\infty = \frac{1}{r^{\beta/2}} (\varepsilon_s - \varepsilon_\infty) \cos(\beta\theta) \tag{3.73}$$

$$\varepsilon'' = \frac{1}{r^{\beta/2}}(\varepsilon_s - \varepsilon_\infty)\sin(\beta\theta) \qquad (3.74)$$

where

$$r = [1 + (\omega\tau)^{1-\alpha}\sin(\frac{\alpha\pi}{2})]^2 + [(\omega\tau)^{1-\alpha}\cos(\frac{\alpha\pi}{2})]^2 \qquad (3.75)$$

$$\theta = arc\tan[\frac{(\omega\tau)^{1-\alpha}\cos(\frac{\alpha\pi}{2})}{1 + (\omega\tau)^{1-\alpha}\sin(\frac{\alpha\pi}{2})}] \qquad (3.76)$$

Equation (3.72) may be examined for extreme values of ω, namely $\omega \to \infty$ and $\omega \to 0$[26].

For the first case, as $\omega \to \infty$, equation (3.72) becomes (for definition of χ'' see section 3.15)

$$\frac{\varepsilon* - \varepsilon_\infty}{\varepsilon_s - \varepsilon_\infty} \simeq (j\omega\tau)^{\beta(\alpha-1)}$$

$$= (\omega\tau)^{\beta(\alpha-1)}\{\cos[\beta(1-\alpha)\pi/2] - j\sin[\beta(1-\alpha)\pi/2]\},$$

$$\tan[\beta(1-\alpha)\pi/2] = \frac{\varepsilon''_{(\omega\to\infty)}}{\varepsilon'_{(\omega\to\infty)} - \varepsilon_\infty}$$

At very high frequencies $\varepsilon'' \propto (\varepsilon' - \varepsilon_\infty) \propto \omega^{\beta(1-\alpha)}$

For the second case, as $\omega \to 0$,

$$\frac{\varepsilon* - \varepsilon_\infty}{\varepsilon_s - \varepsilon_\infty} \simeq 1 - \beta(j\omega\tau)^{(1-\alpha)}$$

$$= 1 - \beta(\omega\tau_H)^{(1-\alpha)}\{\cos[(1-\alpha)\pi/2] + j\sin[(1-\alpha)\pi/2]\},$$

$$\tan\left(\frac{\alpha\pi}{2}\right) = \frac{\varepsilon''_{(\omega\to0)}}{\varepsilon_s - \varepsilon'_{(\omega\to0)}}$$

At very low frequencies, $\varepsilon'' \propto (\varepsilon_s-\varepsilon') \propto \omega^{(1-\alpha)}$. These results have led to the suggestion that the susceptibility functions have slopes as shown in the two extreme cases.

From equations (3.73) and (3.74) we note the following with regard to the dispersion parameter

I.   As $\omega\tau \to \infty$, $\varepsilon' \to \varepsilon_\infty$ and $\varepsilon'' \to 0$. Therefore $\varepsilon^* = \varepsilon_\infty$

II.   As $\omega\tau \to 0$, $\varepsilon' \to \varepsilon_s$ and $\varepsilon'' \to 0$. Therefore $\varepsilon^* \to \varepsilon_s$.

We can therefore evaluate ε_s and ε_∞ from the intercept of the curve with the real axis. To find the parameters α and β we note that equations (3.73) and (3.74) result in the expression

$$\frac{\varepsilon''}{\varepsilon'-\varepsilon_\infty} = \tan \beta\theta = \tan \varphi \tag{3.77}$$

where we have made the substitution $\beta\theta = \phi$. By applying the condition $\omega\tau \to \infty$ to equation (3.76) and denoting the corresponding value of ϕ as ϕ_L (see fig. 3.23) we get

$$\phi_L = (1-\alpha)\beta\pi/2 \tag{3.78}$$

which provides a relation between the graphical parameter ϕ_L and the dispersion parameters, α and β. Again the relaxation time is given by the definition $\omega\tau \equiv 1$, and let us denote all parameters at this frequency by the subscript p. Havriliak and Negami (1966) also prove that the bisector of angle ϕ_L intersects the complex plane plot at ε_p^*. The point of intersection yields the value of α by

$$\frac{1}{\phi_L} \log \frac{\left| \varepsilon_p{}^*-\varepsilon_\infty \right|}{\varepsilon_s - \varepsilon_\infty} = -\frac{1}{\pi(1-\alpha)} \log[2+2\sin\alpha(\pi/2)] \tag{3.79}$$

The analysis of experimental data to evaluate the dispersion parameters is carried out by the following procedure from the complex plane plots:

1. The low frequency measurements are extrapolated to intersect the real axis from which ε_s is obtained.

2. The high frequency measurements are extrapolated to intersect the real axis from which ε_∞ is obtained. If data on refractive index is available then the relation $\varepsilon_\infty = n^2$ may be employed in specific materials.
3. The parameter φ_L is measured using the measurements at high frequencies.
4. The angle φ_L is bisected and extended to intersect the measured curve. From the intersection point the frequency is determined and the corresponding relaxation time is calculated according to $\omega = 1/\tau$. The parameters $\varepsilon_p' - \varepsilon_\infty$ and ε'' are also determined.
5. The parameter α is decided by equation (3.79)
6. The parameter β is calculated by equation (3.78)

Havriliak and Negami analysed the data of several polymers and evaluated the five dispersion parameters (ε_s, ε_∞, α, β, τ) for each one of them (see chapter 5). A more recent list of tabulated values is given in Table 3.3[27]. The Havriliak and Negami function is found to be very useful to describe the relaxation in amorphous polymers which exhibit asymmetrical shape near the glass transition temperature, T_G. In the vicinity of T_G the ε''-$\log\omega$ curves become broader as T is lowered. It has been suggested that the α-parameter represents a quantity that denotes chain connectivity and β is related to the local density fluctuations. Chain connectivity in polymers should decrease as the temperature is lowered. The α-parameter slowly increases above T_G which may be considered as indicative of this[28]. A detailed description of these aspects are treated in ch. 5.

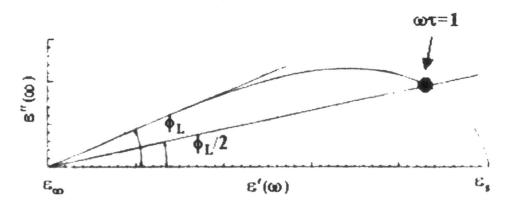

Fig. 3.23 Complex plane plot of ε^* according to H-N function. At high frequencies the plot is linear. At low frequencies the plot is circular (Jonscher, 1999).

A final comment about the influence of conductivity on dielectric loss is appropriate here. As mentioned earlier measurement of ε'' - ω characteristics shows a sudden rise in the loss factor towards the lower frequencies, (e. g., see fig. 3.12) and this increase is due to the dc conductivity of the material. The Havriliak-Negami function is then expressed as

$$\frac{\varepsilon^* - \varepsilon_\infty}{\varepsilon_s - \varepsilon_\infty} = \left[1 + (j\omega\tau_0)^{(1-\alpha)}\right]^\beta - j\frac{\sigma_{dc}}{\omega\varepsilon_o}$$

where σ_{dc} is the dc conductivity.

3.15 DIELECTRIC SUSCEPTIBILITY

In the published literature some authors[29] use the **dielectric susceptibility, $\chi^* = \chi' - j\chi''$** of the material instead of the dielectric constant ε^* and the following relationships hold between the dielectric susceptibility and dielectric constant:

$$\chi^* = \varepsilon^* - \varepsilon_\infty \tag{3.80}$$

$$\chi' = \varepsilon' - \varepsilon_\infty \tag{3.81}$$

$$\chi'' = \varepsilon'' \tag{3.82}$$

The last two quantities are often expressed as normalized quantities,

$$\chi' = \frac{\varepsilon' - \varepsilon_\infty}{\varepsilon_s - \varepsilon_\infty}; \chi'' = \frac{\varepsilon''}{\varepsilon_s - \varepsilon_\infty}$$

Equations (3.81) and (3.82) may also be expressed in a concise form as

$$\chi^* \propto \frac{1}{1 + j\dfrac{\omega}{\omega_p}}$$

where ω_p is the peak at which χ'' is a maximum. Alternately we have

$$\chi^* \propto \frac{1}{1 + j\omega\tau} = \frac{1}{1 + \omega^2\tau^2} - \frac{j\omega\tau}{1 + \omega^2\tau^2} \tag{3.83}$$

Equation (3.83) is known as the **Debye Susceptibility function**. Expressing the dielectric properties in terms of the susceptibility function has the advantage that the slopes of the plots of χ' and χ'' against ω provide a convenient parameter for discussing the possible relaxation mechanisms. Fig. 3.24 (a) shows the variation of χ' and χ'' as a function of frequency[30]. As discussed, in connection with the Debye equations, the decreasing part of $\chi' \propto \omega^{-2}$. The increasing part of χ'' at $\omega \ll \omega_p$ changes in proportion to ω^{+1} and the decreasing part of χ'' at $\omega \gg \omega_p$ changes in proportion to ω^{-1}.

Table 3.4

Selected Dispersion Parameters according to H-N Expression
(Havriliak and Watts, 1986: with permission of Polymer)

Polymer and Temperature	ε_s	ε_∞	$\log f_{max}$	α	β
Poly(carbonate)	3.64	3.12	6.85	0.77	0.29
Polychloroprene (-26°C)	5.85	2.63	7.37	0.57	0.51
Poly(cyclohexyl methacrylate) 121°C	4.33	2.45	5.33	0.71	0.33
Poly(iso-butyl methacrylate) 102.8°C	4.02	2.36	8.28	0.71	0.50
Poly(n-butyl methacrylate) 59°C	4.29	2.44	7.06	0.62	0.60
Poly(n-hexyl methacrylate) 48°C	3.96	2.48	4.40	0.74	0.66
Poly(nonyl methacrylate) 42.8°C	3.51	2.44	8.18	0.73	0.65
Poly(n-octyl methacrylate) 21.5°C	3.88	2.61	9.60	0.73	0.66
Poly(vinyl acetal) 90°C	6.7	2.5	6.43	0.89	0.30
Poly(vinyl acetate) 66°C	8.61	3.02	7.11	0.90	0.51
Syndiotac-Poly(methyl methacrylate)	4.32	2.52	7.96	0.53	0.55
Poly(vinyl formal)	5.85	2.62	7.37	0.56	0.51

The Cole-Cole function has the form:

$$\chi^* \propto \frac{1}{1+(j\omega\tau)^{1-\alpha}} \tag{3.84}$$

which is broader than the Debye function, though both are symmetrical about the frquency at which maximum loss occurs. The Davidson-Cole has the form

$$\chi^* \propto \frac{1}{\left[1+(j\omega\tau)\right]^\beta} \tag{3.85}$$

These susceptibility functions are generalized and the relaxation mechanism is expressed in terms of two independent parameters, α and β, by Havriliak and Negami, their susceptibility function having the form:

$$\chi^* = \frac{1}{\left[1+(j\omega\tau)^{1-\alpha}\right]^\beta} \tag{3.86}$$

Figs. 3.24 (b) and (c) show the variation of these functions as a function of frequency. The similarities and divergences of these functions may be summarized as follows:

1. The Debye function is symmetrical about ω_p and narrower than the Cole-Cole and Davidson-Cole functions. The low frequency region of the dispersion has a slope that is proportional to ω and the high frequency region has a slope that is proportional to $1/\omega$. In this context the low frequency and high frequency regions are defined as $\omega \ll \omega_p$ and $\omega \gg \omega_p$. In the high frequency region χ' has a slope proportional to $1/\omega^2$ (Fig. 3.24 a).
2. The Cole-Cole function (Fig. 3.24b) shows that χ'' has a slope proportional to $\omega^{1-\alpha}$ in the low frequency region and a slope proportional to $\omega^{\alpha-1}$ in the high frequency region. It is also symmetrical about ω_p. The variation of χ' in the high frequency region is also proportional to $\omega^{\alpha-1}$.
3. The Davidson-Cole susceptibility function (Fig. 3.24 c) is asymmetrical about the vertical line drawn at ω_p. In the low frequency region χ'' is proportional to ω which is a behavior similar to that of Debye. In the high frequency region χ'' is proportional to $\omega^{-\beta}$. The real part of the susceptibility function, χ', in the high frequency region is also proportional to $\omega^{-\beta}$. It should be borne in mind that the Cole-Cole and Davidson -Cole susceptibility functions have a single parameter, whereas the Debye function has none.

To demonstrate the applicability of Davidson-Cole function we refer to the measurement of dielectric properties of glycerol by Blochowicz et. al. (1999) over a temperature range of 75-296 K and a frequency range of $10^{-2} \le f \le 3000$ MHz. A plot $\varepsilon''(f)$ as a function of $\log(f)$ is shown in fig. 3.16. At $f < f_{max}$, ε'' behaves similar to a Debye function but at $f > f_{max}$ there are two slopes, $-\beta$ and $-\gamma$ appearing in that order for increasing frequency.

Since Davidson-Cole function uses one parameter only a modification to equation (3.85) is proposed by Blochowicz et. al (1999).

$$\chi^* = \frac{\left(1 + \dfrac{j\omega_o\tau_o}{C_o}\right)^{\beta-\gamma}}{(1 + j\omega\tau_o)^{\beta}}$$

Here $\tau_o = 1/\omega_p$ and C_o is a parameter that controls the frequency of transition from β-power law to γ-power law.

The condition $\beta = \gamma$ signifying a single slope for the log f-ε'' yields Davidson-Cole function because the two slopes merge into one. The mean relaxation time is obtained by the relation

$$\tau = \lim_{\omega \to 0} \frac{\varepsilon''}{\omega} = \beta\tau_o - (\beta - \tau)\frac{\tau_0}{C_0} = \beta\tau_0$$

The temperature dependence of β, γ and C_o is shown in their fig. 4(b). β is weakly dependent on T whereas γ increases rapidly to approach β.

At the glass transition temperature the two power laws merge into one in accordance with equation (3.85). The relaxation in the glass phase (T < T_g) is determined by a single power law over a wide frequency range of $10^{-2} < f < 10^5$ Hz. Below T_g the co-efficient γ is not temperature dependent and very similar for many systems exhibiting this type of behavior. For glycerol $\gamma = 0.07 \pm 0.02$. The behavior below T << T_g occurs according to $\chi'' = \omega^{-\gamma}$. The high frequency contribution of α-relaxation is frozen out at T << T_g. If the data on χ'' is replotted in the T domain at $f = 1$ Hz an exponential relationship is obtained according to

$$\chi'' = e^{T/T_r}$$

where T_r is a constant that is dependent on the material. Their fig. (6a) shows this relationship for many substances, the departure from this equation being due to the onset of the α-process.

4. The Havriliak-Negami function is more general because of the fact that it has two parameters. The comments made with regard to equations(3.73) and (3.74) are equally

applicable to their susceptibility functions; we get the three functions listed above as special cases:

(i) $\alpha = 0$ and $\beta = 1$ gives the Debye equation
(ii) $0 < \alpha < 1$ and $\beta = 1$ gives the Cole-Cole function
(iii) $\alpha = 0$ and $0 < \beta < 1$ gives the Davidson-Cole function

Though this susceptibility function has two parameters which are independently adjustable, the low frequency behavior is not entirely independent of the high frequency behavior. This is due to the fact that that the parameter α appears in both the expressions for χ' and χ''.

The parameters α and β have no physical meaning though these models are successful in fitting the measured data to one of these functions. It has been suggested that a distribution of relaxation times is usually associated with one of the Cole-Cole, Davidson-Cole, and Havriliak-Negami functions. In these models the observed dielectric response is considered as a summation, through a distribution function $G(\tau)$, of individual Debye responses for each group.

3.16 DISTRIBUTION OF RELAXATION TIMES

We have already considered the situations in which there are more than a single relaxation time (section 3.6) and we intend to examine this topic more closely in this section. Polymers generally have a number of relaxation times due to the fact that the long chains may be twisted in a complex array of molecular segments or due to the complicated energy distribution in crystalline regions of the polymer. Since the distribution function is denoted by $G(\tau)$ the fraction of dipoles having relaxation times between τ and $\tau + \Delta\tau$ is given as $G(\tau)\,d\tau$ and hence for all dipoles

$$\int_0^\infty G(\tau)d\tau = 1 \tag{3.87}$$

Assuming that each differential group of dipoles with different relaxation times follows Debye behavior non-interactively, the complex dielectric constant becomes:

$$\varepsilon^* = \varepsilon_\infty + (\varepsilon_s - \varepsilon_\infty)\int_0^\infty \frac{G(\tau)d\tau}{1 + j\omega\tau} \tag{3.88}$$

By separating the real and imaginary parts of equations (3.88), the dielectric constant and the loss factor are expressed as

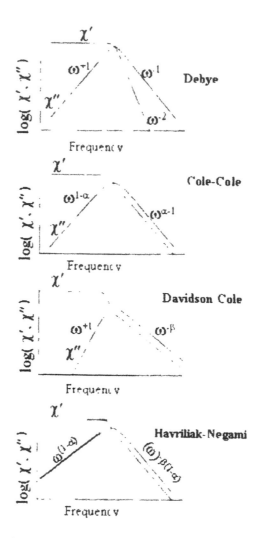

Fig. 3.24 Frequency dependence of susceptibility functions (Das Gupta and Scarpa, ©1999, IEEE). H-N relaxation has been added by the author.

$$\varepsilon' = \varepsilon_\infty + (\varepsilon_s - \varepsilon_\infty) \int_0^\infty \frac{G(\tau)d\tau}{1 + \omega^2 \tau^2} \tag{3.89}$$

$$\varepsilon'' = (\varepsilon_s - \varepsilon_\infty) \int_0^\infty \frac{\omega\tau\, G(\tau)d\tau}{1 + \omega^2 \tau^2} \tag{3.90}$$

The susceptibility function is expressed as:

$$\chi^* = \int_0^\infty \frac{G(\omega)}{1 + j\omega\tau} d\tau \tag{3.91}$$

In some texts (Vera Daniel, 1967) the normalization of equation (3.87) is carried out by expressing the right hand side of equation (3.87) as $(\varepsilon_s - \varepsilon_\infty)$. This results in the absence of this factor in the last term in equations (3.88) - (3.90). The existence of a distributed relaxation time explains the departure from the ideal Debye behavior in the susceptibility functions. However it has been suggested that a distribution of relaxation times cannot be correlated with the existence of relaxing entities in a solid, to justify physical reality.

Equations (3.89) and (3.90) are the basis for determining $G(\tau)$ from ε' and ε'' data though the procedure, as mentioned earlier, is not straight forward. Simple functions of $G(\tau)$ such as Gaussian distribution lead to complicated functions of ε' and ε''. On the other hand, simple functions of ε' and ε'' also lead to complicated functions of $G(\tau)$.

Fig. 3.25 helps to visualize the continuous distribution of relaxation times according to H-N expression for various values of the parameter β and $\alpha\beta=1$ using expressions to follow[31]. The corresponding complex plane plots are also shown.

It is helpful to express equations (3.89) and (3.90) as (Williams, 1963)

$$J_\omega = \frac{\varepsilon' - \varepsilon_\infty}{\varepsilon_s - \varepsilon_\infty} = \int_0^\infty \frac{G(\tau)}{1 + \omega^2 \tau^2} \tag{3.92}$$

$$H_\omega = \frac{\varepsilon''}{\varepsilon_0 - \varepsilon_\infty} = \int_0^\infty \frac{G(\tau)\omega\tau}{1 + \omega^2 \tau^2} d\tau \tag{3.93}$$

To demonstrate the usefulness of equations (3.92) and (3.93) the measured loss factor in amorphous polyacetaldehyde (Williams, 1963) over a temperature range of -9°C to +34.8°C and a frequency range of 25Hz-100kHz is shown in Fig. 3.26. Polyacetaldehyde is a polar polymer with its dielectric moment in the main chain, similar to PVC. Its monomer has a molecular weight of 44 and it belongs to the class of atactic polymers. Its refractive index is 1.437. A single broad peak was observed at all temperatures.

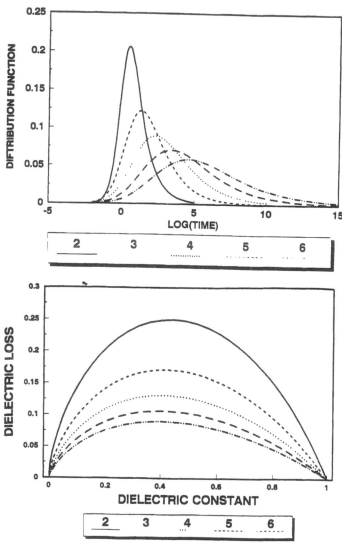

Fig. 3.25 Distribution of relaxation times for various values of β according to H-N dispersion. The corresponding complex plane plots of ε* are also shown for αβ=1 [Runt and Fitzgerald, 1997]. (with permission of Am. Chem. Soc.).

Fig. 3.27 shows these data replotted as J_ω and $H_\omega/H_{\omega p}$ as a function of (ω/ω_p). The experimental points all lie on a master curve indicating that the shape of the distribution of relaxation times is independent of the temperature.

Evaluation of the distribution function from such data is a formidable task requiring a detailed knowledge of Laplace transforms. The relaxation time distribution appropriate to the Cole-Cole equation is[32]

$$G(\tau) = \frac{\sin \alpha\pi}{2\pi \cosh[(1-\alpha)\ln \tau/\tau_0] - \cos\alpha\pi} \qquad (3.94)$$

in which τ_0 is the relaxation time at the center of the distribution.

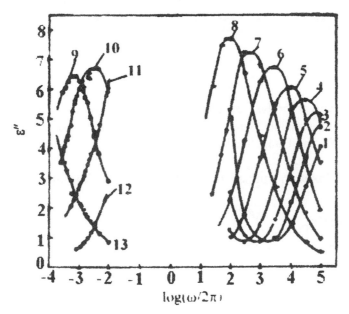

Fig. 3.26 Plot of ε'' against log (w/2p) for 0.598 thick sample. 1-34.8°C, 2-30.5 °C, 3-25 °C, 4-18.5 °C, 5-9.7 °C, 6-3.25 °C, 7—3.5 °C, 8--9 °C, 9—19.2 °C, 10—21.8 °C, 11—24.5 °C, 12—26.4 °C, 13—28.7 °C [Williams, 1963] (with permission of Trans. Farad. Soc.).

As demonstrated earlier (fig. 3-10) the complex plane plot of the Cole-Cole distribution is symmetrical about the mid point and therefore the plot of $G(\tau)$ against log τ or log (τ/τ_{mean}) will be symmetrical about the line. The graphical technique for the analysis of dielectric data makes use of fig. 3.9. The quantity u/v is plotted against log v and the result will be a straight line of slope 1-α. Without this verification the Cole-Cole relationship cannot be established with certainty.

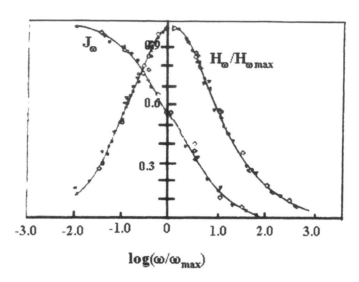

Fig. 3.27 Master curves for J_W , eq. (3.92) and $H\omega/H\omega max$, eq. (3.93) as a function of log ($\omega/\omega max$). 0.598 thick sample. Symbols are the same as in fig. 3-26 [Williams, 1963]. Adopted with permission of Trans. Farad. Soc.]

The distribution of relaxation time according to Davidson-Cole function is

$$G(\tau) = \frac{Sin\beta\pi}{\pi}\left(\frac{\tau}{\tau_0 - \tau}\right)^\beta \qquad \tau < \tau_0 \qquad\qquad (3.95)$$

$$= 0 \qquad\qquad\qquad \tau > \tau_0 \qquad\qquad (3.96)$$

The distribution of relaxation times for the Fuoss-Kirkwood function is a logarithmic function:

$$G(\tau) = \frac{\partial}{\pi} \frac{\cos(\frac{\partial\pi}{2})\cosh(\partial s)}{\cos^2(\frac{\partial\pi}{2}) + \sinh^2(\partial s)} \qquad\qquad (3.97)$$

Where δ is a constant defined in section (3.12) and s= log (ω/ω_p). The distribution of relaxation times for H-N function is given by [Havriliak and Havriliak, 1997]

$$G(\tau) = \left(\frac{1}{\pi}\right) y^{\alpha\beta} (\sin\beta\theta)(y^{2\alpha} + 2y^{\alpha}\cos\pi\alpha + 1)^{-\beta/2} \tag{3.98}$$

In this expression

$$y = \frac{\tau}{\tau_0} \tag{3.99}$$

$$\theta = \arctan\left(\frac{\sin\pi\alpha}{y^{\alpha} + \cos\pi\alpha}\right) \tag{3.100}$$

The distribution of relaxation times may also be represented according to an equation of the form, called Gaussian function (Hasted, 1973) given by

$$G(\tau) = \left(\frac{1}{\sigma\sqrt{2\pi}}\right) \exp\left\{\left(-\frac{1}{2}\right)\left(\frac{y}{\sigma}\right)^2\right\} \tag{3.101}$$

where σ is known as the standard deviation and indicates the breadth of the dispersion. From the form of this function it can be recognized that the distribution, and hence the ε''-ω plot, will be symmetrical about the central or relaxation time (fig. 3.28). As the standard deviation increases the $\log(\varepsilon'') - \log(\omega)$ plots become narrower, and for the case $1/\sigma = 0$, the distribution reduces to a single relaxation time of Debye relaxation. In fig. 3.28 the frequency is shown as the variable on the x-axis instead of the traditional τ/τ_{mean}; conversion to the latter variable is easy because of the relationship $\omega\tau=1$. In almost every case the actual distribution is difficult to determine from the dielectric data whereas its width and symmetry are easier to recognize.

A simple relationship between $(\varepsilon_s-\varepsilon_\infty)$ and ε'' may be derived[33]. The area under the ε''-log ω curve is

$$\int_0^\infty \varepsilon'' d(Ln\omega) = (\varepsilon_s - \varepsilon_\infty) \int_{\tau=0}^\infty \int_{\omega=0}^\infty \frac{G(\tau)\omega\tau}{1+\omega^2\tau^2} d\tau\, d(Ln\omega) \tag{3.102}$$

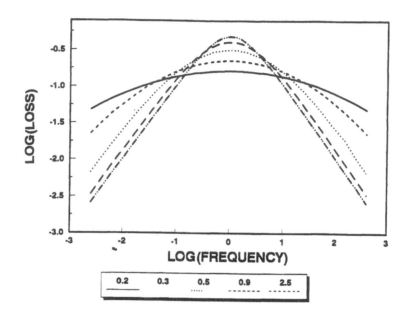

Fig. 3.28 Log ε'' against log(frequency) for Gaussian distribution of relaxation times. The numbers show the standard deviation, σ. Debye relaxation is obtained for $1/s=0$. Note that the slope at high frequency and low frequency tends to +1 and -1 as σ increases.[Havriliak and Havriliak, 1997]. (Permission of Amer. Chem. Soc.)

Using the identity

$$\int\limits_{0}^{\infty} \frac{\omega\tau}{1-\omega^2\tau^2} d(Ln\omega) = \frac{\pi}{2}$$

expression (3.102) simplifies into, because of equation (3.87),

$$\int\limits_{0}^{\infty} \varepsilon'' d(Ln\omega) = \int \frac{\varepsilon'' d\omega}{\omega} = \frac{\pi}{2}(\varepsilon_s - \varepsilon_\infty) \qquad (3.103)$$

The inversion formula corresponding to equation (3.103) is

$$\varepsilon'' \cong -\frac{\pi}{2}\frac{\partial\varepsilon'}{\partial(\ln\omega)} \qquad (3.104)$$

which is useful to calculate ε'' approximately.

Equation (3.103) may be verified in materials that have Debye relaxation or materials that have a peak in the ε'' - log ω characteristic though the peak may be broader than that for Debye relaxation. Such calculations have been employed by Reddish[34] to obtain the dielectric constant of PVc and chlorinated PVc (see Chapter 5). For measurements the frequency range can be extended by making measurements at different temperatures because ω_p, τ and T are related through equations

$$\omega_p \tau = 1 \tag{3.105}$$

$$\tau = \tau_0 \exp \frac{w}{kT} \tag{3.106}$$

Fig. 3.29 (Hasted, 1963) summarizes the dielectric properties ε'- ε'' in the complex plane, the shape of the distribution of relaxation time and the decay function which will be discussed in chapter 6.

3.17 KRAMER-KRONIG RELATIONS

Expressions (3.89) and (3.90) use the same relaxation function $G(\tau)$ and in principle we must be able to calculate one function if the other function is known. This is true only if ε' are related ε'' and these relations are known as Kramer-Konig relations[35]:

$$\varepsilon'(\omega) - \varepsilon_\infty = \frac{2}{\pi} \int_0^\infty \frac{x\varepsilon''(x)}{x^2 - \omega^2} dx \tag{3.107}$$

$$\varepsilon''(\omega) = -\frac{2\omega}{\pi} \int_0^\infty \frac{\varepsilon' - \varepsilon_\infty}{x^2 - \omega^2} dx \tag{3.108}$$

Integration is carried out using an auxiliary variable x which is real. Equations (3.107) and (3.108) imply that at $\omega = \infty$, $\varepsilon' = \varepsilon_\infty$ and $\varepsilon'' = 0$. Daniel (1967) lists the conditions to be satisfied by a system so that these equation are generally applicable. These are:

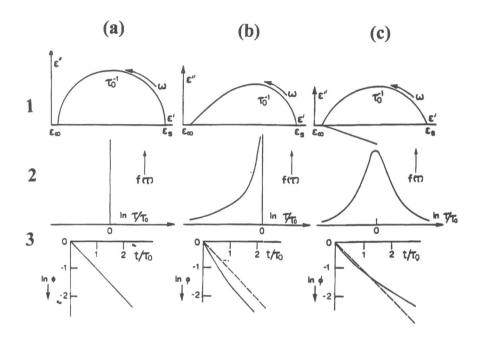

Time Variation of polarization

Fig. 3.29 Graphical depiction of dielectric parameters for the three relaxations shown at top.The applicable equations are also shown (Hasted, 1973). (with permission of Chapman and Hall).Row and column designations are: a1= eq. (3.31), b1 = eq. (3.53), c1 = eq. (3.45), a2 = single value, eq. (3.31), b2 = eq. (3.95), c2 = eq. (3.94), See chaapter 6 for row 3, a3 = eq. (6.46), b3 = eq. (6.49), c3 = eq. (6.47) and (6.48).

1. The system is linear.
2. The constitution of the system does not change during the time interval under consideration.
3. The response of the system is always attributed to a stimulus.

There is no advantage in deriving these equations as we are mostly interested in their application. Application of Kramer-Kronig relations is laborious if the known values are not analytical functions of ω. Jonscher[36] lists a computer program for numerical computation. However the area of the curve $\varepsilon'' - \text{Ln } \omega$ readily gives $(\varepsilon_s - \varepsilon_\infty)$ according to the equation (3.103). Kramer-Konig relations have been used to derive ε' from $\varepsilon'' - \omega$

data and compared with measured ε' as a means of verifying assumed $\varepsilon'' - \omega$ relationship[37].

3.18 LOSS FACTOR AND CONDUCTIVITY

Many dielectrics possess a conductivity due to motion of charges and such conductivity is usually expressed by a volume conductivity. The motion of charges in the dielectric gives rise to the conduction current and additionally polarizes the dielectric. The conductivity may therefore be visualized as contributing to the dielectric loss. Equation (3.8) gives the contribution of conductivity to the dielectric loss. The loss factor is expressed as

$$\varepsilon'' = \varepsilon''(\omega) + \frac{\sigma}{\omega \varepsilon_0} \tag{3.109}$$

Substituting, for $\varepsilon''(\omega)$, from equation (3.29) one gets

$$\sigma = (\varepsilon_s - \varepsilon_\infty) \varepsilon_0 \frac{\omega^2 \tau}{1 + \omega^2 \tau^2} \tag{3.110}$$

If one substitutes

$$\sigma_\infty = \frac{(\varepsilon_s - \varepsilon_\infty) \varepsilon_0}{\tau} \tag{3.111}$$

one obtains

$$\sigma = \frac{\sigma_\infty \omega^2 \tau^2}{1 + \omega^2 \tau^2} \tag{3.112}$$

The conductivity increases from zero at $\omega = 0$ to infinity at $\omega = \infty$ in a manner similar to the mirror image of the decrease of ε' with increasing ω (fig. 3.3). If d.c. conductivity exists then the total conductivity is given by

$$\sigma_{\text{Total}} = \sigma_{\text{dc}} + \frac{(\sigma_\infty - \sigma_{\text{dc}}) \omega^2 \tau^2}{1 + \omega^2 \tau^2} \tag{3.113}$$

Jonscher has compiled the conductivity as a function of frequency in a large number of materials[38] and suggested a "Universal" power law according to

$$\sigma(\omega) = \sigma_{dc} + a\omega^n \qquad (3.114)$$

where the exponent is observed to be within $0.6 \leq n \leq 1$ for most materials. The exponent either remains constant or decreases slightly with increasing temperature and the range mentioned is believed to suggest hopping of charge carriers between traps. The real part of the dielectric constant also increases due to conductivity. A relatively small increase in ε' at low frequencies or high temperatures is possibly due to the hopping charge carriers and a much larger increase is attributed to the interfacial polarization due to space charge, as described in chapter 4.

3.19 REFERENCES

[1] J. F. Mano, J. Phys. D; Appl. Phys., **31** (1998) 2898-2907

[2] Polar Molecules: P. Debye, New York, 1929.

[3] J. B. Hasted: Aqueous Dielectrics, Chapman and Hall, London, 1973, p. 19.

[4] J. Bao, M. L. Swicord and C. C. Davis, J. Chem. Phys., **104** (1996) 4441-4450.

[5] H. Frohlich, Theory of Dielectrics, Oxford University Press, London, 1958
Vera V. Daniel, Dielectric Relaxation, Academic Press, London, 1967, p. 20

[6] K.S. Cole and R. H. Cole, J. Chem. Phys., **9** (1941) 341 – 351.

[7] D. K. Das-Gupta and P. C. N. Scarpa, IEEE Electrical Insulation Magazine, **15** (1999)
23 – 32.

[8] V. V. Daniel, "Dielectric Relaxation", Academic press, London, 1967, p. 97

[9] Y. Ishida, Kolloid-Zeitschrift, **168** (1960) 23-36

[10] A. M. Bottreau, J. M. Moreau, J. M. Laurent and C. Marzat, J. Chem. Phys., **62** (1975)
360-365.

[11] G. P. Johari and S. J. Jones, Proc. Roy. Soc. Lond., **A 349** (1976) 467-495

[12] F. Bruni, G. Consolini and G. Careri, J. Chem. Phys., **99** (1993) 538-547.

[13] G. P. Johari and E. Whalley, J. Chem. Phys., **75** (1981) 1333-1340.

[14] D. W. Davidson and R. H. Cole, J. Chem. Phys., **19** (1951) 1484 – 1490.

[15] T. Blochowitz, A. Kudlik, S. Benkhof, J. Senker and E. Rössler, J. Chem. Phys., **110**
(1999) 12011-12021.

[16] R. P. Auty and R. H. Cole, Jour. Chem. Phys., **20** (1952) 1309-1314.

[17] P. Debye, Polar Molecules (Dover Publications, New York, 1929), p. 84

[18] Dielectric Properties and Molecular behavior, Nora Hill et. al, Van Nostrand, New
York, P. 49

[19] H. Frohlich, "Theory of Dielectrics", Oxford University Press, London, 1986.

[20] G. Williams, Trans. Farad. Soc., **59** (1963) 1397.

[21] J. Melcher, Y. Daben, G. Arlt, Trans. on Elec. Insu. **24** (1989) 31-38. Figure 5 is
misprinted as fig. 8.

[22] R. M. Fuoss and J. G. Kirkwood, J. Am. Chem. Soc., **63** (1941) 385.

[23] K. Mazur, J. Phys. d: Appl. Phys. **30** (1997) 1383-1398.

[24] J. G. Kirkwood and R. M. Fuoss, J. Chem. Phys., **9** (1941) 329.

[25] S. Havriliak and S. Negami, J. Polymer Sci., Part C, **14** (1966) 99-117.

[26] F. Alvarez, A. Alegria and J. Colmenco, Phys. Rev. B., **44** (1991) 7306.

[27] S. Havriliak and D. G. Watts, Polymer, **27** (1986) 1509-1512.

[28] R. Nozaki, J. Chem. Phys., **87** (1987) 2271.

[29] A. K. Jonscher, J. Phys. D., Appl. Phys., **32** (1999) R57-R 70.

[30] D. K. Das Gupta & P. C. N. Scarpa, Electrical Insulation, **15, No. 2** (1999) 23-32

[31] S. Havriliak Jr. and S. J. Havriliak, ch. 6 in "Dielectric Spectroscopy of Polymeric Materials", Ed: J. P. Runt and J. J. Fitzgerald, American Chemical Soc., Washington, D. C., 1977

[32] J. B. Hasted, "Aqueous Dielectrics", Chapman & Hall, London, 1973, p. 24

[33] Daniel (1967). Page 72. Daniel's normalization in equation (3.80) is $\varepsilon_s - \varepsilon_\infty$ and not 1.

[34] W. Reddish, J. Poly. Sci., Part C, (1966) pp. 123-137.

[35] H. A. Kramers, Atti. Congr. Int. Fisici, Como, **2** (1927) 545.
   R. Kronig, J. Opt. Soc. Amer., **12** (1926) 547.

[36] A. K. Jonscher, "Dielectric Relaxation in Solids", Chelsea Dielectric Press, London, 1983.

[37] R. M. Hill, Nature, **275** (1978) 96 .

[38] A. K. Jonscher, "Dielectric Relaxation in solids", Chelsea Dielectric Press, London(1983), p. 214.

There is only a flash between long nights, but this flash is everything.
- Henri Poincaré

4

DIELECTRIC LOSS AND RELAXATION – II

The description of dielectric loss and relaxation with emphasis on materials in the condensed phase is continued in this chapter. We begin with Jonscher's universal law which is claimed to apply to all dielectric materials. Distinction is made here between dielectrics that show negligible conduction currents and those through which appreciable current flows by carrier transport. Formulas for relaxation are given by Jonscher for each case. Again, this is an empirical approach with no fundamental theory to backup the observed frequency dependence of ε* according to a power law. The relatively recent theory of Hill and Dissado, which attempts to overcome this restriction, is described in considerable detail. A dielectric may be visualized as a network of passive elements as far as the external circuit is concerned and the relaxation phenomenon analyzed by using the approach of equivalent circuits is explained. This method, also, does not provide further insight into the physical processes within the dielectric, though by a suitable choice of circuit parameters we can reasonably reproduce the shape of the loss curve. Finally, an analysis of absorption in the optical frequency range is presented both with and without electron damping effects.

4.1 JONSCHER'S UNIVERSAL LAW

On the basis of experimentally observed similarity of the ω-ε'' curves for a large number of polymers, Johnscher[1] has proposed an empirical "Universal Law" which is supposed to apply to all dielectrics in the condensed phase. Let us denote the exponents at low frequency and high frequency as m and n respectively. Here low and high frequency have a different connotation than that used in the previous chapter. Both low and high frequency refer to the post-peak frequency. The loss factor in terms of the susceptibility function is expressed as

$$\frac{1}{\chi''} = (\frac{\omega}{\omega_2})^{-m} + (\frac{\omega}{\omega_1})^{1-n}$$

(4.1)

where $1/\omega_1$ and $1/\omega_2$ are well defined, thermally activated frequency parameters. The empirical exponents m and n are both less than one and m is always greater than $1-n$ by a factor between 2 and 6 depending on the polymer and the temperature, resulting in a pronounced asymmetry in the loss curve. Both m and n decrease with decreasing temperature making the loss curve broader at low temperatures when compared with the loss curve at higher temperatures. In support of his equation Jonscher points out that the low temperature β-relaxation peak in many polymers is much broader and less symmetrical than the high temperature α-relaxation peak.

In addition to polymers the dielectric loss in inorganic materials is associated with hopping of charge carriers, to some extent, and the loss in a wide range of materials is thought to follow relaxation laws of the type:

For $\omega \gg \omega_p$

$$\chi'' = \cot(\frac{n\pi}{2})\chi' \propto \omega^{n-1}$$

(4.2)

For $\omega \ll \omega_p$

$$\chi'' = \tan(\frac{m\pi}{2})[\chi_, - \chi'] \propto \omega^m$$

(4.3)

where the exponents fall within the range

$$0 < m < 1$$
$$0 < n < 1$$

The physical picture associated with hopping charges between two localized sites is explained with the aid of fig. 3-5 of the previous chapter. This picture is an improvement over the bistable model of Debye. A positive charge $+q$ occupying site i can jump to the adjacent site j which is situated at a distance r_{ij}. The frequency of jumps between the two sites is the Debye relaxation frequency $1/\tau_D$ and the loss resulting from this mechanism is given by Debye equation for ε''. τ_D is a thermally activated parameter.

In Jonscher's model some of the localised charge may jump over several consecutive sites leading to a d.c. conduction current and some over a shorter distance; hopping to the adjacent site becomes a limiting case. A charge in a site i is a source of potential. This potential repels charges having the same polarity as the charge in site i and attracts those of opposite polarity. The repulsive force screens partially the charge in question and the result of screening is an effective reduction of the charge under consideration.

In a gas the charges are free and therefore the screening is complete, with the density of charge being zero outside a certain radius which may be of the order of few Debye lengths. In a solid, however, the screening would not be quite as complete as in a gas because the localised charges are not completely free to move. However, Johnscher proposed that the screening would reduce the effective charge to pq where p is necessarily less than one.

Let us now assume that the charge jumps to site j at t=0. The screening charge is still at site i and the initial change of polarization is qr_{ij} The screening readjusts itself over a time period τ, the time required for this adjustment is visualized as a relaxation time, τ. As long as the charge remains in its new position longer than the relaxation time as defined in the above scheme, ($\tau < \tau_D$), there is an energy loss in the system[2].

The situation $\tau > \tau_d$ is likely to occur more often, and presents a qualitatively different picture, though the end result will not be much diffferent. The screening effect can not follow instantly the hopping charge but attains a time averaged occupancy between the two sites. The electric field influences the occupancy rate; down-field rate is enhanced and up-field occupancy rate is decreased. The setting up of the final value of polarization is associated with an energy loss.

According to Jonscher two conditions should be satisfied for a dielectric to obey the universal law of relaxation:
  1. The hopping of charges must occur over a distance of several sites, and not over just adjacent sites.
  2. The presence of screening charge must adjust slowly to the rapid hopping.

In the model proposed by Johnscher screening of charges does not occur in ideal polar substances because there is no net charge transfer. In real solids, however, both crystalline and amorphous, the molecules are not completely free to change their orientations but they must assume a direction dictated by the presence of dipoles in the vicinity. Because the dipoles have finite length in real dielectrics they are more rigidly fixed, as in the case of a side group attached to the main chain of a polymer. The dipoles act as though they are pinned at one end rather than completely free to change orientation by pure orientation. The swing of the dipole about its fixed end is equivalent to the hopping of charge and satisfies condition 1 set above, though less effectively.

The essential feature of the universal law is that the post-peak variation of χ'' is according to eq. (4.2) or superpositions of two such functions with the higher frequency component having a value of n closer to unity. The exponents m and n are weakly dependent on temperature, decreasing with increasing temperature. Many polymeric materials, both polar and non-polar, show very flat losses over many decades of frequency, with superposed very weak peaks. This behavior is consistent with $n \approx 1$, not at all compatible with Debye theory of ω^{-1} dependence. From eq. (4.2) we note that,

$$\frac{\chi''}{\chi'} = \cot\left(\frac{n\pi}{2}\right) \tag{4.4}$$

As a consequence of equation (4.2) the ratio χ''/χ' in the high frequency part of the loss peak remains independent of the frequency. This ratio is quite different from Debye relaxation which gives $\chi''/\chi' = \omega\tau$. Therefore in a log-log presentation χ' - ω and χ'' - ω are parallel.

For the low frequency range of the loss peak, equation (4.3) shows that

$$\frac{\chi''}{\chi_s - \chi'} = \tan\frac{m\pi}{2} \tag{4.5}$$

The denominator on the left side of equation (4.5) is known as the dielectric decrement, a quantity that signifies the decrease of the dielectric constant as a result of the applied frequency. Combining equations (4.2) and (4.3) the susceptibility function given by equation (4.1) is obtained. The range of frequency between low frequency and high frequency regions is narrow and the fit in that range does not significantly influence the representation significantly over the entire frequency range. In any case, as pointed out earlier, these representations lack any physical reality and the approach of Dissado-Hill[3,4] assumes greater significance for their many-body theory which resulted in a relaxation function that has such significance.

Jonscher identifies another form of dielectric relaxation in materials that have considerable conductivity. This kind of behavior is called quasi-dc process (QDC). The frequency dependence of the loss factor does not show a peak and raises steadily towards lower frequencies. For frequencies lower than a critical frequency, $\omega \leq \omega_c$ the complex part of the susceptibility function, χ'', obeys a power law of type ω^{p-1}. Here the real part, χ', also obeys a power law for frequencies $\omega \leq \omega_c$, as shown in fig. 4.1. Here ω_c represents threshold frequency not to be confused with ω_p. In the low frequency region $\chi'' > \chi'$ and in this range of frequencies the material is highly lossy. The curves of χ' and χ'' intersect at ω_c. The characteristics for QDC are represented as:

$$\chi' \propto \chi'' \propto \omega^{m-1} \qquad for\ \omega << \omega_c$$
(4.6)

$$\chi' \propto \chi'' \propto \omega^{n-1} \qquad for\ \omega >> \omega_c \tag{4.7}$$

To overcome the objection that the universal relaxation law, like Cole-Cole and Davidson-Cole, is empirical, Jonscher proposed an energy criterion as a consequence of equation (4.4)

$$\frac{W_L}{W_s} = \cot(\frac{n\pi}{2}) \tag{4.8}$$

in which W_L is the energy lost per radian and W_s is the energy stored. In a field of magnitude E_{rms} the energy lost per radian per unit volume is $\varepsilon_0 \chi'' E^2_{rms}$ and the power lost is σE^2_{rms}. The alternating current (a.c.) conductivity is

$$\sigma_{ac} = \sigma_{dc} + \varepsilon_0 \omega \chi'' \tag{4.9}$$

where σ_{dc} is the d.c. conductivity. This equation defines the relationship between the ac conductivity in terms of χ''. We shall revert to a detailed discussion of conductivity shortly.

The energy criterion of Jonscher is based on two assumptions. The first one is that the dipolar orientation or the charge carrier transition occurs necessarily by discrete movements. Second, every dipolar orientation that contributes to χ' makes a proportionate contribution to χ''. Note that the right sides of equations (4.4) and (4.5) are independent of frequency to provide a basis for the second assumption. Several processes such as the losses in polymers, dipolar relaxation, charge trapping and QDC have been proposed to support the energy criterion. Fig. (4.1c) shows the nearly flat loss in low loss materials. Fig. 4.1(d) applies to H-N equation.

Though we have considered materials that show a peak in ε'' - $\log\omega$ curve the situation shown in fig. 4.1(c) demands some clarification. The presence of a peak implies that at frequencies ($\omega < \omega_p$) the loss becomes smaller and smaller till, at $\omega = 0$, we obtain $\varepsilon'' =$

0, which of course is consistent with the definition of the loss factor (fig. 3.1). There are a number of materials which altogether show a different kind of response; in these materials the loss factor, instead of decreasing with decreasing frequency, shows a trend increasing with lower frequencies due to the presence of dc conductivity which makes a contribution to ε'' according to equation (4.9). The conductivity here is attributed to partially mobile, localised charge carriers.

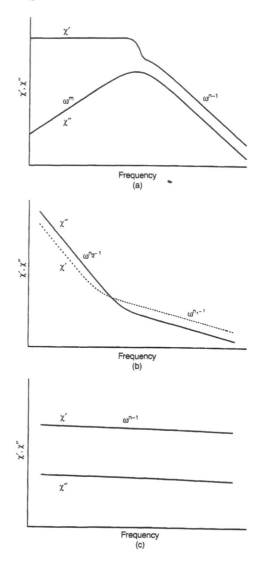

Fig. 4.1 Frequency dependencies of "universal" dielectric response for: (a) dipolar system, (b) quasi-dc (QDC) or low frequency dispersion (LFD) process, and (c) flat loss in low-loss material (Das-Gupta and Scarpa[5] © 1999, IEEE).

As opposed to the small contribution of the free charge carriers to the dielectric loss, localized charge carriers make a contribution to the dielectric loss at low frequencies that must be taken into account. Jonscher[6] discusses two different mechanisms by which localized charges contribute to dielectric relaxation. In the first mechanism, application of a voltage results in a delayed current response which is interpreted in terms of the delayed release of localized charges to the appropriate band where they take part in the conduction process. If the localized charge is an electron it is released to the conduction band. If the localized charge is a hole, then it is released to a valence band.

The second mechanism is that the localized charge may just be transferred by the applied field to another site not involving the conduction band or valence band. This hopping may be according to the two potential well models described earlier in section 3.4. The hopping from site to site may extend throughout the bulk, the sites forming an interconnected net work which the charges may follow. Some jumps are easier because of the small distance between sites. The easier jumps contribute to dielectric relaxation whereas the more difficult jumps contribute to conduction, in the limit the charge transfer to the free band being the most challenging.

This picture of hopping charges contributing both to dielectric relaxation and conduction is considered feasible because of the semi-crystalline and amorphous nature of practical dielectrics. With increasing disorder the density of traps increases and a completely disordered structure may have an unlimited number of localized levels. The essential point is that the dielectric relaxation is not totally isolated from the conductivity.

Dielectric systems that have charge carriers show an ac conductivity that is dependent on frequency. A compilation of conductivity data by Jonscher over 16 decades leads to the conclusion that the conductivity follows the power law

$$\sigma_{ac} = \sigma_{dc} + A\omega^n \tag{4.10}$$

where A is a constant and the exponent n has a range of values between 0.6 and 1 depending on the material. However there are exceptions with n having a value much lower than 0.6 or higher than one.

A further empirical equation due to Hill-Jonscher which has not found wide applicability is[7]:

$$\varepsilon^* = \varepsilon_\infty + (\varepsilon_s - \varepsilon_\infty)_2 F^1(m, n, \omega\tau) \tag{4.11}$$

where

$$F\left(m,n,\omega\tau\right)=\left(1+j\omega\tau\right)^{n-1}{}_{2}F^{1}\left(1-n,1-m;\frac{1}{1+j\omega\tau}\right)$$ (4.12)

and $_{2}F^{1}$ is the Gaussian hypergeometric function.

4.2 CLUSTER APPROACH OF DISSADO-HILL

Dissado-Hill (1983) view matter in the condensed phase as having some structural order and consequently having some locally coupled vibrations. Dielectric relaxation is the reorganizations of the relative orientations and positions of constitutive molecules, atoms or ions. Relaxation is therefore possible only in materials that possess some form of structural disorder.

Under these circumstances relaxation of one entity can not occur without affecting the motion of other entities, though the entire subject of dielectric relaxation was originated by Debye who assumed that each molecule relaxed independent of other molecules. This clarification is not to be taken as criticism or over-stressing the limitation of Debye theory. In view of the inter-relationship of relaxing entities the earlier approach should be viewed as an equivalent instantaneous description of what is essentially a complex dynamic phenomenon. The failure to take into account the local vibrations has been attributed to the incorrect description of the dielectric response in the time domain, as will be discussed later[8].

The theory of Dissado-Hill[9] has basis on a realistic picture of the nature of the structure of a solid that has imperfect order. They pictured that the condensed phase, both solids and liquids, which exhibit position or orientation relaxation, are composed of spatially limited regions over which a partially regular structural order of individual units extends. These regions are called clusters. In any sample of the material many clusters exist and as long as interaction between them exists an array will be formed possessing at least a partial long range regularity. The nature of the long range regularity is bounded by two extremes. A perfectly regular array as in the case of a crystal, and a gas in which there is no coupling, leading to a cluster gas. The clusters may collide without assimilating and dissociating. These are the extremes. Any other structure in between in the condensed phase can be treated without loss of generality with regard to microscopic structure and macroscopic average.

In the model proposed by them, orientation or position changes of individual units such as dipole molecules can be accomplished by the application of electric field. The electric

ld is usually spatially uniform over the material under study and will influence the
ientation or position fluctuations, or both, that are also spatially in phase. When the
sponse is linear, the electric field will only change the population of these fluctuations
d not their nature.

ie displacement fluctuations may be of two kinds, inter-cluster or intra-cluster. Each of
ese interactions makes its own characteristic contribution to the susceptibility function.
ie intra-cluster (within a cluster) movement involves individual dipoles which relaxes
cording to a exponential law ($e^{-t/\tau}$), which is the Debye model. The dipole is linked to
her dipoles through the structure of the material and therefore the relaxing dipole will
fect the field seen by other dipoles of the cluster. The neighboring dipoles may also
lax exponentially affecting the field seen by the first dipole. The overall effect will be
exponential single dipole relaxation.

n the other hand, the inter-cluster (between adjacent clusters) movement will occur
rough dipoles at the edges of neighboring clusters (Fig. 4.2[10]). The inter-cluster motion
s larger range than the intra-cluster motion. The structural change that occurs because
these two types of cluster movements results in a frequency dependent response of the
electric properties. Proceeding from these considerations Dissado and Hill formulate
improved rate equation and determine its solution by quantum mechanical methods.

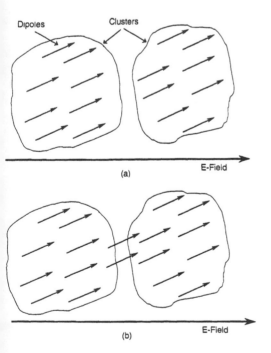

(a)

(b)

Fig. 4.2 Schematic diagram of (a) intra-cluster motion and (b) inter-cluster exchange mechanism in the Dissado-Hill cluster model for dielectric relaxation (Das Gupta and Scarpa, 1999) (with permission of IEEE).

The theory of Dissado and Hill is significant in that the application of their theory provides information on the structure of the material, though on a coarse scale. The inter-cluster displacement arises from non-polar structural fluctuations whereas the intra-cluster motion is necessarily dipolar. Highly ordered structures in which the correlation of clusters is complete can be distinguished from materials with complete disorder.

The range of materials for which relaxations have been observed is extensive, running from covalent, ionic or Van der Waal crystals at one extreme, through glassy or polymer matrices to pure liquids and liquid suspensions at the other. The continued existence of cluster structure in the viscous liquid formed from the glass, to above a glass transition[11] has been demonstrated. Applications to plastic crystal phases[12] and ferroelectrics have also been made. The theory of Dissado-Hill should be considered a major step forward in the development of dielectric theory and has the potential of yielding rich information when applied to polymers.

4.3 EQUIVALENT CIRCUITS

A real dielectric may be represented by a capacitance in series with a resistance, or alternatively a capacitance in parallel with a resistance. We consider that this representation is successful if the frequency response of the equivalent circuit is identical to that of the real dielectric. We shall soon see that a simple equivalency such as a series or parallel combination of resistance and capacitance may not hold true over the entire frequency and temperature domain.

4.3.1 A SERIES EQUIVALENT CIRCUIT

A capacitance C_s in series with a resistance has a series impedance given by

$$Z_s = R_s + \frac{1}{j\omega c_s} \tag{4.13}$$

The impedance of the capacitor with the real dielectric is

$$Z = \frac{1}{j\omega C_0(\varepsilon' - j\varepsilon'')} \tag{4.14}$$

where C_o is the capacitance without the dielectric. Since the two impedances are equal from the external circuit point of view we can equate equations (4.13) and (4.14). To obtain ε' and ε'' as a function of frequency we equate the real and imaginary parts. This gives

$$\varepsilon' = \frac{C_s}{C_0[1+(\omega R_s C_s)^2]} \tag{4.15}$$

$$\varepsilon'' = \frac{\omega R_s C_s^2}{C_0[1+(\omega R_s C_s)^2]} \tag{4.16}$$

$$\tan\delta = \omega R_s C_s \tag{4.17}$$

According to equation (4.16) the ε''-ω characteristics show a broad maximum at the radian frequency corresponding to $\omega C_s R_s = 1$. Substituting $C_s R_s = \tau$ the condition for maximum ε'' translates into $\omega\tau = 1$. τ is the relaxation time which substitutes for the time constant in electrical engineering applications.

Qualitative agreement of the shape of ε''-ω curve with the measured dielectric loss does not justify the conclusion that the series equivalent circuit can be used to represent all polar dielectrics. We therefore consider other equivalent circuits to obtain a comprehensive picture of the scope and limitations of the equivalent circuit approach.

4.3.2 PARALLEL EQUIVALENT CIRCUIT

A capacitance C_p in parallel with a resistance R_p may also be used as a equivalent circuit (fig. 4.3). The admittance of the parallel circuit is given by

$$Y = j\omega C_p + \frac{1}{R_p} \tag{4.18}$$

The admittance of the capacitor with the dielectric is given by

$$Y = j\omega C_0(\varepsilon' - j\varepsilon'') \tag{4.19}$$

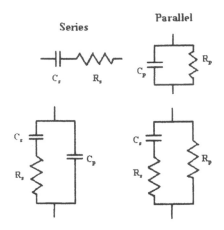

Variations of series-parallel equivalent

Fig. 4.3 Equivalent circuits of a lossy dielectric.

Equating the admittances and separating the real and imaginary parts gives

$$\varepsilon' = \frac{C_p}{C_0}; \qquad \varepsilon'' = \frac{1}{\omega R C_0}; \qquad \tan\delta = \frac{1}{\omega R C_p} \qquad (4.20)$$

Equation (4.20) shows that the ε''-ω curve shows a monotonic decrease. It is clear that a wide range of characteristics can be obtained by combining the series and parallel behavior. Table 4.1 gives the parameters for series and parallel equivalent.

Table 4.1
Equivalent circuit parameters

Series circuit	Parallel equivalent
R_s	$R_p = \dfrac{1+\omega^2 C_s^2 R_s^2}{\omega^2 R_s C_s^2}$
$\tan\delta = \omega C_s R_s$	$\tan\delta = \dfrac{1}{\omega C_p R_p}$
Parallel circuit	Series equivalent
R_p	$R_s = \dfrac{R_p}{1+\omega^2 C_p^2 R_p^2}$

4.3.3 SERIES-PARALLEL CIRCUIT

Fig. 4.3 also shows a series-parallel circuit in which a series branch having a capacitance C_s and a resistance R_s is in parallel with a capacitance C_p. We follow the same procedure to determine the real and imaginary parts of the complex dielectric constant ε^*. The admittance of the equivalent circuit is:

$$Y_{eq} = j\omega C_p + \frac{j\omega C_s}{(1 + j\omega C_s R_s)} \tag{4.21}$$

Substituting again $C_s R_s = \tau$ the above equation becomes

$$Y_{eq} = \frac{\omega^2 C_s R_s}{1 + \omega^2 \tau^2} + j\omega C_p + \frac{j\omega C_s}{1 + \omega^2 \tau^2} \tag{4.22}$$

Equating equations (4.22) and (4.19) we obtain

$$Y = j\omega \varepsilon^* C_0 = j\omega C_0(\varepsilon' - j\varepsilon'') \tag{4.23}$$

Separating the real and imaginary parts yields we obtain the equations:

$$\varepsilon' = \frac{C_p}{C_0} + \frac{C_s}{C_0} \frac{1}{1 + \omega^2 \tau^2} \tag{4.24}$$

$$\varepsilon'' = \frac{C_s}{C_0} \frac{\omega\tau}{1 + \omega^2 \tau^2} \tag{4.25}$$

From these two equations the power factor may be obtained as

$$\tan\delta = \frac{\varepsilon''}{\varepsilon'} = \frac{\omega\tau}{1 + \frac{C_p}{C_s}(1 + \omega^2 \tau^2)} \tag{4.26}$$

These equations may be simplified by substituting conditions that apply at the limiting values of ω.

(a) At $\omega = 0$, $\varepsilon'' = 0$ and ε' has a maximum value given by

$$\varepsilon' = \frac{C_p}{C_0} + \frac{C_s}{C_0} = \varepsilon_s \qquad (4.27)$$

(b) As $\omega \to \infty$, ε' approaches a minimum value given by

$$\varepsilon' = \frac{C_p}{C_0} = \varepsilon_\infty \qquad (4.28)$$

(c) The radian frequency at which ε'' is a maximum is given by

$$\frac{\delta \varepsilon''}{\delta \omega} = C_0(1 + \omega^2 \tau^2)C_s - 2\omega^2 C_0 C_s \tau^2 = 0$$

which yields

$$\omega \tau = 1 \qquad (4.29)$$

This result shows that the equivalent circuit yields ω_{max} that is identical to the Debye criterion.

Substituting Equations (4.27) - (4.28) in equations (4.24) - (4.26) we get

$$\varepsilon' = \varepsilon_\infty + (\varepsilon_s - \varepsilon_\infty)\frac{1}{\omega^2 \tau^2} \qquad (4.30)$$

$$\varepsilon'' = (\varepsilon_s - \varepsilon_\infty)\frac{\omega \tau}{1 + \omega^2 \tau^2} \qquad (4.31)$$

$$\tan \delta = \frac{(\varepsilon_s - \varepsilon_\infty)\omega \tau}{\varepsilon_s + \varepsilon_\infty \omega^2 \tau^2} \qquad (4.32)$$

These are identical with Debye equations providing a basis for the use of the equivalent circuit for polar dielectrics. The parameters of the equivalent circuit may be obtained by the relationships

$$\varepsilon''_{max} = \frac{1}{2}\frac{C_s}{C_0} \qquad (4.33)$$

$$\varepsilon_\infty = \frac{C_p}{C_0} \tag{4.34}$$

$$R_s = \frac{1}{\omega C_s} \tag{4.35}$$

We can also prove, by a similar approach, that the a series branch of C_s and R_s in parallel with R_p yields the Debye equations.

4.3.4 SUMMARY OF SIMPLE EQUIVALENT CIRCUITS

Fig. 4.4 shows a few simple circuits (I column), Z, the impedance (II column), Y, the admittance (III column), C*, the complex capacitance (IV column) defined as

$$C^* = C'' - jC' = \varepsilon_0 \frac{A}{d}(\varepsilon' - j\varepsilon'') \tag{4.36}$$

χ', χ'' in the frequency domain (V column). The real and imaginary parts are plotted for Z, Y, C* as in the complex plane plots[13]. A brief description of each row is given below.

(a) A series circuit with two energy storage elements L and C. The energy is exchanged between the inductive and capacitive elements in a series of periodic oscillations that get damped due to the resistance in the circuit. Resonance occurs in the circuit when the inductive reactance X_L equals the capacitive reactance X_C and the circuit behaves, at the resonance frequency, as though it is entirely resistive. The current in the circuit is then limited only by the resistance. In dielectric studies resonance phenomenon is referred to as absorption and is discussed in greater detail in section 4.6.

(b) A series RC circuit. $X_C = 1/j\omega C$ decreases with increasing frequency. The Cole-Cole plots of Y and C* are semi-circles and Debye relaxation applies. This circuit has been discussed in section 4.3.1.

(c) A parallel combination of R and C, representing a leaky capacitor. The Cole-Cole plot of Z is a semi-circle while ε''-ω plot decreases monotonically. This circuit has been analysed in section 4.3.2.

(d) A series RC circuit in parallel with a capacitance, C_∞. A series RC circuit has been shown to exhibit Debye relaxation. The capacitance in parallel represents any frequency independent process that operates jointly with the Debye process. The Cole-Cole plot of Y shows a upturn due to the additional admittance of the parallel capacitor. The limiting values of capacitance, that is the intercept on the real axis are $C+C_\infty$ and C_∞. The ε''-ω peak is shifted higher and the ε'-ω curve has a limiting value due to ε_∞.

(e) A parallel RC circuit in series with a resistance. An inflection in the loss factor at f requencies lower than ω_p is the dominant characteristic of this circuit.

(f) A parallel RC circuit in series with a capacitance. The response is similar to that shown in (d) with the Cole-Cole plot showing a upturn due to the series capacitance which 'resembles' a series barrier.

(g) Two parallel RC circuits in series. This is known as interfacial polarization and is considered in detail in the next section. The increase in ε'' at lower frequencies is similar to (e) above.

(h) The last entry is in a different category than the lumped elements adopted in the equivalent circuits so far. The so-called transmission line equations are derived using the concept that the circuit parameters, R, L and C, are distributed in practice. In the case of dielectric materials we can use only R and C as distributed, as in the case of a capacitor with electrodes providing a high sheet resistance[14]. In one dimension let r and c be resistance and capacitance per unit length respectively.

The differential equations for voltage and current at a distance x are

$$dV = I(x)rdx \tag{4.37}$$

$$dI = j\omega cV(x)dx \tag{4.38}$$

Equations (4.37) and (4.38) result in differential equations for V:

$$\frac{d^2V}{dx^2} = A^2V(x) \tag{4.39}$$

where

$$A = (1+j)\left(\frac{\omega rc}{2}\right)^{1/2} \tag{4.40}$$

The solution of equation (4.39) is

$$V(x) = V_0(\cosh Ax - B\sinh Ax) \tag{4.41}$$

where V_0 is the input voltage and B is a constant determined by the boundary conditions. For an infinitely long line $B = 1$. The current is given by equation (4.37) as

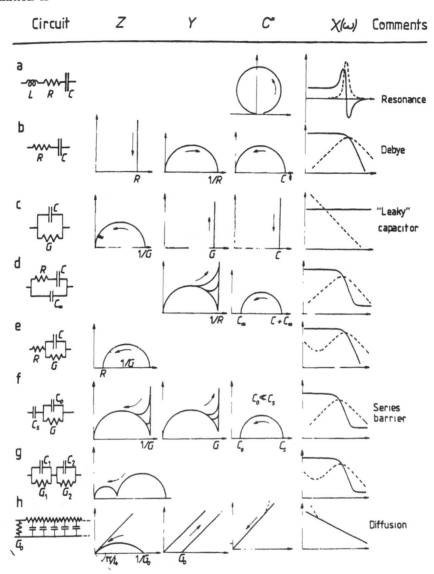

Fig. 4.4 Schematic representations of the equivalent simple circuits, see text (Jonscher, 1983; Chelsea Dielectric).

$$I(x) = \frac{1}{r}\frac{dV}{dx} = \frac{V_0 A}{r}(\sinh Ax - \cosh Ax) \qquad (4.42)$$

The input current I_0 is obtained by substituting $x = 0$, as

$$I_0 = \frac{V_0 A}{r} \tag{4.43}$$

The input admittance of an infinitely long line is the ratio I_0/V_0, giving

$$Y = \frac{A}{r} = \left(\frac{c}{2r}\right)^{1/2} \omega^{1/2}(1+j) \tag{4.44}$$

The Cole-Cole plot of the admittance is a straight line with a slope of one or making an angle of $\pi/4$ with the real axis (see column 3, row h in fig. 4.4). If there is a parallel dc conductance the line through the origin will be displaced to the right. The real and complex part of the dielectric constant in the frequency domain are shown by the same line with slope of $-1/2$ as shown in the last column of row h. The dashed upward tilt at low frequencies is due to the additional parallel conductance, if present. Lack of peak in this situation is particularly interesting.

4.4 INTERFACIAL POLARIZATION

Interfacial polarization, also known as space charge polarization, arises as a result of accumulation of charges locally as they drift through the material. In this respect, this kind of polarization is different from the three previously discussed mechanisms, namely, the electronic, orientational and atomic polarization, all of which are due to displacement of bound charges. The atoms or molecules are subject to a locally distorted electric field that is the sum of the applied field and various distortion mechanisms apply. In the case of interfacial polarization large scale distortions of the field takes place. For example, charges pile up in the volume or on the surface of the dielectric, predominantly due to change in conductivity that occurs at boundaries, imperfections such as cracks and defects, and boundary regions between the crystalline and amorphous regions within the same polymer. Regions of occluded moisture also cause an increase in conductivity locally, leading to accumulation of charges.

We consider the classic example of Maxwell-Wagner to derive the ε'-ω and ε''-ω characteristics due to the interfacial polarization that exists between two layers of dielectric materials that have different conductivity. Let d_1 and d_2 be the thickness of two materials that are in series. Their dielectric constant and resistivity are respectively ε and ρ, with subscripts 1 and 2 denoting each material (Fig. 4.5).

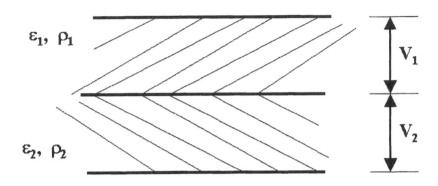

Fig. 4.5 Dielectrics with different conductivities in series.

When a direct voltage, V, is applied across the combination the voltage across each dielectric will be distributed, at $t = 0_+$, according to

$$V_{10} = V \frac{C_2}{C_1 + C_2}$$ (4.45)

$$V_{20} = V \frac{C_1}{C_1 + C_2}$$ (4.46)

When a steady state is reached at $t = \infty$ the voltage across each dielectric will be

$$V_{1\infty} = V \frac{R_1}{R_1 + R_2} \quad ; \quad V_{2\infty} = V \frac{R_2}{R_1 + R_2}$$ (4.47)

The charge stored in each dielectric will change during the transition period. At $t = 0_+$ the charge in each layer will be equal;

$$Q_{10} = Q_{20} = C_1 V_{10} = C_2 V_{20} = \frac{C_1 C_2}{C_1 + C_2} V$$ (4.48)

At $t = \infty$ the charge in each layer will be

$$Q_{1\infty} = C_1 V_{1\infty} = \frac{C_1 R_1}{R_1 + R_2} V ; \quad Q_{2\infty} = C_2 V_{2\infty} = \frac{C_2 R_2}{R_1 + R_2} V$$ (4.49)

During the transition period the change in the stored charge in each layer is:

$$\Delta Q_1 = Q_{1\infty} - Q_{10} = VC_1 \frac{(R_1 C_1 - R_2 C_2)}{(R_1 + R_2)(C_1 + C_2)} \tag{4.50}$$

$$\Delta Q_2 = Q_{2\infty} - Q_{20} = VC_2 \frac{(R_2 C_2 - R_1 C_1)}{(R_1 + R_2)(C_1 + C_2)} \tag{4.51}$$

The redistribution of charges within the layers occurs due to migration of charges and equations (4.50) and (4.51) show that there will be no migration of charges if the condition $C_1 R_1 = C_2 R_2$ is satisfied. Since $C_1 R_1 = \varepsilon_0 \varepsilon_1 \rho_1$ and $C_2 R_2 = \varepsilon_0 \varepsilon_2 \rho_2$ the condition for migration of charges translates into

$$\varepsilon_1 \rho_1 \neq \varepsilon_2 \rho_2 \tag{4.52}$$

Let us suppose that the condition set by expression (4.52) is satisfied by the components of the two layer dielectric. The frequency response of the series combination may be calculated by the method outlined in the previous section. The admittance of the equivalent circuit (fig. 4. 6) is given by

$$Y_{eq} = \frac{Y_1 Y_2}{Y_1 + Y_2} \tag{4.53}$$

where

$$Y_1 = \frac{1}{R_1} + j\omega C_1 \tag{4.54}$$

$$Y_2 = \frac{1}{R_2} + j\omega C_2 \tag{4.55}$$

leading to

$$Y_{eq} = \frac{(1 + j\omega \tau_1)(1 + j\omega \tau_2)}{R_1 + R_2 + j\omega R_1 R_2 (C_1 + C_2)} \tag{4.56}$$

where we have made the substitutions

$$C_1 R_1 = \tau_1; \quad C_2 R_2 = \tau_2 \tag{4.57}$$

We further substitute

$$\tau = \frac{(C_1 + C_2)R_1R_2}{R_1 + R_2} == \frac{R_1\tau_1 + R_2\tau_2}{R_1 + R_2} \tag{4.58}$$

Substituting equations (4.57) and (4.58) into (4.56) and rationalizing the resultant expression yields a rather a long expression:

$$Y_{eq} = \frac{1}{R_1 + R_2} \times$$

$$\frac{[1 - \omega^2\tau_1\tau_2 + \omega^2\tau(\tau_1 + \tau_2) - j\omega\tau(1 - \omega^2\tau_1\tau_2) + j\omega(\tau_1 + \tau_2)]}{1 + \omega^2\tau^2} \tag{4.59}$$

The admittance of the capacitor with the real dielectric is

$$Y = j\omega C_0\varepsilon^* = j\omega C_0(\varepsilon' - j\varepsilon'') \tag{4.60}$$

where ε^* is the complex dielectric constant of the series combination of dielectrics.

Equating the real and imaginary parts of (4.59) and (4.60) we get

$$\varepsilon' = \frac{1}{C_0(R_1 + R_2)} \frac{[(\tau_1 + \tau_2) - \tau(1 - \omega^2\tau_1\tau_2)]}{1 + \omega^2\tau^2} \tag{4.61}$$

$$\varepsilon'' = \frac{1}{\omega C_0(R_1 + R_2)} \frac{[1 - \omega^2\tau_1\tau_2 + \omega^2\tau(\tau_1 + \tau_2)]}{1 + \omega^2\tau^2} \tag{4.62}$$

(a) When $\omega = 0$ equation (4.61) reduces to

$$\varepsilon' = \varepsilon_s = \frac{\tau_1 + \tau_2 - \tau}{C_0(R_1 + R_2)} \tag{4.63}$$

(b) As $\omega \to \infty$

$$\varepsilon' = \varepsilon_\infty = \frac{\tau_1\tau_2}{\tau} \frac{1}{C_0(R_1 + R_2)} \tag{4.64}$$

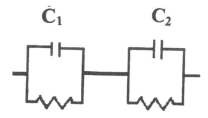

Fig. 4.6 Equivalent circuit for two dielectrics in series for interfacial polarization.

Substituting equations (4.63) and (4.64) in (4.61) we get

$$\varepsilon' = \varepsilon_\infty + \frac{\varepsilon_s - \varepsilon_\infty}{1 + \omega^2 \tau^2} \tag{4.65}$$

A further substitution of equations (4.63) and (4.64) into equation (4.62) gives

$$\varepsilon'' = \frac{1}{\omega C_0 (R_1 + R_2)} + \frac{\omega \tau (\varepsilon_s - \varepsilon_\infty)}{1 + \omega^2 \tau^2} \tag{4.66}$$

Equation (4.65) gives the ε' - ω characteristics for interfacial polarization. It is identical to the Debye equation (3.28), that is, the dispersion for interfacial polarization is identical with dipolar dispersion although the relaxation time for the former could be much longer. It can be as large as a few seconds in some heterogeneous materials. The relaxation spectrum given by equation (4.66) has two terms; the second term is identical to the Debye relaxation (equation 3.29) and at higher frequencies the relaxation for interfacial polarization is indistinguishable from dipolar relaxation. However the first term, due to conductivity, makes an increasing contribution to the dielectric loss as the frequency becomes smaller, Fig. 4.7[15]

The complex dielectric constant of the two layer dielectric including the effects of conductivity is given by

$$\varepsilon^* = \varepsilon_\infty + \frac{\varepsilon_s - \varepsilon_\infty}{1 + j\omega\tau} - j\frac{\sigma_{dc}}{\omega\varepsilon_0} \tag{4.67}$$

The conductivity term (σ/ω) and not the conductivity itself, increases with decreasing ω. An increase in absorbed moisture or in the case of polymers, the onset of d.c. conductivity at higher temperatures, dramatically increases the loss factor at lower

quencies. As stated above, the relaxation time for interfacial polarization can be as
ge as a few seconds in heterogeneous and semi-crystalline polymers. Such behavior is
erved in polyethylene terephthalate (PET), which is semi-crystalline.

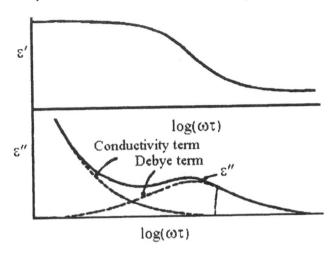

Fig. 4.7 Relaxation spectrum of a two layer dielectric. The conductivity is given by $\sigma =$
$_o/C_o(R_1+R_2)$.

. 4.8 shows the measured ε' and ε'' as a function of frequency at various temperatures
he range 150-190°C[16]. As the frequency is reduced below ~100 Hz both ε' and ε''
rease significantly, with ε' reaching values as high as 1000 at 10^{-2} Hz. This effect is
ibuted to the interfacial polarization that occurs in the boundaries separating the
stalline and non-crystalline regions, the former region having much higher resistivity.
the frequency increases the time available for the drift of charge carriers is reduced
the observed increase in ε' and ε'' is substantially less. Space charge polarization at
ctrodes is also considered to be a contributing factor at low frequencies for the
rease in ε'.

two layer model with each layer having a dielectric constant of ε_1 and ε_2 and direct
rent conductivity of σ_1 and σ_2, in series, has been analyzed by Volger[17]. The
quency dependent behavior of this model is obtained in terms of the following
ations:

$$\varepsilon'(\omega) = \varepsilon_\infty + \frac{(\varepsilon_s - \varepsilon_\infty)}{1 + \omega^2 \tau^2}$$
(4.68)

equation (4.68) the following relationships hold:

$$\varepsilon_s = \frac{(d_1 + d_2)}{(d_1\rho_1 + d_2\rho_2)^2}\left(d_1\varepsilon_1\rho_1^2 + d_2\varepsilon_2\rho_2^2\right) \qquad (4.69)$$

$$\varepsilon_\infty = \frac{d_1 + d_2}{\dfrac{d_1}{\varepsilon_1} + \dfrac{d_2}{\varepsilon_2}} \qquad (4.70)$$

The inset of fig. 4.9 defines the quantities in equation (4.69).

The conductivity and resistivities are also complex quantities and their relationships to the same quantities of the individual dielectrics are given by:

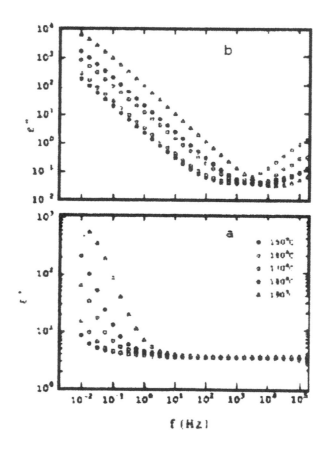

Fig. 4.8 The (a) real and (b) imaginary part of the complex dielectric constant in PET at various temperatures (Neagu et. al., 1997, with permission of the Institute of Phys., UK)

$$\sigma'(\omega) = \frac{\sigma_s + \sigma_\infty \tau_\sigma^2 \omega^2}{1 + \tau_\sigma^2 \omega^2} \tag{4.71}$$

where

$$\sigma_s = \frac{d_1 + d_2}{d_1 \rho_1 + d_2 \rho_2} \tag{4.72}$$

$$\sigma_\infty = \frac{d_1 + d_2}{\left(\dfrac{d_1}{\varepsilon_1} + \dfrac{d_2}{\varepsilon_2}\right)^2} \left(\frac{d_1 \sigma_1}{\varepsilon_1^2} + \frac{d_2 \sigma_2}{\varepsilon_2^2}\right) \tag{4.73}$$

Further we have the following two relationships

$$\rho'(\omega) = \rho_\infty + \frac{(\rho_s - \rho_\infty)}{1 + \omega^2 \tau_\rho^2} \tag{4.74}$$

$$\tan \delta = \frac{\sigma_\infty}{\varepsilon_0 \omega \varepsilon_\infty} \frac{\left(\dfrac{\varepsilon_\infty \sigma_s}{\varepsilon_s \sigma_\infty}\right) + \omega^2 \tau_\delta^2}{1 + \omega^2 \tau_\delta^2} \tag{4.75}$$

Fig. 4.-9 shows these relationships and the similarity between $\varepsilon'(\omega)$ and $\rho'(\omega)$ is evident. The relaxation time for each process is different and the relationship between them is given by the equation

$$\tau = \tau_\sigma = \tau_\rho \left(\frac{\sigma_s}{\sigma_\infty}\right)^{1/2} = \tau_\delta \left(\frac{\varepsilon_s}{\varepsilon_\infty}\right)^{1/2}$$

The increase in dielectric constant, ε', at low frequencies as shown in fig. 4.8 cannot be attributed to conductivity and the observed effect is possibly due to the accumulation of charges at the electrodes. The real part of conductivity, σ', remains constant at low frequencies, increasing to a saturation value at high frequencies. The low frequency flat part is almost equal to dc conductivity and may not be observed except at high temperatures. Fig. 4.10 demonstrates such conductivity behavior.

4.5 The Absorption Phenomenon

So far, our discussion has been restricted to dielectric relaxation that occurs at frequencies generally below 100 GHz. This frequency limit encompasses most of the dipolar and interfacial mechanisms. The frequency dependence of atomic polarizability becomes evident at infrared frequencies ($\sim 10^{12}$ Hz); the frequency dependence of electronic polarizability becomes evident at optical frequencies. Considering the latter, the electron cloud, due to its negligible mass relative to the atom or molecule, is, capable of keeping in phase with the alternating nature of the voltage well into the range of 10^{15} Hz (100 nm).

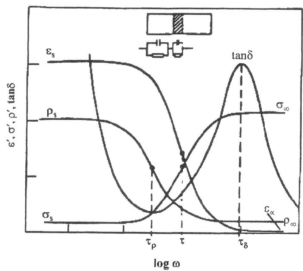

Fig. 4.9 Calculated curves giving the dependence of ε', σ', ρ' and $\tan\delta$ on $\log \omega$ in the two layer model of Volger [17]. Arbitrary units. (with permission of Academic Press).

The oscillating electron bound elastically to an equilibrium position can be viewed as a harmonic oscillator and when the exciting frequency coincides with the natural frequency of the oscillator, resonance occurs. In dielectric studies ε'-ω characteristic at resonance is known as **anomalous dispersion** and ε''-ω characteristic is known as absorption phenomenon. The equation of motion of a linear harmonic oscillator is

$$m\frac{d^2x}{dt^2} + ax = 0 \tag{4.76}$$

where m is the electron mass, x and t are the position and time variables, and a is a constant not to be confused with acceleration.

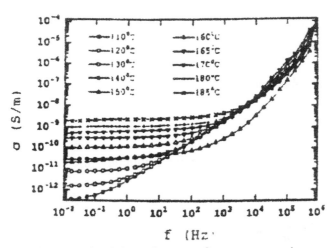

Fig. 4.10 The real part of AC conductivity (σ') versus frequency at various temperatures. The flat region at low frequencies, particularly at high temperatures gives approximate dc conductivity. Compare with theoretical curve shown in fig. 4-9 (Neagu et. al., 1997) (with permission of IEEE)

The solution of equation (4.76) is

$$x = x_0 \cos \omega_0 t \qquad (4.77)$$

where

$$\omega_0 = (a/m)^{1/2} \qquad (4.78)$$

is the characteristic frequency of the harmonic oscillator. If the applied field is represented by

$$E = E_{\max} \cos \omega t \qquad (4.79)$$

where ω is the radian frequency; the force on the electron due to the applied field is

$$F = e E_{\max} \cos \omega t \qquad (4.80)$$

where e is the electronic charge. The equation for the harmonic oscillator with the force acting on the electron is

$$m \frac{d^2 x}{dt^2} + ax = e E_{\max} \cos \omega t \qquad (4.81)$$

The solution of this equation is

$$x = A\cos\omega t \qquad (4.82)$$

where A is a constant, obtained from

$$-\omega^2 mA + a\,A = eE_{max} \qquad (4.83)$$

Expressing m in terms of the characteristic frequency and rearranging equation (4.83) we get

$$A(\omega^2 - \omega_0^2) = \frac{e}{m}E_{max} \qquad (4.84)$$

The solution of equation (4.81) is obtained as

$$x = \frac{eE_{max}}{m(\omega_0^2 - \omega^2)}\cos\omega t \qquad (4.85)$$

Recalling that the induced dipole moment due to electronic polarizability is the product of the electronic charge and the displacement, we get the induced dipole moment as

$$\mu_{ind} = e\,x = \frac{e^2 E_{max}}{m(\omega_0^2 - \omega^2)}\cos\omega t = \frac{e^2}{m(\omega_0^2 - \omega^2)}E \qquad (4.86)$$

The induced electronic polarizability is obtained as

$$\alpha_e = \frac{\mu_{ind}}{E} = \frac{e^2}{m(\omega_0^2 - \omega^2)} \qquad (4.87)$$

By substituting typical values, $e = 1.6 \times 10^{-19}$ C, $m = 9.1 \times 10^{-31}$ kg, $\alpha_e = 5 \times 10^{-40}$ F/m^2, $\omega = 10^9$ s^{-1} we get $\omega_0 \sim 10^{16}$ s^{-1}, which falls in the ultra-violet part of the spectrum.

Using equation (4.87) the variation of α_e with ω may be shown to have the following characteristics: when $\omega_0 >> \omega$, α_e has relatively small value. As ω_0 approaches ω, α_e and therefore ε', increase sharply, reaching theoretically an infinite value at $\omega_0 = \omega$. For a further increase in ω, that is $\omega_0 < \omega$, α_e becomes negative. When $\omega = \omega_0 + \Delta\omega$ an increase in ω results in a decrease of α_e. So this behavior is usually referred to as anomalous dispersion. In practice the α_e-ω characteristics will be modified considerably due to the fact that the motion of electron is subject to a damping effect which was not taken into account in formulating equation (4.76).

The oscillating electron will generate electromagnetic radiation that acts as a damping force. From mechanics we know that the damping factor in a linear harmonic oscillator is proportional to the electron velocity. The equation of motion of the electron is modified as

$$m\frac{d^2x}{dt^2} + b\frac{dx}{dt} + ax = eE_{max}\cos\omega t \tag{4.88}$$

We shall assume that the solution is of the form

$$x = A\cos\omega t = Ae^{j\omega t} \tag{4.89}$$

remembering to consider only the real part of the solution.

Substitution of equation (4.89) in (4.88) gives

$$\left(-\omega^2 A + \frac{aA}{m} + j\frac{\omega bA}{m} - \frac{e}{m}E_{max}\right)e^{j\omega t} = 0 \tag{4.90}$$

Equation (4.90) is identical to equation (4.83) except for the term containing b. Following a procedure similar to the one adopted for the no-damping situation we get

$$A = \frac{e}{m}\frac{E_{max}}{(\omega_0^2 - \omega^2 + j\omega_0/m)} \tag{4.91}$$

where $\omega_0 = \sqrt{\dfrac{a}{m}}$.

Hence

$$x = \frac{e}{m}\frac{E_{max}}{(\omega_0^2 - \omega^2 + j\omega b/m)}e^{j\omega t} \tag{4.92}$$

Equation (4.92) shows that the displacement is out of phase with the applied field and therefore we must introduce a complex polarizability defined by

$$\alpha_e^* = \alpha_e' - j\alpha_e''$$ (4.93)

The induced dipole moment is

$$\mu_e = e\,x = \frac{e^2}{m}\frac{E_{max}}{(\omega_0^2 - \omega^2 + j\omega b/m)}e^{j\omega t}$$ (4.94)

The induced polarizability is

$$\alpha_e' = \frac{e^2}{m}\frac{1}{(\omega_0^2 - \omega^2 + j\omega b/m)}$$ (4.95)

Separating the real and imaginary parts we obtain the pair of equations

$$\alpha_e' = \frac{e^2}{m}\frac{(\omega_0^2 - \omega^2)}{(\omega_0^2 - \omega^2)^2 + (\omega b/m)^2}$$ (4.96)

$$\alpha_e'' = \frac{e^2}{m}\frac{(\omega b/m)}{(\omega_0^2 - \omega^2)^2 + (\omega b/m)^2}$$ (4.97)

From these equations we deduce the following:

(a) At $\omega = 0$ the real part

$$\alpha_e' = \frac{e^2}{m}\frac{1}{\omega_0^2}$$ (4.98)

which is identical to equation (4.87) with $b = 0$. The polarizability is in phase with the applied field. Of course, $\alpha_e'' = 0$.

(b) $\omega \ll \omega_0$, both α_e' and α_e'' are positive.

(c) At $\omega = \omega_0$, α_e' sharply falls to zero and α_e'' has a peak value of $e^2/\omega b$.

(d) For $\omega > \omega_0$, α_e' is negative and α_e'' is positive.

(e) α_e' has a maximum value when $\omega_0 - \omega = \Delta\omega$ is $b/2m$. It has a minimum (negative) value when $\omega - \omega_0 = \Delta\omega$ is $b/2m$. Therefore the distance between the two peaks is b/m.

These characteristics are shown in Fig. 4. 11. It should be emphasized that α_e' - ω and α_e'' - ω characteristics depend on the damping factor. Fig. 4.12[18] shows these characteristics for three values of the parameter $(k=b\omega_0/m)$. For low damping factor, k= 0.1, the characteristics are identical to that shown in Fig. 4.11. With greater damping, k=1, the negative undershoot of α_e' is reduced. With still greater value, k = 10, the negative undershoot is completely suppressed. The characteristics also become broader as the damping increases.

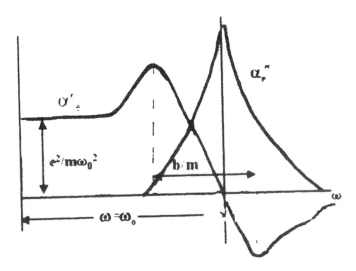

Fig. 4.11 Variation of the real and imaginary parts of the electronic polarizability with frequency.

The theory outlined above has been applied for the study of microwave absorption in gases leading to techniques such as microwave spectroscopy. In crystalline solids the absorption spectra are in the short wavelength regions for electronic transitions within individual atoms or molecules, and in the longer wave length due to vibrations of the crystal as a whole. The lattice vibrations fall into two categories, the optical and acoustical frequency. Acoustical vibrations do not cause a change in polarization and there will be no electromagnetic interaction. The optical vibrations are associated with the oscillatory motion of the charges.

In an ionic crystal the simplest oscillatory motion is due to the vibration of neighboring ions of opposite charges, such as the movement of K^- and Cl^- ions, resembling a dipole of varying moment. The frequency of this oscillation is of the order of 10^{-13} s. There are two basic types of vibrations. One, along the line joining the equilibrium positions of the charges, called **longitudinal optic mode (LO)** and another, in a plane perpendicular to the line joining the equilibrium positions. This mode which is much stronger is called **transverse optic mode** (TO). Evidence for strong resonance absorption in TO mode in KCl at 7K has been reported by Parker et. al.[19].

Extensive studies of infrared spectra have given data of inter-atomic forces and the shape of the electron distribution. A discussion of these are beyond the scope of this book.

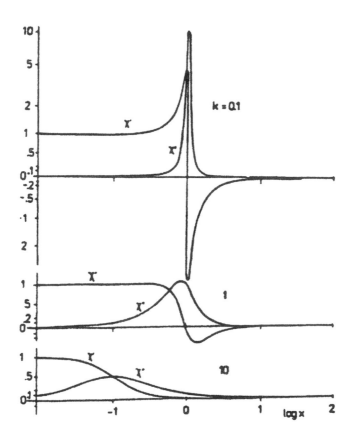

Fig. 4.12 Dispersion (real part) and absorption (imaginary part) of the susceptibility for three values of the damping coefficient as a function of $x = \omega/\omega_0$.

4.6 FREQUENCY DEPENDENCE OF $\varepsilon*$

We are now in a position to represent the variation in the complex dielectric constant as a function of frequency, from $\omega = 0$ to $\omega \to \infty$. Fig. 4-13 shows the contribution of individual polarization mechanisms to the dielectric constant and their relaxation frequencies. As each process relaxes the dielectric constant becomes smaller because the contribution to polarization from that mechanism ceases. Beyond optical frequencies the dielectric constant is given by $\varepsilon_\infty = n^2$.

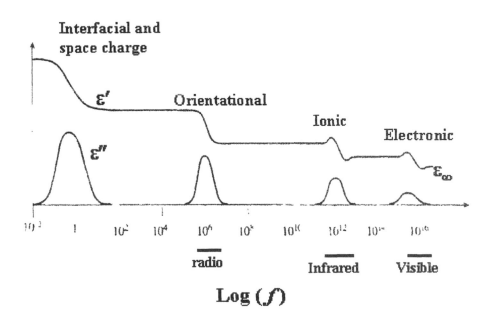

Fig. 4.13 Frequency dependence of the real and imaginary parts of the dielectric constant (schematic).

Though these mechanisms are shown, for the sake of clarity, as distinct and clearly separable, in reality the peaks are broader and often overlap. The space charge polarization may involve several mechanisms of charge build up at the electrode dielectric interface or in the amorphous and crystalline regions of a semi-crystalline polymer. Several types of charges may also be involved depending upon the mechanism of charge generation.

The orientational polarization occurs in the radio frequency to microwave frequency range in dipolar liquids. However in polymers the dipoles may be constrained to rotate or move to a limited extent depending upon whether the dipole is a part of the main chain or side group. Correspondingly the relaxation frequency may be smaller, of the order of a few hundred kHz. Further, in solids there is no single vibration frequency but only a range of allowed frequencies, making the present treatment considerably simplified. The experimental results presented in the next chapter in a number of different polymers will make this evident.

4.7 REFERENCES

[1] A. K. Jonscher, J. Phys. D: Appl. Phys., **32** (1999) R57-R70.

[2] A. K. Jonscher, "Dielectric Relaxation in Solids", Chelsea Dielectrics Press, London, 1983, p. 316.

[3] L. A. Dissado and R. M. Hill: Proc. Roy. Soc., **A 390** (1983) 131-180.

[4] L. A. Dissado and R. M. Hill, Nature, London, **279** (1979) 685.

[5] D. K. Das Gupta and P. C. N. Scarpa, Electrical Insulation, 15 (March-April) (1999) 23-32.

[6] A. K. Jonscher, "Dielectric Relaxation in Solids", Chelsea Dielectrics Press, London, 1983, p. 214.

[7] R. M. Hill and A. K. Jonscher, Contemp. Phys. **24** (1983) 75, also quoted in Y. Wei and S. Sridhar, J. Chem. Phys., **99** (1993) 3119-3125.

[8] G. Williams, Chem. Rev., **72** (1972) 55.

[9] L. A. Dissado and R. M. Hill, Proc. Roy. Soc., A, vol. **390** (1983) 131.

[10] Fig. 3 in Das Gupta and Scarpa (1999).

[11] M. Shablakh, R. M. Hill and L. A. Dissado, J. Chem. Soc. Farad. Trans. II, **78** (1982) 639.

[12] M. Shablakh, L. A. Dissado and R. M. Hill, , J. Chem. Soc. Farad. Trans. II, **79** (1983) p. 369; M. Shablakh, L. A. Dissado and R. M. Hill, , J. Chem. Soc. Farad. Trans. II, **78** (1983) 625.

[13] A. K. Johnscher, (1983), p. 81.

[14] A. K. Johnscher, (1983), p. 80.

[15] A. Von Hippel, Dielectric Materials and Applications, John Wiley & Sons, New York, 1954.

[16] E. Neagu, P. Pissis, L. Apekis and J. L. Gomez Ribelles, J. Phys. D: Appl. Phys., **30** (1997) 1551.

[17] J. Volger, Prog. In Semicond., **4** (1960) 209.

[18] A. K. Johnscher (1983), Figure 4.1.

[19] T. J. Parker, C. L. Mok and W. G. Chambers, 1979, IEE Conference Publication, 177, p. 207.

When you can measure what you are speaking about and express it in numbers you know something about it; but when you cannot measure it, when you cannot express it in numbers, your knowledge is of a meagre and unsatisfactory kind: it may be the beginning of knowledge, but you have scarcely, in your thoughts, advanced to the stage of science.

- Lord Kelvin (William Thomson), 1883.

5

EXPERIMENTAL DATA (FREQUENCY DOMAIN)

We have acquired sufficient theoretical foundation to understand and interpret the results of experimental measurements obtained in various materials. Both the dielectric constant and dielectric relaxation will be considered and results presented will follow, as far as possible, the sequence of treatment in the previous chapters. Anyone familiar with the enormous volume of data available will appreciate the fact that it is impossible to present all of the data due to limitations of space. Moreover, several alternative schemes are possible for the classification of materials for presentation of data. Phase classification as solids, liquids and gases is considered to be too broad to provide a meaningful insight into the complexities of dielectric behavior.

A possible classification is, to deal with polar and non-polar materials as two distinct groups, which is not preferred here because in such an approach we need to go back and forth in theoretical terms. However, considering the condensed phase only has the advantage that we can concentrate on theories of dielectric constant and dielectric loss factor with reference to polymers. In this sense this approach fits well into the scope of the book. So we adopt the scheme of choosing specific materials that permit discussion of dielectric properties in the same order that we have adopted for presenting dielectric relaxation theories. As background information a brief description of polymer materials and their morphology is provided because of the large number of polymer materials cited. We restrict ourselves to experimental data obtained mainly in the frequency domain with temperature as the parameter, though limited studies at various temperatures using constant frequency have been reported in the literature.

Measuring the real part of the dielectric constant centers around the idea that the theories can be verified using molecular properties, particularly the electronic polarizability, and the dipole moment in the case of polar molecules. A review of studies of dielectric loss is published by Jonscher[1] which has been referred to previously. The absorption

197

phenomena in gases and liquids in the microwave region has been has been treated by Illinger[2] and we restrict ourselves to the condensed phases.

The experimental techniques used to measure dispersion and relate it to the morphology, using electrical methods include some of the following:

1. Measurement of ε' and ε'' at various frequencies; each set of frequency measurement is carried out at a constant temperature and the procedure repeated isothermally at other selected temperatures (See fig 5.36 for an example). Plots of ε''- $\log f$ exhibit a more or less sharp peak at the relaxation frequency. In addition the loss factor due to conductivity may exhibit a low frequency peak. The conductivity may be inherent to the polymer, or it may be due to absorbed moisture or deliberately increased in preparing the sample to study the variation of conductivity with temperature or frequency. Fig. 5.1 shows the loss factor in a thin film of amorphous polymer called polypyrrole[3] in which the conductivity could be controlled by electrochemical techniques. The large conductivity contribution at low frequencies can be clearly distinguished. In this particular polymer the conductivity was found to vary according to $T^{-0.25}$. Care should be exercised, particularly in new materials, to distinguish the rise in ε'' due to a hidden relaxation.

2. Same as the above scheme except that the temperature is used as the variable in presenting the data and frequency as the parameter. Availability of computerized data acquisition equipment has made the effort less laborious. Fig. 5.2 shows this type of data for polyamide-4,6 which is a new material introduced under the trade name of Stanyl®[4]. Discussion of the data is given in section 5.4.11.

3. Three dimensional plots of the variation of ε' and ε'' with temperature and frequency as constant contours. This method of data presentation is compact and powerful for quickly evaluating the behavior of the material over the ranges of parameters used; however its usefulness for analysis of data is limited. Fig. 5.2 shows the contour plots of ε' and ε'' in Stanyl ® (Steeman and Maurer, 1992).

4. Measurement of polarization and depolarization currents as a function of time with temperature and the electric field as the parameters. Transformation techniques from time domain to the frequency domain result in data that is complimentary to the method in (1) above; the frequency domain data obtained this way falls in the low frequency region and is very useful in revealing phenomena that occur at low frequencies.

Examples of low frequency phenomena are α-relaxation and interfacial polarization, though care should be exercised to recognize ionic conductivity which is more pronounced at lower frequencies. For example the α-dispersion radian frequency in polystyrene is 3 s^{-1} at its glass transition temperature of 100°C. In this range of frequency, time domain studies appear to be a more desirable choice. This aspect of dielectric study is treated in chapter 6.

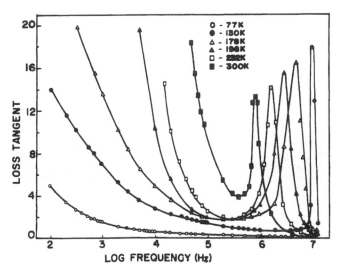

Fig. 5.1 Increase of low frequency loss factor in amorphous polypyrrole film (Singh et. al., 1991, with permission of J. Chem. Phys.)

5. Measurement of the ε''-ω characteristic can be used to obtain the ε'-ω characteristics by evaluation of the dielectric decrement according to eq. (3.103) or Kramer-Kronig equations (equations 3.107 & 3.108). This method is particularly useful in relatively low loss materials in which the dielectric decrement is small and difficult to measure by direct methods. Two variations are available in this technique. In the first, $\varepsilon(t)$ is measured, and by Fourier transformation $\varepsilon^*(\omega)$ is evaluated. In the second method, $I(t)$ is measured and ε'' is then obtained by transformation. Integration according to eq. (3.107) then yields the dielectric decrement[5].

6. Evaluation of the dielectric constant as a function of temperature by methods of (1) or (4) above and determining the slope $d\varepsilon_s/dT$. A change of sign for the slope, from positive to negative as the temperature is increased, indicates a unique temperature, that of order-disorder transition.

7. The dielectric decrement at $\omega = 0$ is defined as $(\varepsilon_s - \varepsilon_\infty)$ and this may be evaluated by finding the area under ε''- $\log\omega$ curve in accordance with eq. (3.103).

8. Presentation of dielectric data in a normalized method is frequently adopted to cover a wide range of parameters. For example the ε''- $\log\omega$ curve is replotted with the x-axis showing values of ω/ω_{max} and y-axis showing $\varepsilon''/\varepsilon''_{max}$. (see fig. 5.3[6] for an example). If the points lie on the same curve, that is the shape of the curve is independent of

temperature, the symmetrical shape represents one of the Debye, Cole-Cole or Fuoss-Kirkwood relaxations.

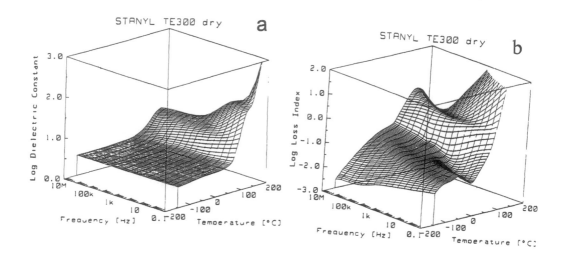

Fig. 5.2 (a) The dielectric constant of dry Stanyl® (aliphatic Polyamide) as a function of frequency and temperature. (b) The dielectric loss factor as a function of frequency and temperature (Steeman and Maurer, 1992, with permission of Polymer).

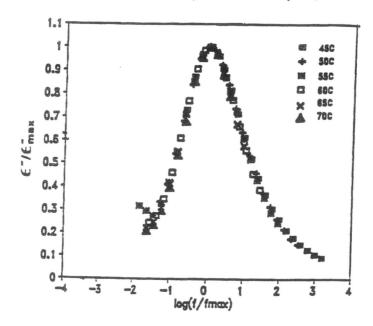

Fig. 5.3 Normalized loss factor in PVAc (Dionisio et. al. 1993, With permission of Polymer).

Another example is due to Jonscher's analysis of the data of Ishida and Yamafuji [7] to discuss relaxation in PEMA as shown in Fig. 5.4. The normalized curves (b) show that the curve becomes broader as the temperature becomes smaller in the pre-peak region indicating strong evidence for overlap of another loss mechanism.

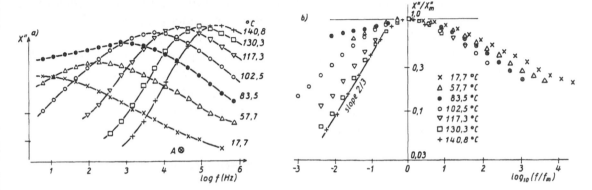

Fig. 5.4 (a) shows the dielectric loss data for poly(ethyl methacrylate) taken from Ishida and Yamafuji (1961). (b) shows the plots using normalized frequency and loss (Jonscher, 1983). (With permission of Chelsea Dielectric Press, London).

The normalization can be carried out using a different procedure on the basis of equations (3.86) as suggested by Havriliak and Negami[8]. In this procedure the co-ordinates are chosen as:

$$x = \frac{\varepsilon' - \varepsilon_\infty}{\varepsilon_s - \varepsilon_\infty}; \qquad y = \frac{\varepsilon''}{\varepsilon_s - \varepsilon_\infty}$$

If the data fall on a single locus then the distribution of relaxation times is independent of temperature.

Williams and Ferry et. al[9] have demonstrated a relatively simple method of finding the most probable relaxation time. According to their suggestion the plots of the parameter $\varepsilon''/(\varepsilon_s - \varepsilon_\infty)$ versus T yields a straight line. The same dependence of reduced loss factor with temperature can exist only over a narrow range because at some low temperature the loss must become zero and it can not decrease further. At the other end the loss can reach a value of 0.5, or approach it, as dictated by Debye equation. A normalized loss factor greater than 0.5 is not observed because it would mean a relaxation narrower than the Debye relaxation.

With this overview we summarize the experimental data in some polymers of practical interest.

5.1   INTRODUCTION TO POLYMER SCIENCE

Polymers are found in nature and made in the laboratory. Rubber and cellulose are the most common example of natural polymers. One of the earliest polymers synthesized as a resin from common chemicals (phenol and formaldehyde) is called phenol formaldehyde (surprise!) commonly known as bakelite. Because of its tough characteristics bakelite found many applications from telephones to transformers. The vast number of polymers available today have a wide range of mechanical, thermal and electrical characteristics; from soft and foamy materials to those that are as strong as steel, from transparent to completely opaque, from highly insulating to conducting. The list is long and the end is not in sight.

5.1.1   CLASSIFICATION OF POLYMERS

Polymers may be classified according to different schemes; natural or synthetic, organic or inorganic, thermoplastic or thermosetting, etc. Organic molecules that make fats (**aliphatic** in Greek) like waxes, soaps, lubricants, detergents, glycerine, etc. have relatively straight chains of carbon atoms. In contrast **aromatic** compounds are those that were originally synthesized by fragrances, spices and herbs. They are volatile and highly reactive. Because they are ready to combine, aromatics outnumber aliphatics. Molecules that have more than six carbon atoms or benzene ring are mostly aromatic. The presence of benzene in the backbone chain makes a polymer more rigid.

Hydrocarbons whose molecules contain a pair of carbon atoms linked together by a double bond are called **olefins** and their polymers are correspondingly called **polyolefins**. Polymers that are flexible at room temperature are called **elastomers.** Natural rubber and synthetic polymers such as polychloroprene and butadiene are examples of **elastomers.** The molecular chains in elastomers are coiled in the absence of external force and the chains are uncoiled when stretched. Removal of the force restores the original positions. If the backbone of the polymer is made of the same atom then the polymer is called a **homochain** polymer, as in polyethylene. In contrast a polymer in which the backbone has different atoms is known as a **heterochain** polymer.

Polymers made out of a single monomer have the same repeating unit throughout the chain while polymers made out of two or more monomers have different molecules along the chain. These are called **homopolymers** and **copolymers** respectively. Polyethylene, polyvinyl chloride (PVC) and polyvinyl acetate (PVAc) are homopolymers. Poly (vinyl chloride-vinyl acetate) is made out of vinyl acetate and vinyl chloride and it is a copolymer.

Copolymers are classified into four categories as follows:
1. **Random copolymer**: In this configuration the molecules of the two comonomers are distributed randomly.
2. **Alternating copolymer:** In this structure the molecules of the comonomers alternate throughout the chain.
3. **Block copolymer:** The molecules of comonomers combine in blocks, the number of molecules in each block generally will not be the same.
4. **Graft copolymer:** The main chain consists of the same monomer while the units of the second monomer are added as branches.

For the ability of a monomer to turn into polymer, that is for polymerization to occur, the monomer should have at least two **reactive sites**. Another molecule attaches to each of the reactive sites and if the molecule has two reactive sites it is said to have **bifunctionality.** A compound becomes reactive because of the presence of reactive functional groups, such as $- OH$, $- COOH$, $- NH_2$, $-NCO$ etc. Some molecules do not contain any reactive functional groups–but the presence of double or triple bonds renders the molecule reactive. Ethylene (C_2H_6) has a double bond and a functionality of two. Depending on the functionality of the monomer the polymer will be linear if bifunctional, branched or cross linked in three dimensions if tri-functional. If we use a mixture of bi-functional and tri-functional monomers the resulting polymer will be branched or cross linked depending on their ratio.

When monomers just add to each other during polymerization the process is called **addition** polymerization. Polyethylene is an example. If the molecules react during the polymerization the process is known as **condensation** polymerization. The reacting molecules may chemically be identical or different. Removal of moisture during polymerization of **hydroxy** acid monomers into polyester is an example. Polymerization of nylon from adipic acid $(C_6 H_{10} O_4)$ and hexamethylenediamine $(C_6 H_{16} N_2)$ is a second example. In addition polymerization, the molecular mass of the polymer is the mass of the monomer multiplied by the number of repeating units. In the case of condensation polymerization, this is not true because condensation or removal of some reaction products reduces the molecular mass of the polymer.

The chemical structure of a polymer depends on the elements in the monomer unit. In polymers we have to distinguish between the chemical structure and the geometrical structure because of the fact that monomers combine in a particular way to yield the polymer. Two polymers having the same chemical formula can have different geometrical arrangement of their molecules. Two terminologies are commonly used; **configuration** and **conformation.** Configuration is the arrangement of atoms in the adjacent monomer units and it is determined by the nature of the chemical bond between adjacent monomer units and between adjacent atoms in the monomer. The configuration of a polymer cannot be changed without breaking the chemical bonds; it is equivalent to changing its finger print, its identity, so to speak.

A conformation is one of several possible arrangements of a chain segment resulting from rotation around a single bond. A change in conformation does not involve breaking or reforming any bond and the rotation of the segment occurs only in space. A polymer of given conformation can assume several different configurations over a period of time depending upon external factors such as thermal energy, mechanical stress, etc.

The conformation assumed by a polymer depends upon whether the polymer is a **flexible chain** type or **rigid chain** type. In a flexible chain type the chain segments have sufficient freedom to rotate about each other. Polymers that have non-polar segments or segments with low dipole moments are flexible chain type. Polyethylene, polystyrene and rubber belong to this class. On the other hand, rigid chain polymers have chain segments in which rotation relative to each other is hindered due to a number of reasons. The presence of bulky side groups or aromatic rings in the back bone acts as hindrance to rotation. Strong forces such as dipole attraction or hydrogen bonding also prevent rotation. Polyimides, aromatic polyesters and cellulose esters belong to this category.

Conformations of polymers in the condensed phase vary from a rigid, linear, rod like structure to random coils that are flexible. In amorphous solids the coils are interpenetrating whereas in crystalline polymers they are neatly folded chains. In dilute solutions molecules of flexible chain, polymers exist as isolated random coils like curly fish in a huge water tank. Molecules of rigid chain polymers in solution exist as isolated stiff rods or helixes.

Stereo-regular polymers, or stereo polymers for short, have the monomers aligned in a regular configuration giving a structural regularity as a whole. The structure resembles cars of the same model, and same color parked one behind the other on a level and straight road. In a non-stereo polymer the molecules are in a random pattern as though identical beads are randomly attached to a piece of flexible material that could be twisted in several different directions.

Chemical compounds that have the same formula but different arrangement of atoms are called **isomers.** The different arrangement may be with respect to space, that is geometry; this property is known as **stereo-isomerism,** sometimes known as geometric isomerism. Stereo-isomerism has a relation to the behavior of light while passing through the material or solution containing the material. It is known in optics that certain crystals, liquids or solutions rotate the plane of plane-polarized light as the light passes through the material. The origin of this behavior is attributed to the fact that the molecule is asymmetric, so that they can exist in two different forms, each being a mirror of the other. The two forms are known as **optical isomers.** If one form rotates the plane of plane-polarized light clockwise, the form is known as **dextro-rotatory (prefix-*d*).** The other form will then rotate the plane in the anti-clockwise direction by exactly the same amount; This form is known as **laevo-rotatory (prefix-*l*).** A mixture of equal molar volume of **D** and **L** forms of the same substance will be optically neutral. If the position of the functional group is different or the functional group is different then the property is called **structural isomerism.** There are a number of naturally occurring isomers but they occur only in *d* or *l* forms, not both.

To describe **isomerism** in polymers we choose polyethylene because of its simple structure. The carbon atoms lie in the plane of the paper, though making an angle with each other, which we shall ignore for the present. The hydrogen atoms attached to carbon, then, lie above or below the plane of the paper. It does not matter which hydrogen atom is above the plane and which below the plane of the paper because, in polyethylene the individual hydrogen atoms attached to each carbon atom are indistinguishable from each other.

Let us suppose that one of the hydrogen atoms in ethylene is replaced by a substituent R (R may be Cl, CN or CH_3). Because of the substitution, the structure of the polymer changes depending upon the location of R with regard to the carbon atoms in the plane of the paper. Three different structures have been identified as below (Fig. 5.5[10]).

1.   R lies on one side of the plane and this structure is known as **isotactic configuration.** This is shown in Fig. (5.5 a).
2.   R lies alternately at the top and bottom of the plane and this structure is known as **syndiotactic** configuration (Fig. 5.5 b).
3.   R lies randomly on either side of the plane and this structure is known as the **atactic** or **heterotactic** configuration (Fig. 5.5 c).

Though the chemical formula of the three structures shown are the same, the geometric structures are different, changing some of its physical characteristics. Atactic polymers have generally low melting points and are easily soluble while isotactic and syndiotactic polymers have high melting points and are less soluble.

5.1.2 MOLECULAR WEIGHT AND SIZE

The number of repeating units of a molecule of polymer is not constant due to the fact that the termination of polymerization of each unit is a random process. The molecular mass is therefore expressed as an average based on the number of molecules or the mass of the molecules. The **number average** molecular mass is given by

$$M_{av.n} = \sum \frac{n_i m_i}{n_i} \tag{5.1}$$

Where n_i and m_i are the number and mass of the ith repeating unit respectively. The mass average molecular mass is given by

$$M_{av.m} = \frac{\sum n_i m_i^2}{\sum n_i m_i} \tag{5.2}$$

For synthetic polymers $M_{av,m}$ is always greater than $M_{av\,n}$. For these two quantities to be equal requires that the polymer should be homogenous, which does not happen. The mechanical strength of a polymer is dependent upon the number of repeating units or the degree of polymerization.

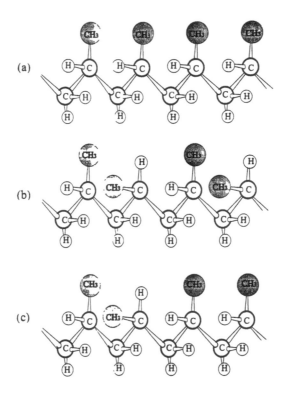

Fig. 5.5 Three different stereoregular structures of polypropylene: (a) isotactic, CH3 groups are on the same side of the plane C=C bond (b) Syndiotactic, CH3 groups alternate on the opposite of the plane (c) atactic, CH$_3$ groups are randomly distributed (Kim and Yoshino, 2000) (with permission of J. Phys. D: Appl. Phys.)

A substance that has the same mass for each molecule is called a **mono-dispersed** system. Water is a simple example. In contrast, polymers have molecules each with a different mass. Such substances are called poly-dispersed systems. However, certain polymers can be polymerized in such a way that they are mono-dispersed. Polystyrene is an example of this kind of polymer.

A polymer chain can assume different shapes. At one end of the spectrum the molecule may be fully extended while at the other end it may be tightly coiled. Random walk theory shows that these extreme shapes occur with very low probability while a randomly coiled shape is more common.

5.1.3 GLASS TRANSITION TEMPERATURE

Solids and liquids are phase separated at the melting point. Polymers have an intermediate boundary called the **glass transition temperature** at which there are remarkable changes in the properties of the polymers. In a simplistic view this is the temperature at which a needle can be inserted into the otherwise hard polymer, such as polystyrene. Of course this method of determining the glass transition temperature is not scientifically accurate and other methods must be employed. The transition does not have latent heat and does not show change in thermodynamic parameters or x-ray diffraction pattern. However the **specific volume**, defined as the inverse of specific density, shows an abrupt increase with increasing temperature. This method of determining the transition temperature by various experimenters gives results within a degree.

In a crystalline solid, at low temperatures, the molecules occupy well defined positions within the crystal lattice in three dimensions, though they vibrate about a mean position. Long range order is said to exist within the crystal and there is no **Brownian motion**. In view of the rigidity of the solid any applied external force, below yield stress, will be transferred without disruption of the lattice structure.

As the temperature is increased kinetic energy is added to the system and the molecules vibrate more energetically. A point is reached eventually when the vibrational motion of the molecule can exceed the energy holding the molecule in its position within the lattice. The molecule is now free to float about and this is the onset of Brownian motion. At a higher temperature still, the motion of molecules spreads in the lattice and the long range order is lost. At this temperature, called the **melting point**, the system will yield to any external force and flow.

In a polymer with large molecules, such as polyethylene, an intermediate state between solid and liquid phases exists, with segments of a chain moving. The large molecule is still immobile with motion of chain segments confined to local regions. This state is referred to as the **rubbery state**. Though the long range order of the solid is lost there is still some order. Some segments of a long chain molecule may have freedom of movement while the molecule itself is not free to move. Only when the temperature is increased still higher does the polymer melt into a highly viscous fluid with the entire chain moving. From the point of view of dielectric studies, our interest lies in the temperature range below and just above the glass transition temperature, unless of course the polymer is a liquid at operating temperature.

5.1.4 CRYSTALLINITY OF POLYMERS

As explained above crystalline solids possess long range order. At low temperatures the molecules are immobile due to intermolecular forces. At high temperatures the energy required to impart mobility is almost equal to the melting point and therefore the rubbery state exists over a very narrow temperature range. In polymers the situation is quite different due to the fact that most polymers are not usually entirely crystalline or amorphous. They are partially crystalline and contain regions that are both crystalline and amorphous. For example in polyethylene prepared by the high pressure method crystallinity is about 50% with both crystalline and amorphous material present in equal amounts. The region of crystallinity is about 10-20 nm[11].

In high polymers measurement of conductivity on seemingly identically prepared specimens show considerable scatter in the activation energy and conductivity. It has been realized that the morphology of the polymer is the influencing factor even though all the experimental conditions are kept identical. While the chemical structure can be maintained the same, the crystallinity of a sample is not easy to reproduce except for single crystals. Fig. 5.6[12] shows the dependence of crystallinity on the rate of cooling of polyethylene produced from melt. The width of each square is the maximum uncertainty. Both crystallinity and density increases with decreasing cooling rate.

Partially crystalline polymers possess both a glass transition temperature and a melting point. If the temperature of the polymer $T < T_g$, the amorphous regions exist in the glassy state and the crystalline regions remain crystalline. Molecular motion in this temperature region is limited to rotation of side groups or parts of side group. Another kind of motion called 'crankshaft' motion of the main chain is also possible below the glass transition temperature; 'crankshaft' motion is the rotation of the four inner carbon atoms even though the outer atoms remain stationary [Bueche, 1962]. At $T \sim T_g$ the amorphous regions become rubbery with no appreciable change in the crystalline regions. The space between molecules or free volume must increase to allow for motion of molecular chain. At $T > T_m$ the distinction between the amorphous and crystalline regions disappears because the polymer melts.

Crystalline polymers obtained from melts do not show anything extraordinary when viewed in a microscope with unpolarized light. However when a polarizing microscope is used complex polycrystalline regions are observed. The regions have high geometrical symmetry, roughly 1 mm in diameter, circular in shape. They are regions of birefringence. The regions resemble a four sector ceiling fan and are called **spherulites** (fig. 5.7). They are believed to be made of several inter-connected lamallae which are stacked one upon the other like the pages of a book. Their presence is an evidence of crystallinity of the polymer. The spherulites keep growing spherically till they collide with another spherulite and the growth stops. It is generally believed that spherical spherulites are formed as the polymer crystalizes in the bulk [Bueche, 1962].

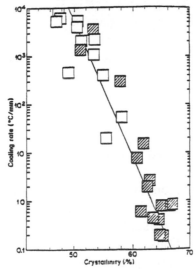

Fig. 5.6 Rate of cooling from the melt versus percent crystallinity in 3% carbon filled polyethylene (H. St. Onge, 1980, with permission of IEEE ©).

Single crystals of some polymers can be obtained from very dilute solutions by crystallization. When examined under electron microscope the minute crystals revealed thin slabs or lamellae. X-ray studies of the **lamallae** revealed, quite unexpectedly that the molecular chain was perpendicular to the plane of the slab. Since the length of a chain (100-1000 nm) was several times the transverse thickness of the slab (10 nm), the molecular chain must fold making a round about turn at the edges. The spacing between folds and therefore the thickness of the lamallae is extremely regular. At the fold surface the long molecule does not fold neatly but makes distorted ' U ' type loops or leave loose ends. This is a region of disorder even in a single crystal.

The configuration of the chain of the polymer determines whether the polymer is crystalline. Table 5.1 lists the glass transition temperature of some crystalline and amorphous polymers.

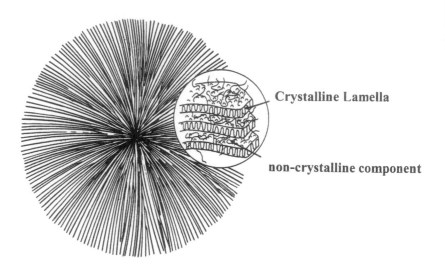

Fig. 5.7. Schematic Spherulite structure in semi crystalline polymers. Molecular chain axes are approximately normal to the surfaces of the Lamellar platelets which grow radially from the center of the structure. (Broadhurst and Davis, 1980, with permission of Springer Verlag, Berlin).

Accepting the view that crystallization in the bulk occurs via spherulites and single crystallization occurs via lamallae, the relation between spherulites and lamallae needs to be clarified. We recall that bulk crystallization is obtained from melts and single crystallization from very dilute solutions. It is reasonably certain that the growth begins at a single nucleus and ribbon like units propagate developing twists and branches. A spherical pattern is quickly established and folded chains grow at right angles to the direction of growth. Spherulites are complicated assemblies of lamallae and amorphous materials exist between fibrous crystals. The region between spherulites themselves is also filled with amorphous materials. Growth of a Spherulite ceases when its boundy overlaps that of the next [Bueche, 1962].

Table 5.1 leads to the question: What is the relationship of T_g to T_m? From a number of experimental results the relation

$$\frac{1}{2} < \frac{T_g}{T_m} < \frac{2}{3}$$

(5.3)

has been found useful to estimate the glass transition temperature.

Table 5.1

Glass Transition Temperature of Polymers

Polymer	T_g (°C)	T_m (°C)
polyisoprene (Natural Rubber)	-73	36
Nylon 6 (Polyamide 6)	50	250
Nylon (6,6)	50	270
Polyvinyl chloride	81	310
Polytetrafluoroethylene		
Polybutadiene (trans-)	-58	100
Polybutyl acrylate	-54	47
Polycarbonate (bisphenol-A)	145	265
Polychlorotrifluoroethylene (PCTFE)	40-59	185-218
Polyethylene (high density)	-125	146
Polyethylene terephthalate	69-75	264
polyamide	90	295
Polyimide	260-320	
Polymethyl methacrylate		
atactic	105	-
isotactic	38	160
syndiotactic	105	>200
Polypropylene (PP)		
atactic	105	-
isotactic	-8	208
Polystyrene	100	250
Polyvinyl acetate	32	-
Polyvinyl alcohol		
Atactic	85	-
Isotactic		212
syndiotactic		267

Amorphous polymers generally exhibit three different relaxations, called α-, β- and γ-relaxation in order of decreasing temperature or increasing frequency. At the glass transition temperature there is a rapid increase in viscosity if the polymer is being cooled and the main relaxation, α-relaxation, slows down. The faster relaxation, β-relaxation, also occurs below T_g and has a much weaker temperature dependence. The splitting of the two relaxations has been observed to occur in three different ways, as shown in fig. 5.8[13].

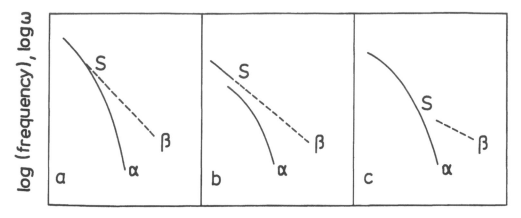

Fig. 5.8 Splitting of the α-, and β-relaxation in amorphous polymers, from high to low temperature (a) true merging of the two relaxations leading to a single high temperature process, (b) Separate onset of the α- transition, (c) Classical β-relaxation stimulated by the development of the α-process. S is the splitting region (Garwe et. al. 1994, with permission of Inst. of Phy., England).

The β-relaxation follows Arrhenius law (log $\tau \propto 1/T$) and the ε''-log(ω) curve is symmetrical on either side at the maximum. On the other hand, the ε''-log(ω) curve is asymmetric with a high frequency broadening. The α-relaxation time follows a non-Arrhenius behavior that is normally described by the Vogel-Fulcher (VF) equation (5.18) or the WLF equation (5.19), presented in sections (5.4.5) and (5.4.7) respectively. A detailed discussion of the relaxation laws is deferred to maintain continuity.

5.1.5 THERMALLY STABLE GROUPS

Certain chemical groups are known to increase thermal resistance, and by increasing these groups, polymers with high thermal stability have been synthesized. Some of these groups are shown below.

5.1.6 POLYMER DEGRADATION AND DEFECTS

Degradation and defects are topics of vital interest in engineering applications. A Polymer degrades basically by two methods: (1) **Chain end degradation** (2) **Random degradation**. Chain degradation consists of the last monomer in the chain dropping out and progressively the chain gets shorter. This is the inverse process of polymerization. This mode of degradation is often termed as **depolymerization** or **unzipping,** the latter term having the connotation that polymerization is a zipping process. Often the degradation is accelerated by higher temperature, presence of moisture, oxygen and carbon dioxide. For example poly(methyl methacrylate) degrades at 300°C releasing an almost equal amount of monomer. Recycling polymers involves recovering the monomer and polymerizing again.

imide   ether   ketone   amide   ester   sulfone

sulfide   siloxane

Thermally stable groups (R. Wicks, "High Temperature Electrical Insulation", (unpublished) Electrical Insulation Conference/ International Coil Winders Association, 1991, with permission of IEEE ©).

Random degradation is initiated at any point along the chain and is the reverse process of polymerization by poly-condensation process. This kind of degradation can occur in almost all polymers. In random degradation of polyethylene a hydrogen atom may migrate from one carbon atom to another elsewhere along the chain and cause chain scission yielding two fragments. Polyesters absorb moisture and chain scission occurs. The causes for degradation may be classified as follows:

1. Thermal degradation
2. Mechanical degradation
3. Photodegradation
4. Degradation due to radiation
5. Degradation due to oxidation and water absorption.

The stability of bonds in the chain is a vital parameter in determining the degree of thermal degradation. For C-C bonds a simple rule is enough for our purpose: more bonds give less stability. A single bond is stronger than a double bond or a triple bond. Increasing the number of constituents in the backbone also decreases stability. Polyethylene is thermally more stable than polypropylene; in turn polypropylene is more stable than polyisobutylene (see the Table of formulas 5.3). Bond dissociation energies for C—C are shown in Table 5.2 and the simple ideas stated above are found to be generally true for molecules containing a H atom. If the molecule does not contain H at all (Teflon is an example) then other considerations such as electronegativity enter the picture. The last entry in Table 5.2 is included to illustrate this point.

Polymers with aromatic groups in the backbone are generally more stable thermally. Poly(phenylene) which has only aromatic rings in the backbone also has high thermal stability. Poly(tetrafluorophenylene) which combines the characteristics of PTFE and poly(phenylene) is even more stable, up to 500°C. Fig. 5.9[14] (Sessler 1980) shows the types of defects that occur in polymers.

5.1.7 DIPOLE MOMENT OF POLYMERS

We are now ready to consider the dipole moment of polymer molecules. We have already shown (Ch. 2) that the dielectric constant of a polar liquid is higher than a non-polar liquid because the dipoles align themselves in an electric field. The same concept can be extended to polymers with polar molecules, in the molten state. In the condensed phase the molecules arrange themselves in a given configuration and eq. (2.54) shows that the polarization due to dipoles is proportional to $N\mu^2$ where N is the number of molecules per unit volume and μ the dipole moment.

For the purpose of continuity it is convenient to repeat the Clausius-Mosotti ratio

$$\frac{\varepsilon - 1}{\varepsilon + 2} = \frac{N\alpha_e}{3\varepsilon_0} + \frac{N\mu^2}{9\varepsilon_0 kT} \tag{2.54}$$

Table 5.2
Dissociation Energies of C—C, C-F Bonds

Molecule	Energy (eV)
CH_3—CH_3	3.82
CH_3CH_2—CH_3	3.68
$(CH_3)_3C$—CH_3	3.47
$C_6C_5CH_2$—CH_3	3.04
CF_4	4.68

Let us consider the variation of the term in brackets on the right side in the case of a polymer that has been polymerized by Z monomer units, each having a dipole moment of μ. We assume that μ is unaffected by the process of polymerization. Before polymerization the aggregate of Z monomers make a contribution of $Z\mu^2$ to the second term in equation (5.1).

After polymerization the contribution of this term depends upon one of the three possibilities:

(1) The dipoles are rigidly fixed aligned in the same direction. The contribution of Z units is $Z^2\mu^2$.

(2) The dipoles are aligned in such a way that their dipole moments cancel; the contribution is zero.

(3) The dipoles are completely free to rotate, each dipole making a contribution of μ^2; the contribution of Z units is $Z\,\mu^2$, (see box below).

Thus the dipole moment contribution can vary from zero to $Z^2\mu^2$. An infinitely large variation is possible depending upon the configuration of the polymer. To put it another way the dipole moment of a polymer provides information as to the structure of the chain. The average dipole moment of a polymer containing Z number of units, each possessing a dipole moment μ, is given by [Bueche, 1962] as

$$\mu_{av}^2 = \left\langle \sum_{n=1}^{z}\sum_{m=1}^{z} \mu_n \cdot \mu_m \right\rangle_{av} \tag{5.4}$$

where the symbolism $<\mu_n \bullet \mu_m>_{av}$ is the average value of the product $\mu_n\,\mu_m\,cos\theta$ over all chain configurations, where θ is the angle between dipoles with index n and m.

For a freely jointed molecule it is easy to see that $\mu^2_{av} = Z\,\mu^2$, that is the contribution of the monomer unit in the polymer, is equal to the dipole moment of the monomer unit in the un-polymerized state. For other configurations the average dipole moment contribution is found to be $\mu^2_{av} = k\,Z\,\mu^2$ where k is a constant. Actual measurements show that $k = 0.75$ for poly (vinyl chloride).

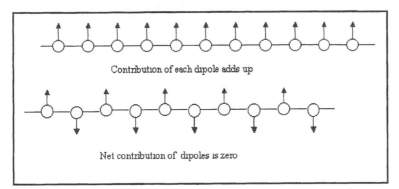

Contribution of each dipole adds up

Net contribution of dipoles is zero

A few general comments with regard to the dielectric loss factor of polymers is appropriate here. We have already explained that the glass transition temperature induces segmental motion in the chain and a jump frequency ϕ may be defined as the number of jumps per second a segment translates from one equilibrium position to the other. ϕ depends on the molecular mass, in the range of 10^{-3}-10 /s. The jump distance δ is the average distance of the jump. δ will not vary much from material to material, 10-100 nm.

As an approximation we can assume that the dipoles are rigidly attached to the chain segment, as in poly(vinyl chloride) for example. The implication of this assumption is that the dipoles move with essentially the same rate as the segment. Further, in accordance with Frohlich's theory (fig. 3-5) it is assumed that the dipoles are oriented parallel or anti-parallel to the applied field. The dielectric loss is a maximum at the radian frequency

$$\omega_{max} = 2\phi \qquad\qquad\qquad\qquad\qquad\qquad (5.5)$$

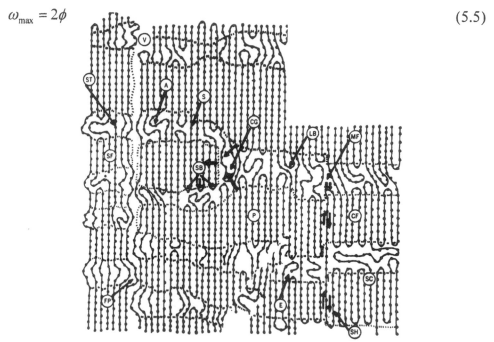

Fig. 5.9 A amorphous phase; CF clustered Fibrils; CG crystal growth in bulk; E end of a chain; FP four point diaagram; LB long backfolding; MF migrating fold; P paracrystalline lattice; S straight chains; SB short backfolding; SC single crystals; SF single fibrils (cold stretched); SM shearing region; ST Statton model; V voids (Van Turnhout, 1980). (with permission of Springer Verlag).

Equation (5.5) provides a good method for determining the jump frequency by measurement of dielectric loss. Such a technique has been used by Bueche (1962).

The assumption that the dipole is rigidly attached to the segment is not so restrictive because equation (5.5) still holds true, though the jump frequency now applies to the dipole rather than the segment. The jump frequency of the segment will be different. We consider the molecule of poly-*m*-chlorostyrene (Bueche, 1962). The bond C-Cl is the dipole attached to the ring. The dipole moment which is directed from Cl to C can be resolved in two mutually perpendicular directions; one parallel to the chain and the other, perpendicular to it, along AB as shown in fig. (5.10). This latter component acts as though it is rigidly attached to the main chain. For this component to move the entire segment must move and equation (5.5) gives the frequency of jump. In other words the loss maximum observed at the glass transition temperature is related to the segmental motion. This kind of maximum is known as α- **loss peak** or α-**dispersion.**

Now consider the component perpendicular to AB. It can rotate more freely with the aromatic ring and this movement occurs more frequently than the rotation of a main chain segment. The loss maximum frequency will be correspondingly higher than that for the α-dispersion. This loss region is known as the β-**dispersion**.

The description presented so far assumes that the chain is made of segments that are freely jointed. If the segment is rigid or the molecule is short the molecular mass enters the picture. The α-dispersion frequency decreases with increasing molecular mass. In the case of stiff cellulose molecules, the α-dispersion frequency is found to depend upon the molecular mass. The rotation of units smaller than a side chain occurs even more frequently than the α or β dispersions, and the associated dielectric loss leads to so called γ **dispersion,** which occurs at a higher frequency than α- or β- dispersions. It is useful to recall that temperature may be used as a variable at constant frequency for the measurement of dielectric loss.

In view of eqs. (3.32) and (3.40) a dispersion that is observed at lower frequency with constant temperature corresponds to that at higher temperature at constant frequency. Hence, at constant frequency α-dispersion occurs at the highest temperature, β- and γ-dispersions occur at lower temperatures, in that order. δ-relaxation process occurs at even higher temperatures or lower f.

5.1.8 MOLECULAR STRUCTURE

In the following sections the dielectric properties of several polymers are discussed and it is more convenient to collect the molecular structure, as shown in Table 5.3.

5.2 NOMENCLATURE OF RELAXATION PROCESSES

The importance of morphology in interpreting dielectric data cannot be overstressed. For the sake of continuity we recall that measured dielectric loss as a function of temperature at constant frequency reveals the relaxation processes. Fig. 5.11[15] is a further description of the relation between the dielectric loss and morphology. Though the nomenclature described here was adopted by Hoffmann et. al. (1966) in describing the relaxation processes in poly(chlorotrifluoroethylene) (PCTFE), it is applicable for other polymers as well. For convenience sake, the temperature scale is normalized with respect to the melting point T_m. The polymer is assumed to be free of independent rotatable side groups and the dipoles are at right angles to the main chain.

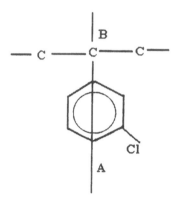

Fig. 5.10. Molecule of poly(m-chlorostyrene) used to define α- and β- dispersion (Bueche. 1962). (with permission of Inter science).

Further, the measurement frequency is assumed to be 1 Hz. The choice of this frequency as a reference is dictated by the fact that the frequency of molecular motion is 1 per second at T_m. As the temperature increases the peak shifts to higher frequencies for the same relaxation mechanism (Fig. 5.16 for example).

The highest temperature of interest is the melting point and at this temperature the relaxation is designated as α-relaxation. The processes at glass transition temperature in semi-crystalline and amorphous materials are designated as β-relaxation. Processes that occur at lower temperatures, possibly due to side groups or segmental polar molecules, are designated as γ- and δ-relaxations in order of decreasing temperature.

1. Single crystal slabs: The highest temperature peak in single crystals is the α_c-relaxation that occurs close to the melting point ($T/T_M \sim 0.9$) as shown in fig. (5.11 C). The relative heights of peaks shown are arbitrary. If the density of defects is low the peak will be higher than if the density is high. The subscript c merely reminds us that we are considering a crystalline phase. Since a single crystal is assumed to have no glass transition the β-relaxation is non-existent. The γ_c peak occurs at temperatures lower than T_G and may have one of the two components depending upon the defect density. The origin of the γ_c peak is possibly the defects wherein chains reorient themselves.

Table 5.3 Selected Molecular structure[16] (Dissado and Fothergill, 1992).

Generic structure		Name (abbreviation)
$\begin{array}{cc} X & X \\ | & | \\ -C-C- \\ | & | \\ X & X \end{array}$	X = H X = F	polyethylene (PE) polytetrafluoroethylene (PTFE)
$\begin{array}{cc} H & X \\ | & | \\ -C-C- \\ | & | \\ H & H \end{array}$	X = CH$_3$ X = Cl X = C$_6$H$_5$ X = OCOCH$_3$	polypropylene (PP) poly(vinyl chloride) (PVC) polystyrene (PS) poly(vinyl acetate) (PVA)
$\begin{array}{cc} H & X \\ | & | \\ -C-C- \\ | & | \\ H & X \end{array}$	X = Cl X = F X = CH$_3$	poly(vinylidine chloride) (PVDC) poly(vinylidine fluoride) (PVDF) polyisobutylene (butyl rubber)
$\begin{array}{cc} H & X \\ | & | \\ -C-C- \\ | & | \\ H & Y \end{array}$	X = CH$_3$ Y = COOCH$_3$	poly(methyl methacrylate) (PMMA)

$$\begin{array}{ccc} H & & X \\ & \diagdown \quad \diagup & \\ & C=C & \\ & \diagup \quad \diagdown & \\ -CH_2 & & CH_2- \end{array}$$

X = H   polybutadiene (BR)
X = CH$_3$ polyisoprene (natural rubber)

$$-(CH_2)_n-O-\overset{\overset{\displaystyle O}{\|}}{C}-\!\!\bigcirc\!\!-\overset{\overset{\displaystyle O}{\|}}{C}-C-$$

n = 2 poly(ethylene terephthalate) (PET)

$$-(CH_2)_n-\overset{\overset{\displaystyle O}{\|}}{C}-\underset{\underset{\displaystyle H}{|}}{N}-$$

n = 5 polyamide 6 (PA6, nylon 6)
n = 10 polyamide 10 (PA10, nylon 10)

m = 4, n = 6, polyamide 6.6 (PA6.6, nylon 6.6)

$$-(CH_2)_n-\underset{\underset{\displaystyle H}{|}}{N}-\overset{\overset{\displaystyle O}{\|}}{C}-(CH_2)_m-\overset{\overset{\displaystyle O}{\|}}{C}-\underset{\underset{\displaystyle H}{|}}{N}-$$

$$-\!\!\bigcirc\!\!-\underset{\underset{\displaystyle CH_3}{|}}{\overset{\overset{\displaystyle CH_3}{|}}{C}}-\!\!\bigcirc\!\!-O-\overset{\overset{\displaystyle O}{\|}}{C}-O-$$

polycarbonate (PC)

$$-\!\!\bigcirc\!\!-O-\!\!\bigcirc\!\!-O-\!\!\bigcirc\!\!-\underset{\underset{\displaystyle O}{\|}}{C}-$$

poly(ether ether ketone) PEEK

2. Completely amorphous phase (Fig. 5-11A): β-relaxation occurs above T_G ; the polymer is in a super cooled phase. In the glassy phase γ_a peaks are observed at temperatures lower than T_G. If the chemical structure is the same then γ_c and γ_a are likely to occur at the same temperature for moderately high frequencies. α-relaxation does not arise in the amorphous state while β-relaxation does not occur in single crystals.

3. Semi-crystalline (Fig. 5-11B): All the three relaxations may be observed; α_c peak occurs in the crystalline regions and β peak in amorphous regions. The γ peak is quite complicated and may occur both in the crystalline and amorphous regions.

5.3 NON-POLAR POLYMERS

We first consider the measured dielectric properties of selected non-polar materials.

5.3.1 POLYETHYLENE

Polyethylene is non-polar polymer that has a simple molecular structure. It is a thermoplastic, polyolefin with physical characteristics that can be controlled to a limited extent. The monomer is ethylene gas (C_2H_4) and additional polymerization yields a polymer that is linear with a C-C carbon chain as the backbone. The carbon atoms make an angle of about 107° with each other, the entire carbon chain lying in the same plane. There are no independently rotatable side groups.

Low density polythene (LDPE) is produced at pressures as high as 150 MPa (1500 atmospheres) with appropriate safe guards to prevent explosion due to exothermic reactions. Recent advances are to lower the pressure to 600 atmospheres. The process yields a single chain polymer with short branches of a few carbon atoms along the main chain. Molecular weight is typically 20000-50000 and the number of monomers in a chain can be as large as 10,000. It is highly resistant to alkalis and acids. It does not have any solvent at room temperature but at 100°C there are a number of organic solvents in which it dissolves. As the solution cools polythene precipitates out.

LDPE is susceptible to sunlight and cannot be used out doors unless additives are used. It has moderate mechanical strength, but its cheapness is an attractive feature for low voltage cable insulation. To improve the electrical characteristics, copolymers with 5% of 1-butene ($CH_2=CH-CH_2-CH_3$) or cross linked polyethylene is used for electrical cables.

High density polythene (HDPE) is more crystalline, has a higher density, has a higher melting point, is more resistant to chemicals and is mechanically stronger. The physical characteristics are shown in Table 5.4.

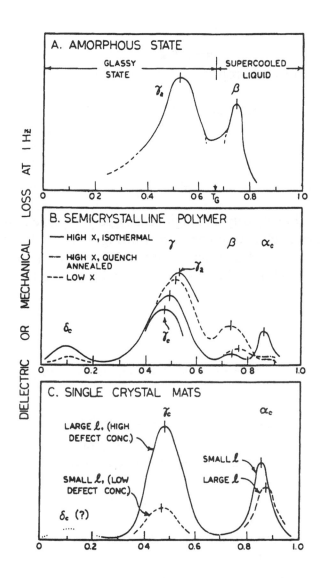

Fig. 5.11 Nomenclature for relaxation processes. T$_m$ is the melting temperature, T$_G$ the glass transition temperature, l the lamella thickness, x the mass fraction of crystallization (Hoffmann et. al., 1966, with permission of J. Poly. Sci.)

The glass transition temperature is -125°C, quite low because (i) strong intermolecular cohesive forces are absent (ii) the atom bonding with carbon is hydrogen which has the lowest mass. Linear polyethylene is highly crystalline (90%) though branching reduces it to 40%. Branching introduces irregularity to the molecular structure and reduces the ability of the molecules to pack closely together. The ability to crystallize is correspondingly reduced.

Table 5.4
Physical Properties of Polyethylene.

Property	Low density PE	High Density PE
Melting point (°C)	110-125	144-150
Density (kg/m³)	910-920	965
T_g (°C)	-	-125
Crystallinity (%)	40	90

The dielectric constants of low density and high density polyethylene at low frequencies are 2.286 and 2.355 respectively. A comparison with Fig. 2.4[17] with calculated values shows a reasonable agreement. The Clausius-Mosotti factor is a linear function of density and Barrie et. al.[18] find from their measurements that the expression

$$\frac{\varepsilon - 1}{\varepsilon + 2} = k\rho \tag{5.6}$$

where k is a constant having a value of 3.24×10^3 (kg/m³)⁻¹ and ρ the density holds true.

Buckingham and Reddish[19], Barrie, Buckingham and Reddish (1966) and Reddish[20] have measured the loss factor. The loss angle of about 30 μ rad at 1 kHz and 12 μ rad at 5 kHz, at very low temperatures is independent of frequency from 4.24 K to 80 K.

Fig. 5.12 shows the measured loss data in medium density polyethylene at various temperatures and separation of the loss into two nearly equal mechanisms, β and γ mechanisms. The high temperature β-process (peak 10°C) and the low temperature γ-process (peak -90°C) is evident in all cases; the peak of the β dispersion is clearly seen at 10°C (fig. d). There is evidence that the α- process has an onset temperature at 30-40°C (fig. e) and from other studies it is known that this process occurs at lower frequencies at 80°C.

Polyethylene is a dielectric with low loss angle covering a five decade frequency scale as measured by Reddish at room temperature. In this frequency range the dielectric loss is reasonably independent of temperature. Jonscher considers this as evidence for the fact that the index *n*, in his empirical loss equation (4.1), is approximately equal to one, independent of frequency. In contrast the loss angle varies with temperature more vigorously at lower temperatures (fig. 5.12 b) due to change in structure. We recall that the glass transition temperature of polyethylene is 160 K, which lies in the region where the fluctuations are observed. We should not overlook the fact that this is the region of interest from the behavior point of view, and Jonscher's equation does not throw much light on this aspect.

A brief explanation of the reasons for observing three dispersions in non-polar polyethylene is in order. Theoretically a non-polar polymer with linear chain should show only the α-dispersion since the only motion that can occur is due to the main chain at the glass transition temperature. However the LDPE has branches of carbon atoms, about 50 per molecule, distributed along the chain at intervals of about 100 main units. Branches of approximately the main chain length may also exist at intervals of a few main chain lengths. The degree of chain branching changes rapidly with density and comparing results obtained by different experiments is difficult because the sample used may not have exactly the same morphology. It has also been suggested that the β-dispersion is not due to the branches but due to the glass transition of the amorphous regions. The γ-process involves at least four CH_2 molecules that may participate in a crankshaft movement in the amorphous region.

During the 1960's the use of polyethylene in submarine cables spurred research into dielectric loss mechanisms, in particular on the effects of moisture and oxidation. Micro-droplets of water in PE cause a dielectric loss in proportion to the amount of water absorbed. With a water content of 190 ppm the dissipation factor (tan δ) increases to 1.5 $\times$ 10^{-4} at 10 MHZ[21]. The origin of the loss is due to Maxwell-Wagner dispersion which occurs due to the motion of impurity ions in the water molecule.

Dielectric loss in oxidized PE occurs at a frequency of ~10 GHz and this is considered to be an intrinsic property of PE, due to localized dipolar motion in the amorphous phase. The oxidation of PE introduces polar carbonyl groups which are short and rigidly attached to the main chain. Any movement of the dipole is associated with the motion of the chain and the polar molecule acts as a probe to observe the natural motion of the chain.

The results in polyethylene may be summarized as follows:

1. Three loss regions are observed in polyethylene; the exact temperature and frequency depends upon the morphology and impurity in the samples.
2. The α-dispersion occurs at high temperature ~ 80°C at 100 HZ, these values changing with crystallinity. It is probably due to chain motion in the crystalline region.

3. The β-dispersion occurs at ~10°C and 1 kHz, probably due to branches attached to the main chain. The magnitude of the loss depends upon the number of side groups attached. The temperature at which relaxation occurs is not affected, however, by the length or number of side groups since the side groups themselves do not interact.

4. The γ-dispersion occurs at low temperatures, possibly due to the crankshaft motion in the amorphous regions.

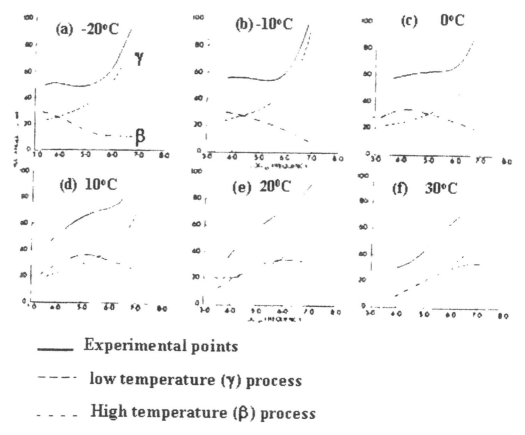

_____ **Experimental points**

− − − − **low temperature (γ) process**

. − − − **High temperature (β) process**

Fig. 5.12 Relative contributions to loss angle of the two relaxation processes in medium density polyethylene (Barrie et. al., 1966, with permission of IEE).

5.3.2 POLY(TETRAFLUOROETHYLENE)

Poly(tetrafluoroethylene), abbreviated as PTFE, is a non-polar polymer that has a chemical structure as simple as polyethylene except that all the hydrogen atoms are replaced with fluorine atoms (Table 5.3). The monomer is a gas with a boiling point of − 76°C. The polymer has a highly regular helical structure with a linear main chain and no branches. This is due to the fact that the C-F bonds are quite strong and they are unlikely to be broken during polymerization. It is also highly crystalline (upto 93-98%) and has a melting point of 330°C. It is highly resistant to chemicals and has a low coefficient of friction. It has a low dielectric constant, large acoustic absorption, very good compatibility with biological tissues and high vapor transmission rate. Highly porous PTFE[22] (fig. 5-13) is commercially available in various grades and has demonstrated high surface charge stability even up to 300°C. Recent progress includes development of amorphous grades (Teflon® AF) that have 95% optical transmission in the range of 400-nm to 2000 nm. Since this grade does not absorb light it does not deteriorate when exposed to light. It has a low dielectric constant of ~1.9.

The early results of Krum and Muller[23] over a wide temperature range of 83-623 K using frequencies of 1-316kHZ, also exhibited the three dispersions like polyethylene. However more recent investigations[24] have led to the conclusion that the α- and β-relaxations are possibly due to polar groups as impurities and not representative of the polymer. The low temperature γ-relaxation at ~ 194 K is the characteristic of the material, attributed to motions of short chain segments in the amorphous regions. The Arrhenius relationship

$$\omega_{max} = A\exp-\frac{\varepsilon_a}{kT} \tag{5.7}$$

is found to hold true for this relaxation in the temperature range of 160-225 K with an activation energy of 0.7 eV.

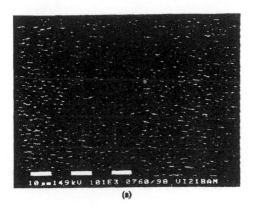 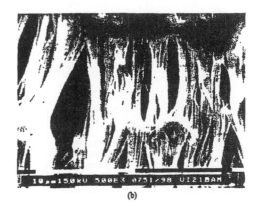

(a)   (b)

Fig 5.13 Scanning electron micrograph of (a) non-porous and (b) porous PTFE films taken at magnifications of 1010 and 5000, respectively (Xia et. al. 1999, with permission of J. Phys. D: Appl. Phys.)

High pressure has a pronounced influence on the complex dielectric constant, ε^*, due to the fact that internal motions are restricted due to decreased volume. Fig. 5-14 shows the dissipation factor (tanδ) at constant temperature with pressure as the parameter[25]. With increasing pressure the activation energy increases from 0.7 eV at 1 bar to 0.92 eV at 2500 bar.

Though pure Teflon shows only γ-relaxation at low temperatures the effect of impurities has been demonstrated by immersing PTFE in both polar and non-polar liquids. While non-polar carbon tetrachloride did not introduce any significant change, polar liquids like chloroform and fluorocarbon-113 increased the maximum for tan δ at 1 kHz. Further the polar molecules introduced an additional large loss peak at ~ 50 K. These changes were also associated with the shift of the γ-peak to lower temperatures[26].

5.4 POLAR POLYMERS

Replacement of one or several hydrogen atoms in CH_4 renders the molecule polar. As a rule the various characteristics like thermal stability, mechanical strength, etc. are achieved by suitable substitutions and the process inevitably imparts a dipole moment to the molecule. There are many more polar polymers than non-polar. Polymers possess a dipole moment due to the presence of one or several types of polar bonds located in the polymer structure in such a way that the dipole moment of the bonds do not neutralize each other. The polar group may be situated in the main chain, a side group or attached to an aromatic ring. In the following we discuss the relaxation phenomena in selected polymers.

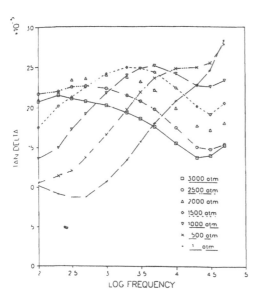

Fig. 5.14 The effect of pressure on the γ- relaxation in PTFE at a constant temperature of 212 K (Starkweather et. al. 1992). (with permission of A. Chem. Soc.)

5.4.1 POLYPROPYLENE

Polypropylene is a mildly polar hydrocarbon that has a low density of 910-920 kg/m^3. Structurally it is related to polyethylene with the hydrogen atom linked to alternate carbon atoms replaced with CH$_3$. It can be prepared in isotactic, syndiotactic or atactic forms (fig. 5-5). The isotactic form melts at 208°C and has high crystallinity. The molecules are essentially linear and form a helical configuration. It has high stiffness and tensile strength due to its high crystallinity. Though it is highly resistant to chemicals at room temperatures it dissolves in aromatic and chlorinated hydrocarbons near its melting point. The evolution of spherulites in i-PP from the melt is shown in fig. 5-15 (Kim, 2000). Crystallization starts at 130°C and ends at 118°C. In the s-PP, on the other hand, crystallization begins only at 110° C and saturates at 100°C.

Polymer crystallization is generally affected by the probability that a molecule is able to form a crystalline arrangement with the neighbors. This probability depends upon the length of the repeating unit. A long and uniform length of main chain acts as a generation center of a spherulite. The i-PP has a ratio of Molecular weight to molecular number of 4-12, whereas s-PP has a ratio of 2. This makes the packing efficiency in s-PP greater giving it a good stereochemically symmetrical molecular structure, with a zig-zag configuration of the methyl groups (CH$_3$) along the main chain (Kim, 2000).

Kramer and Heif[27] have measured ε' and ε" in polypropylene over a temperature range of -75°C-140°C and a frequency range of 0.15 kHZ-140°C. The samples were mostly isotactic with 2-3% atactic, having a crystallinity of 5%. The molecular mass is 30000. The dielectric constant increases as the temperature is increased from -75°C and reaches a maximum at 25°C and for further increase, there is change of slope of ε'- 1/T curve. This aspect has been briefly touched upon earlier in this chapter. We will consider it in greater detail in connection with PVC.

The loss angle shows both α- and β-relaxations. The β-relaxation occurs 20°C (T$_G$ ~0°C) at 0.15kHz and the relaxation temperature increases to 60°C at 300 kHz as the theory requires. The α-relaxation occurs in the 100-140°C range. The maximum frequency for both absorptions follows the **Arrhenius law**

$$f_{max} = A\exp(-\frac{W}{kT})$$
(5.8)

where the activation energy W is 6.5eV and 1.21eV for α- and β-relaxation respectively. The relatively large energy observed is not unusual for polymers with a molecular mass of several thousand. A simplified explanation of this phenomenon is given below [Daniel, 1967].

Proceeding on the assumption that the jump frequency ϕ is thermally activated we can express that

$$\phi = \phi_0 \exp(-\frac{W}{kT}) \tag{5.9}$$

where the pre-exponential factor ϕ_0 is much less sensitive to the temperature than the exponential that contains the critical energy for a jump, W. If the temperature is very high the exponential factor will almost be equal to unity and ϕ_0 is approximately equal to ϕ.

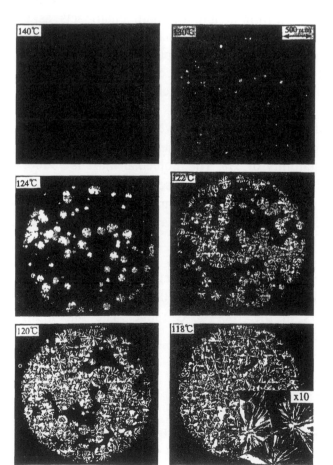

Fig. 5.15   Growth of Spherulites in i-PP during cooling process (Kim, 2000, with permission of J. Phys. D: Appl. Phys.)

Now, the viscosity of a polymer depends upon many factors including the jump frequency and only φ appears to be temperature sensitive. Since the viscosity is high when φ is low, the consequence is that W is large near α-relaxation temperature. For most polymers the energy is close to 0.5 eV and at glass transition temperature it could be as large as 10 eV. In PMMA, for example, a critical energy as high as 12 eV has been reported.

The glass transition temperature of polypropylene is -18°C and melting temperature is ~170°C. It is a slightly polar semicrystalline polymer. The α-relaxation occurs at a much higher temperature relative to the glass transition temperature. In an attempt to explain the discrepancy it has been suggested (Work, et. al., 1964) that the observed high temperature relaxation is possibly due to oxidation products and impurities.

Work, McCammon and Saba[28] have measured ε'' in highly purified atactic polypropylene as shown in fig. 5.16 in the temperature range of 4-300 K at each of several frequencies in the range 100-20 kHz. The ε'' peak occurs in the temperature range of 275-295K which is much closer to the glass transition temperature of their sample, 267K. Using the Cole-Cole plot and from the dielectric decrement obtained, the dipole moment of a polypropylene chain unit has been determined as 0.05 D. Work, et. al have not commented on the origin of this dipole moment.

Jonscher has evaluated values for the index in his formula (4.1) as *m = 0.37 and n = 0.24* from the loss data of Work et. al. Low frequency loss factors in isotactic PP has been reported by Banford et. al. [29]

The relaxation properties of this polymer may be summarized as follows:
1. A high temperature α-relaxation around 100°C due to reorientation of main chain in the crystalline regions
2. A lower temperature β-relaxation at ~10°C in the amorphous regions.

5.4.2 POLY(VINYL CHLORIDE)

Poly(vinyl chloride), abbreviated as PVC, is one of the cheapest and extensively used polymers. The monomer, vinyl chloride, boils at 259 K and therefore it is a gas at room temperature. Hence polymerization is carried out in pressure reactors. PVC has low crystallinity (~10%) and the molecules are either linear or slightly branched. Normal PVC has a molecular weight of 60,000-150,000. The molecular structure is only partially syndiotactic and lacks complete regularity with low crystallinity. The material is hard to process unless **plasticizers** are used; plasticizers lower the glass transition temperature from a value of 81°C for the unplasticized material. PVC has a melting point of 310°C decomposing before the melting point is reached and it is not thermally stable even at moderately high temperature. Stabilizers such as Zinc octoate, sodium carbonate, certain salts of calcium, barium or lead, organo-tin compounds, epoxidised vegetable oils are used to minimize decomposition at high temperature and exposure to sun light.

The chlorine content of PVC is 56.7% by weight and can be further increased by dissolving it in a solvent such as chlorobenzene and chlorinating at temperature ~100°C. The resulting polymer is called **chlorinated PVC** (CPVC) with chlorine content increasing in the range 60-65%. During chlorination, the chlorine replaces the hydrogen atoms in CH_2 units rather than in the CH—Cl units. Chlorination increases the chemical resistance but thermal resistance is reduced.

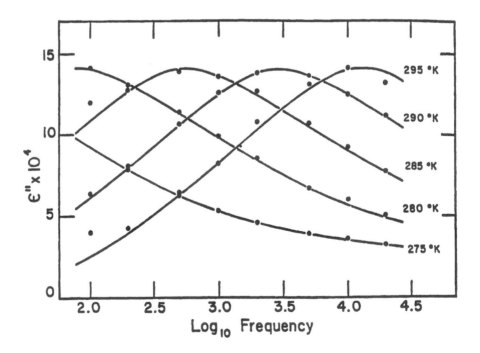

Fig 5.16 Dielectric loss factor of polypropylene (works et. Al., 1964). (with permission of A. Chem. Soc.).

Ishida[30] has measured the dielectric constant and loss in PVC and Deutsch et. al.[31] have further extended the study to chlorinated PVC using both electrical bridges and polarization techniques. The latter consists of poling the material at an appropriate electric field and measuring both polarization and depolarization currents as a function of time (see ch. 6). These measurements are carried out isothermally and the correction f ε'' due to contribution from conduction currents (eq. 3.8) applied to the polarization urrents. The time domain current is transformed to frequency domain to yield the loss actor at lower frequencies.

ig. (5.17) shows the dielectric properties of PVC at higher temperatures. The observed naximum in ε'' is due to the β- relaxation. In this temperature range the absorption ecomes stronger as the temperature increases, the frequency of the maximum ε'' nifting to higher values.

ig. (5.18) shows similar data at lower temperatures and the peak here is attributed to ne γ-relaxation. It is noted that the nomenclature which is adopted here is due to Bur 1984) because the α-absorption, in fact corresponds to the melting point. We will refer o this value while another polymer of the same family, poly(vinyl fluoride) is discussed. ig. 5.19 shows ε' presented as a function of temperature at various constant requencies[32]. The dielectric constant at various temperatures is shown in Table 5.5. The epeating unit of PVC has a dipole moment of 1.1 D. The sharp rise in ε_s at 354 K is dentified as a transition temperature at the Curie point as discussed in section 2.7. At emperatures higher than T_c, the dielectric constant decreases according to $1+3T_c/(T-T_c)$ s the Debye theory requires.

he transition becomes steeper and reaches higher values as the frequency is reduced. A tructural change could explain the existence of a critical temperature. The implication is nat the polymer has structural order above T_c and disorder below T_c , the transition ccurring due to the motion of rod-like short segments. In the disordered state that exists elow T_c alternate chlorine sides are directed in opposite directions, the dipoles being at ight angles to the rod segment. At higher temperatures the rods rotate co-operatively to lign the dipoles. The rotational motion increases the dipole moment leading to a higher ielectric constant. The number of repeating units N_u in each moving segment depends n the temperature as shown in Table 5.5. With increasing temperature, the number of nits decreases explaining the peak of the dielectric constant at T_c.

he number of repeating units in a moving segment is calculated, for a first pproximation, as follows. The Onsager equation for the dielectric constant is given as:

$$(\varepsilon_s - \varepsilon_\infty) = \frac{3\varepsilon_s}{2\varepsilon_s + \varepsilon_\infty}\left(\frac{\varepsilon_\infty + 2}{3}\right)^2 \frac{N\mu_v^2}{3\varepsilon_0 kT} \tag{2.84}$$

vhere N is the number of repeat units per m^3 and μ_v is the moment of a dipole in acuum. Applying the relationship:

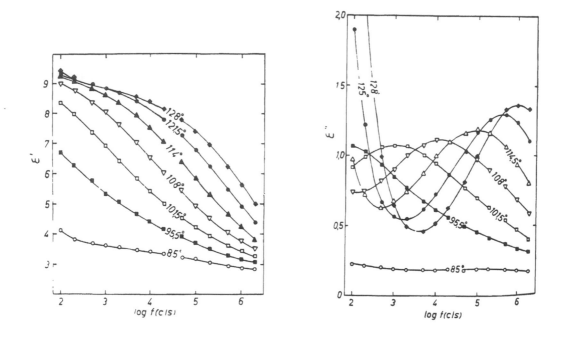

Fig. 5.17 Dielectric constant and loss factor of PVC at high temperatures showing β-absorption.(Ishida, 1960, with permission of Dr. Dietrich Steinkopff Verlag, Darmstadt, Germany)

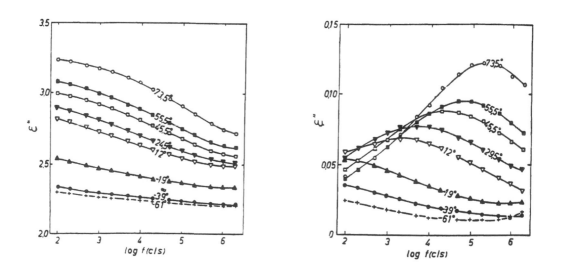

Fig. 5.18 low temperature γ- relaxation in PVC (Ishida, 1960). (With permission of Dr. Dietrich Steinkopff Verlag, Darmstadt, Germany)

$$N = N_A \frac{\rho}{M}$$

where N_A is the Avagadro number (6.06×10^{23}/mole), ρ the density (1400 kg/m^3), M the molecular mass (62.5×10^{-3} kg/mole). Substitution gives $N = 1.36 \times 10^{28}$ /m^3. Several dipoles move co-operatively and let *n* be the number of repeat units that move during relaxation.

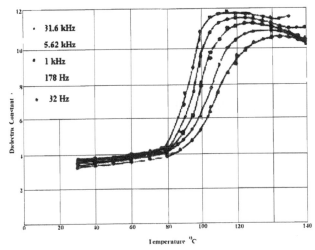

Fig. 5.19 Dielectric constant of chlorinated PVC in alternating fields, 56% Cl. (Reddish, 1966). (with permission of J. Poly. Sci.)

Table 5.5

Dielectric properties of Chlorinated PVC (Reddish, 1966)

T (K)	ε_s	No. of repeat units (N_u)	$\mu^2 N_u$ (D^2)*
425	11	2.38	3.14
400	12	2.48	3.27
375	15	3.02	3.98
370	16.2	3.23	4.26
365	18.5	3.68	4.38
360	22.2	4.44	5.85
354.3	37	7.52	9.91

* D=1 Debye unit=3.33×10^{-30} C m

Then *N/n* is the number of motional units and μ_v *n* is the dipole moment of a motional unit. Substituting these values, the Onsager's equation now becomes:

$$n\mu_v^2 = \frac{(\varepsilon_s - \varepsilon_\infty)(2\varepsilon_s + \varepsilon_\infty)}{\varepsilon_s}\left(\frac{3}{\varepsilon_\infty + 2}\right)^2 \frac{\varepsilon_0 kT}{N}$$

(5.10)

The dipole moment in a vacuum is related to the moment of the dipole (1.1D) according to

$$\mu^2 = \left(\frac{\varepsilon_{\infty+} 2}{3}\right)^2 \mu_v^2$$

(5.11)

Substituting equation (5.11) into (5.10) results in

$$n = \frac{(\varepsilon_s - \varepsilon_\infty)}{3\varepsilon_s}\frac{(2\varepsilon_s + \varepsilon_\infty)}{\mu^2}\frac{3\varepsilon_0 kT}{N}$$

(5.12)

We collect the values to calculate **n**:
T = 425 K, ε_s = 11 (from top row of Table 5.5), μ = 1.1×3.3×10^{-30} C m, N = 1.36× 10^{28} /m^3, ε_0 = 8.85×10^{-12} F/m, k = 1.38×10^{-23} J/K. We get n = 2.38. From Table 5.5 we see that the number of co-operating units increase as T_c is approached.

The motion of the dipoles changes the internal field and at each temperature a finite time is required for the equilibrium state to be established. At T_c this time is of the order of several thousand seconds, which explains the increase of the dielectric constant with decreasing frequency at T_c (fig. 5.19).

Increasing the chlorine content of PVC has been found to decrease the dielectric constant and loss factor. The second chlorine atom attaches to the same carbon atom as the first, neutralizing the dipole moment, and this has the effect of neutralizing some of the dipole moment. Consequently both ε' and ε'' are reduced, the reduction being largest for the highest chlorine content (69.1%). The peak is also shifted to higher temperatures.

We summarize the results in PVC as follows;
1. At T_g the dielectric constant exhibits a disorder-order transition involving unrestricted rotation, in place of restricted rotation, of rod like segments of chain units. The dipoles are perpendicular to the main chain (Bur, 1985).
2. The dielectric constant at this temperature shows a peak at audio frequencies. The change of dielectric constant is time dependent, reaching a discontinuity at 10^{-4} Hz.
3. Chlorination of PVC results in reduced maximum of ε'' and ε' at 1 kHZ.

5.4.3 POLYCHLOROTRIFLUOROETHYLENE

Polytrifluoroethylene, abbreviated as PCTFE, with the trade name of KEL-F 300 has density in the range of 2090-2160 kg/m^3. The number average molecular weight is approximately 415,000 corresponding to a number average chain length of 900 nm. The glass transition temperature is in the range of 40-59°C and crystals have a melting point of ~200°C.

PCTFE, like polyethylene, does not possess any independently rotatable side groups and any observed relaxation process is, in some way, related to the motion of parts or of the entire chain. It has a dipole unit on each monomer chain and the dielectric properties have been studied quite extensively. PCTFE exhibits a wide range of crystallinity (12-80 %) depending upon the method of preparation. By changing the crystallinity and measuring the dielectric properties, Hoffmann et. al. (1966) identified the relaxation mechanisms with the crystalline or amorphous regions.

A. MORPHOLOGICAL DETAILS

Scott, et. al.[33] have measured the dielectric properties of this polymer in the frequency range of 0.1-10^{10} Hz. and an extensive discussion of these studies have been published by Hoffmann et. al (1966). We follow the presentation of these authors because of the thorough treatment. The data were obtained in three different samples which had varying degrees of crystallinity.

1.   Bulk polymer crystallized from the melt with slow cooling: The sample had 80% crystallinity. The growth of spherulites in bulk crystallized polymers has already been discussed and PCTFE behaves entirely typically in this respect. Each spherulite starts from a nucleation center with the plate like lamallae having a thickness of 30-80 nm depending on the temperature of crystallization and time of annealing. Higher temperatures and longer annealing result in thicker samples. The lamallae has a chain folded structure with nearly regular folds and adjacent reentry.

These characteristics mostly, though not entirely, determine the dielectric properties. The inter-lamaellar region is filled with low molecular weight material that has failed to crystallize. It has been established that only molecular chains that have a length greater than a critical length undergo chain folding; shorter molecules are excluded from the crystallization mechanism. The short molecules grow outside the lamallae as extended chains. The crystalline region therefore has two separate regions: 1. Chain folded lamallae. 2. Extended-chain region. The existence of these two regions has also been confirmed in polyethylene (Hoffmann et. al., 1966).

2.   Bulk polymer crystallized from melt by quench cooling: This method yields bulk polymer having low crystallinity of 12% and 44%. Since a large number of nucleation centers are active only very small crystallites are formed in the majority explaining the low crystallinity. Larger crystallites give the polymer higher crystallinity. It is noted that small crystallites are not formed in polyethylene due to crystallization kinetics. The amorphous regions consist of numerous chains interlinked like cooked spaghetti.

3.   Quench crystallized and annealed samples have intermediate crystallinity of 73%. Both the samples, 73% and 80% crystalline, have inter-lamallar links. Such links exist in crystalized polyethylene too. The inter-lamallar links decrease crystallinity and therefore in highly crystallizd samples the number of these links is small. Annealing seems to decrease the number of these links.

B. THE α-RELAXATION

Fig. 5.20 shows the dielectric loss factor plotted as a function of frequency at various temperatures for a sample with 80% crystallinity. The figure also includes the ε'-log f data. The details at low frequency and higher temperatures ($\geq 150°C$) are shown in fig. 5-21. The α_c-relaxation peak at 150°C *(T = 0.86 T_M)* can be seen clearly at 1 Hz. Note the shift of the peak to a higher frequency at 175°C. The α_c-relaxation is not evident at the highest temperature of 200°C because the peak has moved beyond the highest frequency available. Another distinct behavior observed at this temperature is the onset of dc conductivity at the lowest frequencies. Hoffmann et. al. (1966) provide evidence to suggest that the loss is associated with the surface of the crystal rather than the interior. Using equation (5.8) the activation energy is determined as 3.47 ± 0.43 eV. The relaxation is attributed to overlapping mechanisms; motion of chain folds and translations of chains in the interior. For long chains, twisting of chains is also a contributing factor.

C. THE β-RELAXATION

The β-relaxation occurs in samples with lower crastallinity (44% and 73%). A peak is observed in the ε''-T curve at a temperature of 125°C. This temperature is much higher than T_G (~50-59°C) for both samples, but lower than T_M.

Do we assign the peak to α_c-loss in the crystalline region, or to β-loss? Such questions frequently arise in dielectric studies in interpreting experimental data. Since the peak value of ε'' increases with decreasing crystallinity it is reasonable to assume that the loss is related to the amorphous regions. Further the α-, β-, γ-relaxations are closely interlinked with the distribution of relaxation times discussed in ch. 4 and the shape of the ε"- log ω curves, which is dependent on this distribution, provides complementary data about the nature of the loss mechanisms.

By plotting f_{max} as a function of $1/T$ the activation energy may be determined according to equation (5.8). According to Davidson and Cole[34] the mean relaxation time in amorphous materials is given by

$$\tau_m = \frac{1}{\omega_p} = \tau_0 e^{\frac{w_t}{k(T-T_c)}} \qquad (5.13)$$

where τ_0 and T_c are constants. The viscosity of liquids is given by the equation of the same type

$$\eta_g = \eta_0 e^{\frac{w_g}{k(T-T_c)}} \qquad (5.14)$$

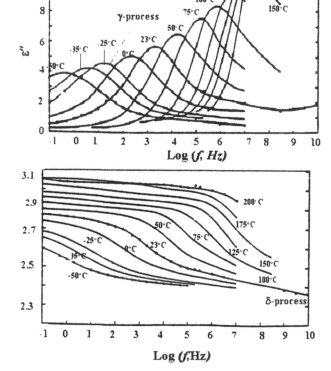

Fig. 5.20 Loss factor and dielectric constant of 80% crystalline PCTFE at various temperatures (Scott et al., 1962). (With permission of National Bureau of Standards, Washington D. C.)

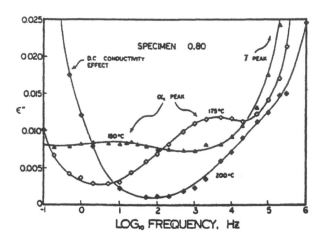

Fig. 5.21 α-relaxation in 80% crystalline PCTFE, (Hoffmann et. al. 1966). (with permission of J. Poly. Sci.)

Equations (5.13) and (5.14) are equally applicable to polymers at the glass transition temperature. William, Landel and Ferry[35] have assigned approximate values $W_n = 0.18$ eV, $T_c \cong T_G - 51.6$ to cover many amorphous polymers. The essential idea is that the activation energy for the β-relaxation is temperature dependent according to, after substituting the WLF constants,

$$\varepsilon_\beta = \frac{0.18\, T^2}{(51.6 + T - T_g)^2} \tag{5.15}$$

where W_β is the energy for the β- process. There is good agreement between the energy calculated using equation (5.15) and that using the ε''-log f curves. A temperature dependent energy is evidence of the β-process.

Before we proceed further a comment about the application of WLF equation is appropriate. The general equation that applies for the activation energy is:

$$W_\beta = \frac{C_1 C_2 T^2}{(C_2 + T - T_0)^2} \tag{5.16}$$

where C1 and C_2 are called the WLF parameters and T_0 is a reference temperature. The equations (5.15) and (5.16) are valid at $T = T_g$ or $T = T_0$ as appropriate.

D. THE γ-RELAXATION

The method of analysis for γ-relaxation runs parallel to that for β-relaxation though the details are different. The γ-relaxation occurs in both 100% amorphous and semi-crystalline material. Denote the relaxation as γ_a and γ_c in the amorphous and crystalline materials, respectively. The area under the curve of ε''-$log\,f$ curve indicates that γ_a is twice as intense as γ_c. Moreover two other major differences are discernible. They are (a) γ_a is broader than γ_c and (b) γ_a is symmetrical and γ_c is asymmetrical about the peak. Hoffman, et. al. suggest that chain ends induce rows of vacancy in the crystal and the molecular chains reorient themselves in the vacancy so induced.

Evidence for a δ-relaxation is also obtained in the frequency range of $10^8 - 10^{10}$ Hz. though it was not observed in the temperature variable mode because the sample temperature could not be lowered without obtaining a change of phase.

We have followed the analysis of Hoffman et. al. in considerable detail for the purpose of elucidating the line of reasoning that is necessary to interpret the experimental measurements. The material should be viewed as an aggregate of its electrical, optical, mechanical and thermal properties for extracting a consistent picture of the relation between structure and properties. A piece-meal evaluation based on a narrow segment of data should always be treated with caution.

5.4.4 POLYCARBONATE

Polycarbonate is transparent and is as strong as steel. The aromatic rings in the back-bone give it a high melting point of ~265°C.

The dielectric properties of polycarbonate is interesting from several points of view[36]. The polymer has high impact strength below the glass transition temperature and it has been suggested that the large free volume trapped between segments in the amorphous state may account for this behavior. Alternately the high impact strength has been linked, in an empirical way, to the low temperature β-loss.

In most polymers this relaxation has been attributed to the motion of side groups which can occur at temperatures well below T_G. If both relaxations occur in the same main chain then the relationship between the two transitions can be obtained with reasonable certainty. This is thought to occur in polycarbonate in the group –O–CO– O– along the main chain. Polycarbonate is also interesting from the point of view of studying the relationship between the degree of crystallinity and dielectric loss, since this polymer can be prepared in both amorphous and semi-crystalline forms, similar to PCTFE.

The dielectric constant of polycarbonate in the proximity of β-relaxation is shown in fig. 5-22[37]. The loss factor as a function of frequency is shown in fig. (5.23)[38] for the amorphous polymer which has T_G = 145°C. The shapes of the ε''-$log\ f$ curves are essentially unaltered which is typical of amorphous polymers for α-relaxation[39]. The relaxation time decreases with increasing temperature, but the distribution about a mean remains unaltered. The loss curves have sharp maximum and the activation energy calculated from f_{max}-$1/T$ plots gives an energy of ~ 6.50 eV.

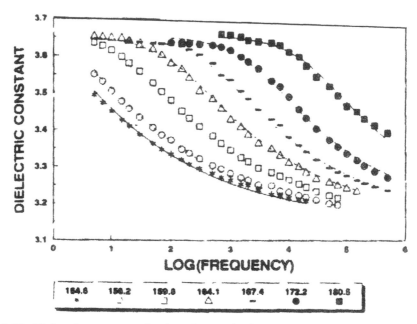

Fig. 5.22 Dielectric constant of polycarbonate in the β-relaxation region. Temperature range as shown (Havriliak and Havriliak, 1997, with permission of A. Chem. Soc.)

We have already seen that such high energy is characteristic of β-transition of amorphous polymers. The shift in the peak with temperature can be used to calculate the specific volume using theory developed by William, Landel and Ferry. The amorphous polymer is shown to have large specific volume compared with the dimensions of a unit cell in the crystalline structure. Matsuoka and Ishida calculate the occupied volume of the T_G state which shows that it is lower than the specific volume. These data show that amorphous polycarbonate has enough free space to allow molecular motion.

The shape of the α-relaxation of the amorphous polymer, while independent of temperature, is dependent upon the degree of crystallinity. The width of the peak of the α-relaxation increases as the crystallinity is introduced. The maximum value does not appear to be affected. Such behavior has also been observed in other polymers, for example in polyethylene terephthalate [PET][40]. Increasing the crystallinity makes the peak broader. It is argued that the introduction of crystallinity imposes tension on the amorphous polymer chains and such tension results in a broader ε''-$log\,f$ curve.

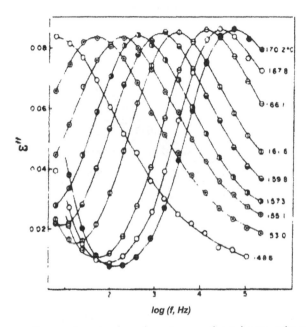

Fig. 5.23 Loss factor of amorphous polycarbonate near the primary relaxation region. (S. Matsuoka and Y. Ishida, 1966, with permission of J. Poly. Sci.).

The low temperature (~150 K) relaxation shows characteristics which are counter to the α-relaxation, namely, they are not sensitive to the degree of crystallinity but change with temperature. With increasing temperature above 161 K the loss curves become less broad as shown in fig. (5.24). Increasing width means that there is resistance to motion and this resistance probably comes from the chains in the local regions surrounding the moving part. In the terminology of relaxation studies a broad distribution means a distribution of relaxation times and the mean relaxation time is higher.

Table 5.6 summarizes the glass transition and β-relaxation temperature for several polymers obtained from both electrical studies. An empirical relation that the ratio T_β / T_G has a value of 0.5 - 0.75 has been suggested by Matsuoka and Ishida and polycarbonate has a much lower value of 0.36, meaning that the γ-transition occurs at an unusually low temperature. The temperature difference (T_β - T_γ) for amorphous polycarbonate is much higher than that for Polyethylene terephthalate (PET). This fact is interpreted as favorable to transfer mechanical energy, by allowing motion down to very low temperatures resulting in the high impact strength of this polymer.

A more recent study of polycarbonate is due to Pratt and Smith[41] who observed an activation energy of 0.28-0.56 eV for γ-relaxation and 4.97-8.65 eV β-relaxation for polycarbonate. Both relaxations are observed to follow Arrhenius law (5.8).

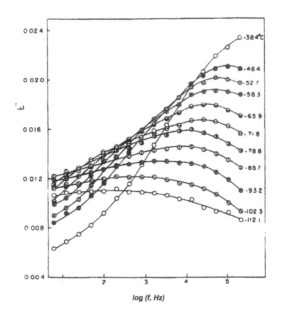

Fig. 5.24 The loss factor in amorphous polycarbonate near secondary relaxation region. The curves become broader as the temperature is lowered. This is attributed to the local motions of the segments of main chain and these motions persist even at lower temperatures (Matsuoka and Ishida (1966). (with permission of J. Poly. Sci.)

Table 5.6

Glass transition and β-relaxation temperature (Runt and Fitzgerald, 1997).

Polymer	T_G (°C)	T_β (°C)	ω_p (Hz)	T_β / T_G	Ref.
Electrical					
Poly(vinyl chloride)	100	0	100	0.73	42
Polychloroprene	-30	-85	100	0.77	43
Poly(vinyl acetate)	30	-80	100	0.57	44
Poly(vinyl benzoate)	100	-70	100	0.55	42
PMMA, atactic	120	40	100	0.80	45
Poly(ethylene terephthalate)	90	-50	100	0.61	46
Poly(ethylene adipate)	-40	-120	100	0.66	44
Polycarbonate	150	-120	100	0.36	37

5.4.5 POLY(METHYL METHACRYLATE)

Poly(methyl methacrylate), abbreviated as PMMA, is a transparent polymer with tough outdoor characteristics. Its optical clarity and easy machinability combined with high mechanical strength makes it an excellent substitute for glass. It has bulky side group giving it an amorphous structure. Such polymers are generally known as glassy polymers, not to be confused as polymers not possessing a glass transition temperature. It is completely recyclable. When heated at 300°C it yields almost entirely the original quantity of the monomer, the depolymerization taking place due to a weak chain breaking under the influence of heat. It is resistant to many chemicals but has many solvents such as acetone. The commercial form is atactic, though isotactic and syndiotactic forms which are crystallible have been prepared. The refractive index measured at room temperature is $n = 1.47$.

The dielectric properties of PMMA has been studied by Bergman et. al.[47] in its s-form, and by Ishida and Yamafuji[48] in its a-form. The latter study is referred to here to provide a necessary background, and the more recent literature will be described subsequently.

There is some confusion in the literature with regard to designation of peaks in PMMA and other acrylates. According to Hoffmann et. al (1966) amorphous polymers do not exhibit α-peak and only the peak above T_G should be designated as β-relaxation. The next lower temperature peak should be designated as γ-peak. Bur (1984) also adopts this nomenclature. To avoid confusion we follow the nomenclature used by several authors which is due to Ishida and Yamafuji.

Fig. (5.25) shows the dielectric constant and loss factor in atactic PMMA with $T_G = 105°C$ over a wide range of frequencies and temperatures (Ishida & Yamafuji, 1961). At 35°C there is a single relaxation at about 150 Hz. This is β-relaxation. Broadhurst and Bur (1984) have observed two relaxations, one at 30 Hz which is identified with this relaxation, and a second smaller δ-relaxation at 1 MHz due to absorbed moisture. As the temperature increases towards T_G the relaxation frequency shifts to higher values and ε''_{max} also becomes larger.

The higher temperature, low frequency (~30 Hz) relaxation in fig. (5.25) is the α-relaxation. This relaxation is difficult to see at the onset temperature of 137°C because the ionic conductivity is very dominant. However a close examination of the ε'-$log\ f$ curve shows that the dielectric constant at this temperature decreases in two distinct steps. The first decrease is due to the α-relaxation and the second due to the β-relaxation. A plot of *log f_m* versus 1/T for the β-peaks yields an activation energy of ~ 0.8 eV.

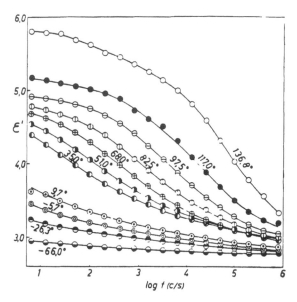

Fig. 5.25 (a). Dielectric constant of atactic PMMA (Ishida & Yamafuji, 1961, with permission of Dr. Dietrich Steinkopff verlag, Darmstadt, Germany).

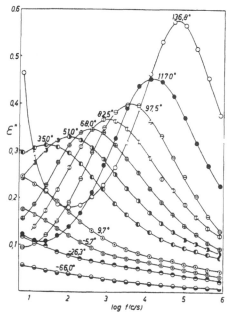

Fig. 5.25 (b). Loss factor in atactic PMMA (Ishida & Yamafuji, 1961). (with permission of Dr. Dietrich Steinkopff verlag, Darmstadt, Germany).

PMMA has two kinds of dipoles; the ester group (O=C-O) and the methyl group (CH₃). While the methyl group with a dipole moment of 0.4 D is short and fixed to the main chain rigidly, the ester groups are long and loosely attached to the main chain. The main dipole in PMMA is the O=C-OCH₃ group in the side chain of PMMA molecule, and it has a dipole moment of ~1.8 D. In this respect PMMA differs from PVC which has the dipole in the main chain and relaxes in the mode of crank shaft motion. At temperatures much lower than T_G the motion of the main chain is frozen and the ester group is active with micro-Brownian motion resulting in γ-peaks. The γ-relaxation is more dominant than the β-relaxation because the latter involves the main chain.

We shall now consider the s-PMMA which has the glass transition temperature at 131°C. The merging of the two relaxations occurs at some temperature above T_G. As the temperature is lowered towards T_G the β-relaxation slows down and merges with the α-relaxation that occurs below T_G. The merging of the two relaxations is important to understand the transitions that occur at T_G.

Bergman, et. al (1998) have used measurement frequency in the range of 10^{-2} Hz-1000 MHz over a temperature range of 220 K-500 K. They devote attention to the merging of α- and β-relaxation in the vicinity of T_G, though the latter is pronounced as discussed above. Fig. 5.26 shows the loss factor of syndiotactic PMMA. The loss curves show the familiar peaks and as the temperature is increased f_{max} becomes higher. Moreover the peaks become more narrow as the temperature is increased. At high temperature, T~400 K, conductivity contributes to ε″ at lower frequencies, < 1 Hz. As the inset of Fig. 5.26 shows the merging of α- and β-relaxation is initiated at 410 K (T_G = 404 K). At 370 K, which is below T_G, only the β-relaxation occurs. 420 K is close to, but slightly higher than T_G, and the merging of the two relaxations dominates the loss characteristics. 470K is well beyond T_G and presumably the contribution due to β-relaxation is negligible, if not totally absent.

The relaxation parameters can also be found by using the so called KWW equation (Kohlrausch William-Watts) and transforming data from the time domain to the frequency domain. The frequency domain data may also be transformed to the time domain using Laplace transforms and evaluating the distribution of relaxation times. This aspect of the analysis is deferred till ch. 6. The curves are fitted to the Havriliak-Negami function (eq. 3.72)

$$\frac{\varepsilon^* - \varepsilon_\infty}{\varepsilon_s - \varepsilon_\infty} = [1 + (j\omega\tau_0)^{1-\alpha}]^{-\beta} \tag{3.65}$$

where τ_0 is the mean relaxation time and, α and β are empirical constants applicable to the material.

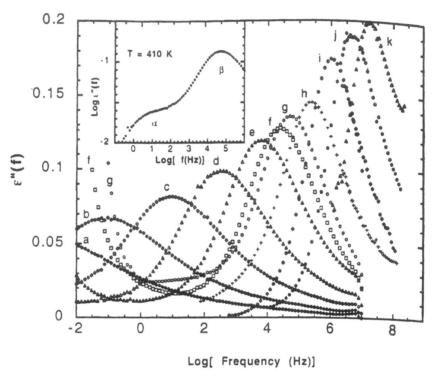

Fig. 5.26 Loss factor in s-PMMA as a function of frequency at various temperatures. Curves (a-e) correspond to temperatures below T_g: 220, 260, 300, 340, and 340 K. Curve (f) to 400K (close to T_g). (g) to (k) to temperatures above T_g: 410, 430, 450, 470 and 490 K. Curves (f) and (g) also show the contribution from conductivity at low frequencies. Curve g (T=410K) is also shown in the inset in a log-log representation with the conductivity subtracted, in order to facilitate recognition of the a-process (Bergman et. al. 1998, with permission of J. Chem. Phys.)

A word of caution is appropriate here to avoid confusion over the choice of symbols for α-relaxation of the polymer and α-parameter for the dispersion. The two alphas are entirely unrelated. α-relaxation is a property that depends on the order-disorder in the polymer whereas α-parameter is a proposed empirical constant. To avoid confusion we always associate α- with its noun such as α-parameter or α-relaxation. Further the α-parameter will be underlined in this chapter only to emphasize that it has meaning only with reference to the dielectric data in the complex plane. Similar comments apply to β-relaxation.

The loss curves at 370 K (well below T_G) and 470 K (well above T_G) show good fit with the Havriliak-Negami function as shown by the full curves, but in the merging region the fit is poor. However by choosing two such functions, each having its own parameters, and adding the superposition principle, results in good agreement. The β-parameter is approximately constant from 280 K-450K and the α-parameter increases linearly. This is equivalent to the qualitative observation that the α-relaxation gets narrower as the temperature increases beyond T_G and the β-relaxation retains the shape as the temperature is lowered below T_G.

Fig. (5.27) shows the relaxation time τ evaluated using several methods of analyses referred to above. These are: (1) H-N function fit (2) KWW transformation (3) Williams' method of transformation, (see Ch. 6). However above this temperature there is some deviation which is attributed to the merger phenomenon.

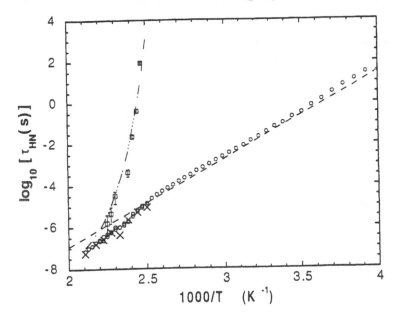

Fig. 5.27 Relaxation times obtained from fitting ε″(f) to a sum of two HN equations. The open square symbolizes $τ_a$ and open circles the effective $τ_b$ as obtained by H-N fits. The x-symbols stand for the $τ_b$ obtained from KWW fits (ch. 6). The dashed and the dash-dotted lines represent the Arrhenius and VF behavior of the β- and α-relaxations respectively (Bergman et. al. 1998). (with permission of J. Chem. Phys.)

The magnitude of the β-relaxation process, below T < T_G, increases with temperature according to

$$\varepsilon_s - \varepsilon_\infty = 0.373 + 7.17 \times 10^{-3} T; \quad T < T_G \tag{5.17}$$

Many polymers show this increase for β-relaxation that may be attributed to increased number of dipoles relaxing or the amplitude of motion increasing because of an increase in free volume. The linear dependence is terminated at T_G. The relaxation time for the α-relaxation τ_α is also temperature dependent; τ_α decreases with increasing temperature changing very sharply at T_G. This kind of behavior makes the polymer very fragile. The relaxation time follows the Vogel-Fulcher equation:

$$\tau_\alpha = \tau_0 \exp(\frac{DT_0}{T - T_0})$$

(5.18)

Where D is a constant and T_0 is the temperature at which the plot of the $1/T$-ln τ line departs from linearity. T_0 is usually 50K lower than T_G in many polymers. The values for α-relaxation may be approximaed to: τ_0 = 14.8 ps, T_0 = 371 K and C = 2.27. Note the very high α-relaxation time of 100s at a temperature of ~415 K.

Selected activation energies for the α- and β- relaxations in PMMA are shown in Table 5.7[49]. The relaxation mechanism in s-PMMA may be summarized as follows:
1. Both α- and β- relaxation are asymmetric, though the latter is the dominant process.
2. There is evidence for the fact that the two processes are statistically independent and there is no need to invoke any change of relaxation mechanism.
3. In thin samples of isotactic PMMA (10 nm thick) the β-relaxation (local) is observed to be independent of thickness whereas the α-relaxation (dynamic glass transition) is thickness dependent. With decreasing thickness the dielectric decrement ($\varepsilon'-\varepsilon_\infty$) decreases. (Ref: Dielectric News letter, F. Kramer and L. Hartmann, Sept. 2001).

5.4.6 POLY(VINYL ACETATE)

Poly(vinyl acetate), abbreviated as PVAc not to be confusd with poly(vinyl acetal), is a polar polymer; the monomer has ~2D dipole moment and is soluble in aromatic solvents, alcohols and esters. It falls into a category known as poly(vinyl esters). It is amorphous with glass transition temperature of ~31°C. Though its dielectric applications are limited it is interesting from two points of view. First, its low glass transition temperature makes it more convenient for study in the region of glass transition temperature. Second it has been quite extensively studied in the time domain and comparing the measured and calculated results often checks the suitability of the applied transformation techniques to the frequency domain. Poly(vinyl acetate) exhibits both α- and β-relaxations. The more recent measurement of the complex dielectric constant of

PVAc is due to Mashimo, et. al.[50], Nozaki and Mashimo[51] and Dionisio, et.al.[52] in the frequency domain.

Table 5.7

Selected values of activation energies for α- and β-relaxations in PMMA.

α-relaxation (eV)	β-relaxation (eV)	Reference
	0.92	Mead and Fuoss (1942)[53]
	0.91	Bröns and Müller (1950)[54]
	0.79, 0.87	Deutsch, et. al., (1954)[55]
4.35		Heijboer (1956)[56]
3.16	0.61	de Brouckere & Offergeld (1958)[57]
	0.91	Mikhailov (1958)[58]
	0.83	(Ishida and Yamafuji)[48]
	0.79	Reddish (1962)[59]
5.25		Lewis (1963)[60]
	0.74, 0.76	McCrum and Morris (1964)[61]
3.47 to 4.30	0.83 to 0.99	McCrum et. al. (1967)[62]
	0.78	Thompson (1968)[63]
10.62	0.91	Sasabe and Saito (1968)[64]
	0.93	Kawamura, et. al. (1969)[65]
	1.15	Solunov and Ponevsky (1977)[66]
4.35	0.86	Hedvig (1977)[67]
4.35 ± 1.04	0.87 ± 0.06	Gilbert, et. al. (1977)[68]
	0.74 – 1.09	Vanderschuren and Linkens (1977)[69]
	0.83	Mashimo et. al. (1978)[70]
7.25 ± 1.04	0.97 ± 0.07	Pratt and Smith (1986)[49]

(with permission of Polymer)

A. β-RELAXATION

The low temperature ($T < T_G$) loss factors, due to Ishida et. al. (1962), are shown in figs. 5.28 covering a temperature range of -61°C to –6.8°C and a frequency range of 100-1MHz. PVAc clearly exhibits the relaxation process at each temperature and as the temperature is increased f_{max} shifts to higher values. The low temperature relaxations are due to β-relaxation which is attributed to the micro-Brownian motions of the side groups accompanied with the local distortions of the main chains. The low temperature β-relaxation curves are symmetrical about f_{max} as shown in fig. 5-28. Broadening of the β-relaxation occurs as the temperature is increased towards the glass transition, probably due to the overlap of the two relaxations. At sufficiently low temperatures only, we get the pure β-relaxation and from an analysis the dipole moment of the side group is obtained as 1.7 D and the number of side groups as 4.9×10^{27} m $^{-3}$. The activation energy for the α- and β-relaxations are 2.6 eV and 0.43 eV respectively. At $T \ll T_G$, $(\varepsilon_s - \varepsilon_\infty) = 0.35$. We shall refer to these values again in Ch. 6.

B. α-RELAXATION

Fig. 5.29 shows the loss factor data for temperatures in the region of the glass transition (~30°C). Again a relaxation peak is observed at each temperature. A plot of the normalized ε" ($\varepsilon''/\varepsilon_{max}$) versus the normalized log f (f / f_{max}) shows that they lie on a single curve and therefore there is no change in the shapes of loss factor curves with temperature. The high temperature plots exhibit Cole-Cole arc in the complex plane plots of ε* as shown though H-N function is more appropriate as has been discovered recently. The α-relaxation is attributed to the re-orientations of the dipoles due to the segmental motions of the main chains.

The β-parameter together with other parameters are shown in Table 5.8. The constant value of the β-parameter shows that the shape of the loss factor curve remains nearly the same with change in temperature. We recall that β = 1 yields Debye relaxation. The dielectric decrement decreases with increasing temperature as shown in fig. 5.29 and Table 5.8 This does not imply that the dipole, in the relaxed state, has a lower dipole moment. The dipole moment calculated at the measured temperatures (see chapter 2 for the method) shows that the dipole moment is ~2 D at all temperatures, higher than that of the monomer, 1.6 D.

The data of Dionisio et. al (1993) shows that WLF equation is satisfied for the activation energy in the temperature range of α-relaxation. The activation energy obtained from the plot is 2.35 eV which is in the expected range. The results in PVAc may be concluded as follows:

1. There are two relaxations, the main relaxation occurring at ~29°C. The main relaxation is much stronger and it is attributed to the re-orientations of the dipoles due to the segmental motions of the main chains.
2. Almost all of the mean square dipole moment $\langle \mu^2 \rangle$ is carried into the α-relaxation.
3. The β-relaxation occurs at -61°C at 1 kHz.

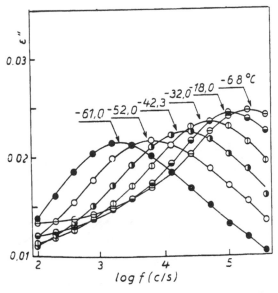

Fig. 5.28 Dielectric loss factor in PVAc at low temperatures showing β-relaxation. The curves become broader as the temperature is increased due to the over-lapping α-relaxation (Ishida et. al, 1962, with permission of Dr. Dietrich Steinkopff Verlag, Darmstadt, Germany).

Table 5.8

α-relaxation parameters in Poly(vinyl acetate), (Dionisio, et. al., 1993).

$T°(C)$	f_{max} (kHZ)	τ_0 (s)	$\varepsilon''_{max} C_0{}^a$ (pF)	$(\varepsilon_s - \varepsilon_\infty) C_0$ (pF)	β-parameter
45	0.0343	4.64×10^{-3}	14.90	54.9	0.56
50	0.1658	9.60×10^{-4}	14.69	50.2	0.58
55	0.6036	2.64×10^{-4}	14.33	47.9	0.59
60	2.0484	7.77×10^{-5}	13.87	46.2	0.60
65	5.8080	2.74×10^{-5}	13.68	45.0	0.61
70	15.8742	1.00×10^{-5}	13.39	43.0	0.61

(a) C_0 is the capacitance of the measuring cell, 7.9 pF

5.4.7 POLYSTYRENE

Polystyrene, abbreviated as PS, consists of linear chain and is chemically inert. It has many organic solvents. It cannot be used above 85°C due to heat distortion and is unsuited for outdoor applications. The glass transition temperature is ~ 100°C and melting point is 250°C. The molecular mass is between 200,000-300,000. Addition of plasticizers lowers T_G depending upon the quantity and the type of compound used. For example T_G of PS can be lowered to as low as -150°C by adding just 1% of methyl acetate.

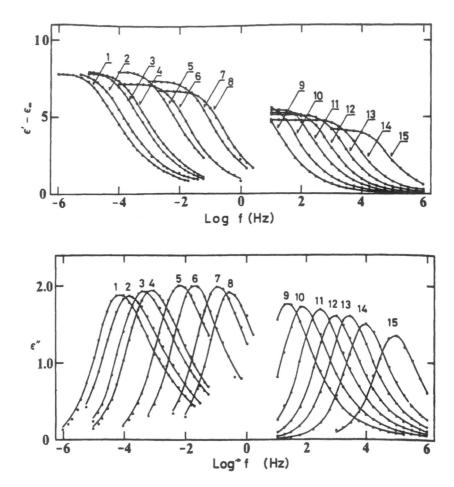

Fig. 5.29 Dielectric constant and loss factor in PVAc in the region of T_G. Temperatures are: 1-301.25K, 2-301.71K, 3-302.90K, 4-303.60K, 305.80K, 6-309.16K, 7-311.15K, 8-315.65K, 9-326.25K, 10-330.47K, 11-335.05K, 12-339.30K, 13-343.75K, 14-349.80K, 15-365.95K. Solid curves will be referred to in ch. 6. (Mashimo, et. al., 1982, with permission of J. Chem. Phys.)

Plasticizers are usually low molecular weight non-volatile compounds that increase the flexibility and processibility of polymers. The plasticizing molecules, being much smaller than a polymer chain, are capable of penetrating the chain like bugs in woodwork. They establish attractive polar forces with chain segments there by reducing the cohesive forces between chains. This increases chain mobility and correspondingly, lowers T_G. The commercial PS is atactic and therefore amorphous though an isotactic form with about 30% crystallinity can be prepared. PS has low water absorption, ~0.03% though moisture attacks it more vigorously than polyethylene. It has a small dipole moment of 0.26D due to the asymmetry at the phenyl side group (C_6H_5).

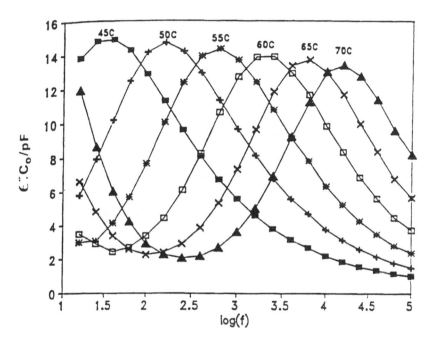

Fig. 5.30 The dielectric loss factor in PVAc above T_G showing α-relaxation. With increasing temperature the peak is marginally reduced. The area under the peak shows the number of relaxing dipoles and the decrease in height, without change in shape, indicates a decrease in the total amount of relaxed dipoles with increasing temperature (Dionisio et. al., 1993, with permission of Polymer).

In certain applications such as encapsulation and cable termination, the procedure involves *in-situ* polymerization. Unmodified PS can be made into films and used in capacitors and radio frequency cables because of low loss. Its high insulation resistance and ability to retain charge for a long period is also an advantage in capacitor applications. Glass reinforced PS has been successfully used as base for printed circuit boards and for microwave strip line circuits because of a low dielectric constant and high dimensional stability.

We refer to the results of Yano and Wada[71] who studied three forms of PS. The types and some of their physical properties are given in Table 5.9. The measurements were performed on PS films *150 μm-250 μm* thick in the frequency range of 30 Hz-100kHz covering a temperature range of 250 K- 400 K.

Table 5.9
Polystyrene samples used by Yano and Wada (1971)

Symbol	Type	Average M	Density Kg/m³
APS	Atactic	2.4×10^5	1050
MPS	Monodisperse	4.1×10^5	1051
IPS	Isotactic	9.2×10^5	1071

Fig. 5.31 shows the complex dielectric constant of atactic polystyrene. Six relaxations are observed in PS, when several techniques (mechanical, NMR) are employed. Of these only three α, β, γ', are observed in dielectric measurements.

The α-relaxation is due to segmental motion and follows the WLF equation for the temperature dependence of relaxation time:

$$\ln \tau = \ln \tau_0 - \frac{C_1(T - T_0)}{C_2 + (T - T_0)} \tag{5.19}$$

where τ is the relaxation time which is a function of temperature, τ_0 is the relaxation time at an arbitrarily chosen reference temperature $T_0 < T_G$, C_1 and C_2 are constants known as the WLF parameters.

Activation energies obtained from the plots of log f_m-1/T are, 1.30 eV, 0.39 eV, 0.12 eV, 0.07 eV for β, γ, γ' and δ processes respectively. Not all the six processes were observed by dielectric measurements. The γ relaxation was not observed at all in this study. The γ' relaxation which was observed clearly in APS is attributed to the introduction of polar impurities during bulk polymerization. This peak appears when the sample is heated at 280°C for 30 minutes providing support for this reasoning. Table 5.10 summarizes the results in PS.

5.4.8 POLY(ETHYLENE TEREPHTHALATE)

Polyethylene terephthalate, abbreviated as PET, has an aromatic ring in the main chain (Table 5.3) and possesses a higher T_G of ~ 69° -75°C. It melts around 265°C and is resistant to heat and moisture. It is extensively used in textile fibres due to its resistance for wrinkles. Electrical industry uses PET as high performance capacitor foils and polymer electrets. It is available in unoriented, uniaxially and biaxially oriented form. Its crystallinity can be controlled in the range of 0-60%. The dielectric properties depend on the degree of crystallinity and the method adopted to achieve crystallinity.

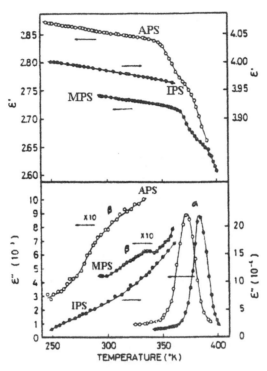

Fig. 5.31 Dielectric constant and loss factor of polystyrene: APS, 110 Hz; MPS, 110 Hz; IPS, 300 Hz (Yano and Wada, 1971, with permission of Polymer).

Table 5.10
Relaxation processes in polystyrene (Yano and Wada, 1971)
(Adapted with permission from J. Poly. Sci.)

Symbol	Temp. (K)	Energy (eV)	Mechanism
α	400	WLF type	Segmental motion of backbone chain
β	350	1.3	Local oscillation of backbone chain
γ	180	0.37	Rotation of pheyl group (not observed electrically)
γ'	97	0.12	Polar defects such as oxygen bonds
δ	55	.07	Defects in tacticity (not observed electrically)
ϵ	(25)	-	Not observed electrically

Being a main chain polymer it possesses both α- and β-relaxations. The α-relaxation is associated with the glass-rubber transition in amorphous regions. The origin of β-relaxation is subject to various interpretations. Recent studies are due to Coburn, et.al.[72], Miyairi[73], Adamec and Calderwood[74], Tatsumi, et. al.[75], Dargent, et. al.[76], Osaki[77] and Neagu, et. al[78].

Fig. 5.32 shows the real part ε′ and the loss factor ε″ above T_G. The α-relaxation is clearly seen, the relaxation frequency shifting to higher values as the temperature is increased from 100° – 150°C. At lower frequencies and higher temperatures the α-relaxation is overlapped with the increase of conductivity as shown in fig. 4.8.

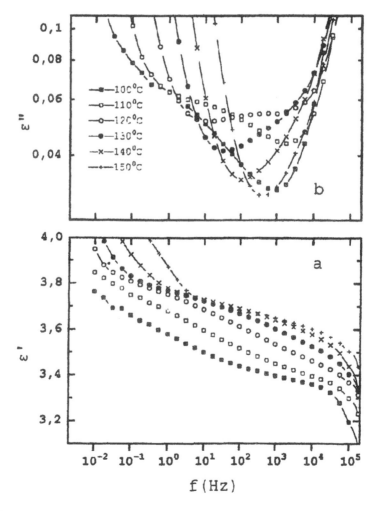

Fig. 5.32 Real part ε′ and the imaginary part ε″ of the dielectric constant of PET exhibiting α-relaxation (Neagu et. al. 1997, with permission of J. Phys. D: Appl. Phys.)

The following conclusions may be drawn with regard to relaxation in PET.

(1) The α-relaxation is associated with the glass-rubber transition in the amorphous regions. The peak shifts to higher temperatures or lower temperatures with increasing degree of crystallinity. Increasing crystallinity also renders the loss peak weaker and broader possibly due to the constraint it imposes on molecular movement in the amorphous regions.

(2) The β-relaxation is due to the local movement of the chain. The α-relaxation is much weaker than the β-relaxation.

(3) At low frequencies and high temperatures both ε' and ε'' increase due to conductivity effects; the bulk interfacial polarization between the amorphous and crystalline regions possibly causes this increase.

5.4.9 POLYISOPRENE

The degree of crystallinity of a polymer is closely related to the long range order of the molecules, that is, the configuration of the chain. Stereo-regular polymers, isotactic and syndiotactic forms, are crystalline; atactics are amorphous. This is the reason for the high density of linear polymers like HDPE exhibiting crystallinity up to 90% while LDPE, which is branched, has only 40% crystallinity. Branched polymers do not crystallize as easily as linear polymers do. Polyisoprene is another example. It has two possible geometrical arrangement of the molecules. In the first, the molecules of isoprene bend back giving the chain a coiled structure. This type of isomer is called *cis*-1,4 polyisoprene and it is the synthetic version of natural rubber. On the other hand, in the trans-isomer, isoprene molecules are packed more straightly, giving it a rod-like structure, and this form is known as *trans*-1, 4 polyisoprene which is the synthetic version of gutta percha. Gutta percha has, therefore, more crystallinity.

Polyisoprene, like polystyrene undergoes depolymerization to yield the monomer to a limited extent. Under pressure scission of monomer connecting bonds ($CH_2 - CH_2$) occurs and big molecules are broken into smaller units. The structure of polyisoprene is:

The polymer bears components of the dipole moment both parallel and perpendicular to the chain backbone. Pressure exerts a greater influence on the segmental motion when compared with the longest normal mode. Above T_G global chain motion and local segmental motion occurs and the global chain motion can be activated by lowering the temperature at constant pressure or by increasing pressure at constant temperature. Recent dielectric studies of Floudas et. al.[79, 80] in polyisoprene are directed towards obtaining the relaxation times under the combined effects of temperature and pressure.

Floudas et. al studied five different samples of *cis*-isoprene with number averaged molecular mass of 1200, 2500, 3500, 10600 and 26000 and identified by these numbers. The glass transition temperatures are 191, 199, 200, 204 and 208K respectively. The samples had a polydispersity of less than 1.1. The molecular mass indicates the degree of entanglement as follows: 20%, 50%, 67%, 200% and 500%. The samples vary from no entanglement to well entangled. The interest of the study lies in understanding the effect of pressure on entanglement as this is of technological interest. Fig. 5.33 shows the loss factor for the PI-1200 under isothermal (top) and isobaric conditions (bottom). Two modes of relaxations are identified as normal (*n*) and segmental (*s*). Both modes become slower with increasing pressure and decreasing temperature. The fastest of the modes is due to segmental motion and the spectra of the slower modes are due to normal chain motion. The pressure is changed from 1 bar to 3 bar and the temperature from 205 K to 243 K.

It is easily seen that the temperature is more effective in slowing the normal mode when compared with pressure. The shift in the peak of the normal mode due to temperature change is five decades, whereas the shift due to pressure change is only three decades. Normalized plots of ε'' at a reference temperature of 295K or 1 bar shows that the normal mode is independent of temperature or pressure while the segmental mode is dependent on these two factors. The temperature change produces greater deviation than the change in pressure.

The dielectric spectra are analyzed using the Havriliak and Negami function (eq. 3.71) which is generally more successful in the analysis of amorphous polymers near T_G. The rise of the loss curves at $f < 100$ Hz is due to conductivity (eq. 3.8), and the contribution due to conductivity is subtracted from measured ε''.

In all systems the main and secondary relaxations exhibit a temperature dependence according to two laws:

Main relaxation: The Vogel-Fulcher Law.

The V-F law is expressed as

$$\log \tau = \log \tau_0 + \frac{DT_0}{(T - T_0)}$$

(5.20)

where $\log \tau_0$ is the limiting value at high T, T_0 the apparent activation temperature of the α-relaxation, sometimes called **Kauzmann temperature** and D the so called fragility parameter. Lower values of D denote increased fragility in the glassy phase.

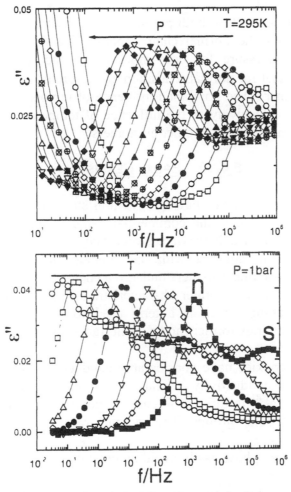

Fig. 5.33 Dielectric loss for PI-1200 plotted under isothermal conditions (top) at 295 K, and under isobaric conditions (bottom) at 1 bar. The corresponding pressures to the isothermal spectra are: (□): 1 bar, (○): 0.3 kbar, (●): 0 6 kbar, (◇): 0 9 kbar, (⊕): 1.2 kbar, (⊗): 1.5 kbar, (▲): 1.8 kbar, (△): 2.1 kbar, (▼): 2.4 kbar, (▽): 2.7 kbar, (♦): 3.0 kbar. The temperatures for the isobaric spectra are (○): 205 K, (□): 208 K, (△): 213 K, (●): 218 K, (▽): 225 K, (◇): 233 K, (■): 243 K. The symbols "s" and "n" indicate the segmental and normal modes, respectively. (with permission of AIP)

Table 5.11

Relaxation parameters in polyisoprene-1200 (Floudas & Gravalides, 1999)

Mode	τ_0 (s)	D	T_0 (K)
segmental	9.2×10^{-7}	4.34	153 ± 1
Normal	1.96×10^{-5}	4.34	149 ± 1

The relaxation time is dependent on both pressure and temperature, though it is for the segmental motion than for the normal chain motion. The relaxation time increases linearly with increasing pressure for both modes. The fact that T_0 for segmental relaxation is higher than that for normal relaxation is interpreted to mean that the log τ - 1/T curves cross each other. The combined influence of temperature and pressure has not been investigated as extensively as the influence of temperature alone and has the potential of yielding valuable results in other elastomers.

5.4.10   EPOXY RESINS

Epoxy resins are excellent adhesives having structural strength and they are used in a variety of applications. Araldite is the commercial name of one of the most popular epoxies that can be bought in a hard ware store. They come under the category of **thermo-setting** polymers in contrast with the **thermo-plastic** polymers we have considered so far. Thermo-setting polymers undergo a chemical change when heated and upon cooling do not regain their former characteristics. They are extensively used in electrical industry for potting which is a term used for encapsulation of electrical equipment such as potential transformers and current transformers.

Commercially available epoxies have considerable conductivity and advantage can be taken to study the role of conductivity in relaxation studies. Such studies have been carried out by Corezzi et. al.[81] in three different commercial epoxy compounds, having an increasing number of polar groups. The samples were chosen for the following reasons:
1. The samples attain glassy phase easily avoiding crystalline phase even at the slowest cooling rate.
2. Their dipole moment is due to the same polar group i.e. the epoxy ring, having a moment of 2.1 D
3. They possess appreciable conductivity due to the presence of ionic impurities.

Fig. 5.34 shows the spectra of the loss factor for one sample at various temperatures. Below the glass transition temperature (204K) there is only one small peak lasting over several decades of the frequency due to the β-relaxation. By increasing the temperature the main α-relaxation is observed and the peak shifts to higher frequencies as the temperature is increased, merging with the β-peak. At lower frequencies ε'' increases due to conductivity and the loss factor in this region increases proportional to ω^{-1} (fig. 4.1). In view of the almost perfect linearity observed it is clear that conductivity dominates the dielectric loss in regions where the α-relaxation may be neglected.

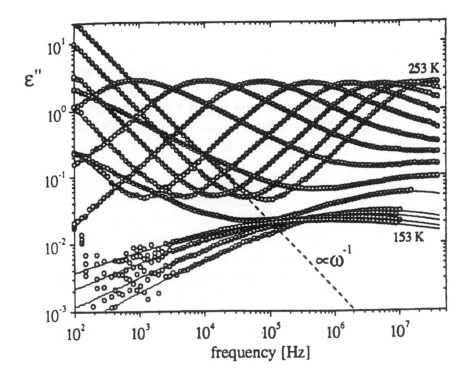

Fig. 5.34 Dielectric loss factor in epoxy resin above and below glass transition temperature. The solid lines represent the fitting equation built up by the superposition of H-N functions and the conductivity term ew_0. The dashed line shows the dc conductivity contribution (Corezzi, et. al., 1999, with permission of J. Chem. Phys.)

The loss curves are fitted to Havriliak-Negami function taking into account the conductivity, (eq. 3.109) and the quality of the fit is shown in fig. (5.34) by solid lines. The relaxation time varies in the range of $10^{-10} - 10^{2}$s for both relaxations.

5.4.11 POLYAMIDES

Polyamides are either aliphatic or aromatic depending on whether they have CH_2 or benzene in the main chain. Aliphatic polyamides are called **nylons** commercially and come in a number of different varieties. Nylons are designated according to a numbering system; a single number designates the number of carbon atoms in the monomer, as in nylon 6. Two numbers designate the number of carbon atoms in each of the constituent chemicals (diamine and dicarboxylic acid) as in nylon 6,6 and nylon 6,10. Aliphatic nylons have a melting point in the range of 250°C-300°C. More recently nylons having melting points of 450°C-500°C have become available. Aromatic polyamides are called **aramids** and they have properties that make them suitable for high temperature applications.

More recently polyamide-4,6 with the trade name Stanyl® has become available commercially. The polymer has improved characteristics making it suitable for electrical insulation at elevated temperatures. The improved characteristics arise out of the high number of amide groups per unit chain length when compared with other polyamides. Stanyl® is highly crystalline and has a melting point of 295°C. Its density is 1180 kg/m^3 and has a molar mass of $M_W = (4-4.5) \times 10^3$ g/mol.

Fig. 5.35 (Steeman and Maurer, 1992) shows the dielectric constant and the loss factor of dry Stanyl® as a function of temperature at various frequencies. The weak γ-relaxation is due to the rotation of the non-polar CH_2 groups as in other nylons and the β-relaxation is much stronger because the amide bonds are polar. The glass transition temperature is ~90°C and the loss factor curves show the α-relaxation at high frequencies. At very low frequencies, for example at 0.1 Hz this peak is clearly missing, the ε''-T curve rising monotonously. This is due to the interfacial polarization which occurs predominantly at high temperatures and low frequencies. A comparison with fig. 5.2 shows similarities when data are presented in the frequency domain.

The relaxation processes in nylons show three distinct features (fig. 5.35, Steeman and Maurer, 1992).
(1) A low temperature γ relaxation in the range 125K-175K, probably due to local motion of CH_2 segments in the inter-chain.
(2) An intermediate β relaxation at ~ 225 K which is related to the rotation of amide bonds (O = C—N— H) together with water molecules that are bonded to them.
(3) A strong α-relaxation near the glass transition temperature (~325 K).
(4) Above T_G the polymer becomes highly conductive with the loss factor increasing rapidly. The mechanism is possibly the interfacial polarization (section 4.4) with the charge carriers accumulating at the boundaries between the crystalline and amorphous regions. This kind of relaxation due to space charges is sometimes called δ-relaxation, not to be confused with the low temperature relaxation such as the one observed with polystyrene (Yano and Wada, 1971).

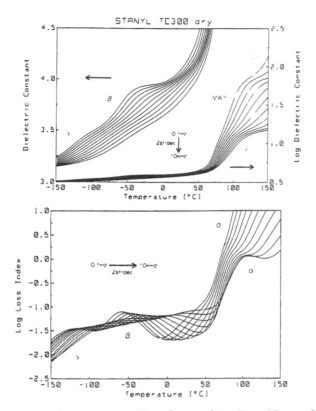

Fig. 5.35 The dielectric constant and loss factor of dry Stanyl® as a function of temperature, for several frequencies. EP stands for electrode polarization which increases the dielectric constant at the highest temperature. MWS stands for Maxwell Wagner Sillars' interfacial polarization which is dominant at higher temperatures and low frequencies (Steeman and Maurer. 1992. with permission of Polymer).

There are two alternative methods to obtain the loss factor due to dipolar processes only. One is to apply equation (3.8) provided σ_{dc} is known. An alternative is to use the approximate equation

$$\varepsilon'' \cong -\frac{\pi}{2}\frac{\partial \varepsilon'}{\partial \ln(\omega)} \qquad (5.21)$$

which has the advantage that ε' is not influenced by the dc conductivity. Steeman et. al. have used this method and obtained ε'' due to dipolar processes only.

The space charge polarization occurs above the glass transition and it is observed only by dielectric measurements. Space charge accumulation usually leads to electrode polarization and an associated increase in the dielectric constant. As expected the activation energy increases for the relaxation processes in the order γ-, β-, α-process. In engineering applications moisture content is of concern and the effect of moisture is also shown in the relaxation map. With increasing moisture the activation energy appears to decrease indicating that the water molecules assist the relaxation processes. Moisture increases the electrical conductivity which also lowers the temperature at which interfacial polarization shows up.

5.4.12 POLYIMIDES

Polyimide (PI) is a high temperature dielectric with a long chemical name that keeps chemists happy: poly-4,4′-oxydiphenylene-pyromelitimide. However the repeat unit is simpler to understand :

It is commercially known as H-film or Kapton® and is amorphous. It has a glass transition temperature between 260°C - 320°C. Kapton is available in several grades such as corona resistant, antistatic, thermally conductive etc. The polymer has the interesting property that conducting lines can be drawn by irradiation with many pulses of ultraviolet laser radiation, and with many possible practical applications[82].

Amborski[83] has measured the dielectric loss in H-film over the temperature range of - 60°C-220°C and Wrasidlo[84] has extended the higher temperature range to 550°C. Three dielectric loss regions are identified. The dielectric constant and tan δ as a function of frequency at various temperatures are shown are shown in fig. 5.43. The width of the loss factor at half peak according to Debye theory is 3.46 (section 3.2) and applies reasonably well to the polymer. The maximum dissipation factor according to Debye theory is

$$\tan \delta_{max} = \frac{\Delta \varepsilon}{\varepsilon_\infty} \frac{1}{\sqrt{1 + \dfrac{\Delta \varepsilon}{\varepsilon_\infty}}} \tag{5.22}$$

where $\Delta\varepsilon = (\varepsilon_s - \varepsilon_\infty)$ is the dielectric decrement. It is easy to prove that Equation (5.22) simplifies to (3.37). Substituting $\tan\delta_{max} = 0.4$ the dielectric decrement is obtained as $\sim$ 6 which agrees with experimental measurements. Since the repeat unit is non-polar except for the diphenyl ether segment which has only 1.15 D, such large dielectric decrement is probably due to amide side group. Since the polymer is amorphous the possibility of glass transition for the peak was considered and found not feasible. The method of calculation is instructive and shown below:

From measurements the aromatic polyimide has the following parameters. $\varepsilon_\infty = 2.4$ (fig. 5.36), $\varepsilon''_{max} = 2.0$, molecular weight of repeat unit = 364 gm/mole and density of 1400 kg/m^3. Substituting in

$$\varepsilon''_{max} = \frac{\varepsilon_s - \varepsilon_\infty}{2} \tag{3.34}$$

$\varepsilon_s = 6.4$. Onsager's equation (2.77) can be applied at the glass transition temperature. We first calculate the number of repeat units as

$$N = 6.02 \times 10^{23} \times \frac{1400}{342 \times 10^{-3}} = 2.46 \times 10^{27} / m^3$$

Substituting in eq. (2.79) the dipole moment at 573 K is found to be 3.6 D which is much greater than the dipole moment of the repeat unit ($\sim$1.15 D). Further substitution of $T_g = 573$ K in the WLF equation (5.12a) gives an activation energy of $\sim$21 eV which appears to be too high by a factor of ten. Hence the relaxation is not due to the glass transition but occurs in the solid state.

A. INFLUENCE OF MOISTURE

Polyimide films absorb considerable moisture, $\sim$3.3% in an atmosphere of 100% humidity (2500 Pa at 21°C). In fact this property has been utilized in building humidity sensors. The absorbed moisture amounts to two molecules per PI repeat units. The dielectric constant increases upto 20% due to absorption of water. The increase occurs even below room temperature, in the range of 80 K-270 K.

Fig. 5.37[85] shows the loss factor of 'dry' film and a single peak is observed at each frequency. In the dry film the familiar characteristic of peak shifting to higher temperature as the frequency is increased is seen. Arrhenius plots give an activation energy of 0.53 eV and a relaxation time of 10^{-16} s. This peak could be eliminated by drying the film for at least 250 days; this suggests that the observed relaxation is possibly due to the presence of the moisture and not due to the intrinsic property of the polymer.

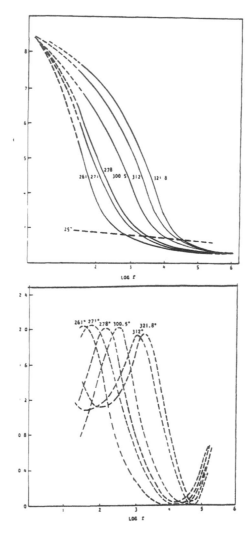

Fig. 5.36 ε' and loss tangent versus log (*f*) at various temperatures in PI. (1) 261.5°C, (2) 271°C, (3) 278°C, (4) 282°C, (5) 300.5°C, (6) 312°C, (7) 321.8°C (Wrasidlo, 1972). (with permission of J. Poly. Sci.)

In contrast to the dry film, films that have absorbed moisture show two distinct peaks, the sharpness of the peak increasing with the moisture content. The peak exhibited by the dry film also increases in magnitude. When normalized plots of ε" - T show that the high temperature peaks merge into a single curve suggesting that the peak height is independent of water content.   The lower temperature peaks do not merge into a single curve. The peaks are attributed to water residing in two different sites. One is the oxygen of the ether linkage and the other is one of the four carbonyl groups. At low humidity the latter is more likely to be the host. Since the activation energy for the high energy peak is determibed as 0.53 eV, which is twice that of a hydrogen bond, both bonds of a water molecule are likely to be involved in the relaxation process.

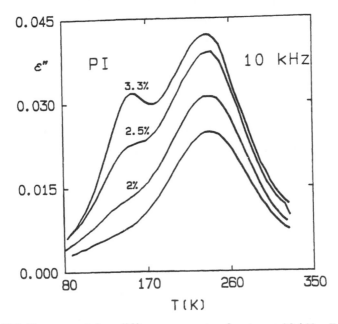

Fig. 5.37 ε'' of PI films containing different amounts of water at 10 kHz. For the lowest curve the film was dried for two days in vacuum, for the other curves the water content is given in wt% (Melcher et. al. 1989, with permission of Trans. Diel. Elec. Insul.)

5.5 SCALING METHODS

We have seen that a large number of liquids could be cooled very slowly, that is super cooled, into an amorphous solid called glass. As the liquid cools its relaxation time increases and at the glass transition temperature the relaxation time τ increases even more rapidly. The Arrhenius law does not hold any more and the relaxation time at the transition region is given by the Vogel-Fulcher law, equation (5.20). We have also seen that many polymers show agreement with different expressions over a limited range of temperatures, but no general expression appears to have been successful in covering the entire range of temperature in the super cooled state. Since τ is a strong function of T and hence a strong function of $1 / \omega$, a large range of frequency needs to be considered before deciding whether any particular scheme is general and applicable to a large number of materials.

Moreover the shape of the complex plane plot changes from a near Debye semi-circular shape to that described by Cole-Cole arc, Davidson-Cole's skewed arc and Havriliak and Negami function. The latter has been particularly successful in the vicinity of the transition temperature. In an attempt to find a general expression, several scaling methods have been suggested and we describe the method adopted by Nixon et. al[86] as it extends over a wider range of frequency from 10^{-3} Hz $\leq f \leq 10^{+10}$ Hz. Many samples were measured including glycerol which we have discussed in connection with Davidson-Cole expression. A difficulty usually encountered in designing a suitable scaling law is that the primary relaxation overlaps a secondary relaxation, and in such cases the frequency range needs to be chosen appropriately to make sure that only primary relaxation is covered.

Both ε' and ε'' were measured by Nixon et. al. (1990) combining a number of different techniques to cover a wide range of frequency. At low frequencies below 10^{-4} time domain measurements were carried out. In the range 10 k HZ < f < 10 MHz four probe impedance measurements were adopted. For frequencies greater than 50 MHz, data were taken in both transmission and reflection with a 50 Ω co-axial transmission line and network analyzer.

The factors involved in the scaling law of Nixon et. al. are the loss factor ε'', f, f_{max}. the dielectric decrement $\Delta\varepsilon$, and the width of the loss factor curve at half peak W. For Debye relaxation which has a single relaxation time this width, W_D is $\cong 1.14$ decades. The width is normalized according to $w \equiv W/W_D$. This is acceptable because the narrowest width for any material is the Debye width. We recall that as the temperature increases, f_{max} also increases and the width of the loss factor curves becomes smaller.

The familiar $\log_{10}(f_{max}$-Hz) versus 1/T curves show similar behavior though displaced from each other. However the increase occurs in different ways resulting in a curve for each material. The purpose of scaling is to find a scheme in which all of the data fall on a single curve.

The scaling law proposed consists of plotting

$$x = \frac{1}{w}(1+\frac{1}{w})\log_{10}(\frac{f}{f_{max}}); \; y = \frac{1}{w}\log_{10}(\frac{\varepsilon'' f_{max}}{\Delta\varepsilon f}) \qquad (5.23)$$

The factor f_{max}/f for the y-axis aligns the peaks of the loss factor curves and the entire term normalizes the curves to the same half width. Fig. (5.38) shows that data for all seven materials investigated by Nixon, et al. fall on a single curve verifying the applicability of the proposed scheme. Though the scaling methods provide a compact way of providing data for a large number of polymers its use in interpreting the measured values in terms of morphology is limited.

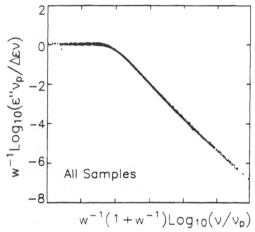

Fig. 5.38 Scaling law for several polymers proposed by Nixon et. al. (1990). (with permission of AIP)

5.6 CONCLUDING REMARKS

An attempt, however concise, has been made in the previous pages to describe the relaxation processes that occur in polymers. The present summary includes only the polymers which we have discussed except where a reference is required for other polymers. The classification of relaxation peaks is mainly according to Hoffmann, et al. (1966). The relation between temperature and frequency for the same peak is such that lowest frequency correspond to the highest temperature. In crystalline polymers α_c-relaxation occurs at the lowest frequency and the highest temperature, $T/T_m \sim 0.9$-1.0. With decreasing temperature and increasing frequency γ_c and δ_c relaxations occur in that order. The β-relaxation is not observed in single crystals.

Semi-crystalline polymers exhibit α_c-, β_a-, γ_c-, γ_a-, and δ_c-relaxations. α_c-mechanism occurs close to T_m and β-mechanism just above T_G. The γ-peak occurs just below T_G. One of the methods used to ascertain the region in which relaxation occurs is to use samples of varying crystallinity, if that is possible at all, and relate the magnitude of the loss peak with the degree of crystallinity. In highly crystalline polymers having crystallinity of ~90%, such as high density polyethylene, it is difficult to visualize a distinct amorphous phase; it may be more appropriate to think in terms of motion in disordered region and in-between crystallites[87].

In wholly amorphous polymers α-relaxation is absent, β-relaxation occurs at ~T_G and γ_a-relaxation at lower temperatures. This designation, though not practiced uniformly, makes the nomenclature for crystalline and amorphous polymers uniform (Bur, 1984).

For each relaxation process, increase of temperature increases the frequency at which the loss peak is observed. Plotting f_{max} versus 1/T yields a relaxation map and from these plots the activation energy for each process in obtained. The energy increases generally in the order of δ-peak to α-peak . For most polymers the activation energy ε_β / ε_α << 1. This means that the β-peak is less pronounced than the α-peak. This is true in the case of isotactic PMMA (Fig. 5.40) but the syndiotactic PMMA exhibits a contrary behavior. The inset of fig. 5.26 may be referred to, to see this. In fact the ratio of ε_β / ε_α has been shown to gradually increase and exceed a value of one as the tacticity of PMMA is systematically increased from isotactic to syndiotactic polymer.

Recent advances in the study of PMMA are measurements carried out on free standing very thin films of polymers, in the range of ~ 10 nm[88]. Thin polymer layers exhibit a glass transition temperature that is dependent on the thickness and the explanations for this observed phenomenon are not yet clear. Further, nanometer thick polymers play an important role in coatings and knowledge of molecular dynamics is highly desired.

Fig. 5.39 shows the dielectric loss ε'' versus frequency for isotactic PMMA having a sample thickness of 21 nm (molecular weight 44900 g mol^{-1}) at various temperatures. The bulk properties are reproduced well in thin films.   At lower temperatures, in the range 305 K-325 K, only the β-relaxation is present. From the top figure, it is noted that at 325 K, the high frequency wing of the α-relaxation appears. The higher temperature dielectric loss in the region of 329 - 357 K shows the region of merging of α- and β-relaxations. The inset shows the separation of the two relaxations.

Fig. 5.40 shows that the β-relaxation has a lower activation energy when compared with the α-relaxation. Further the β-relaxation follows the Arrhenius law and the α-relaxation follows the Volger-Fulcher-Tamann law. The inset shows the shift of the α-relaxation to lower temperatures as the thickness increases with the bulk having the lowest T_g. T_g is determined by the intersection of the relaxation curve with the line $1/\tau_{max} = 10^{-2}$ Hz, as shown by the dashed line in fig. 5.40.

The dielectric properties of polyimide has received considerable attention in recent years, due to the possibility of modifying its dielectric properties by various means. A low dielectric constant is preferred for several applications, such as an inter layer medium in multilayer integrated circuits. Addition of flourine appears to lower the dielectric constant to a value close to 1.9 when compared with the value of ~ 2.6-3.1

without this addition. There is experimental evidence to show that irradiation by high energy beams also reduces the dielectric constant.

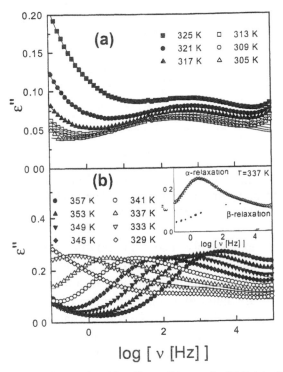

Fig. 5.39 Dielectric loss factor in thin film of isotactic PMMA. The top figure shows the relaxation in at lower temperatures. The bottom figure shows the merger of both α- and β-relaxation at higher temperatures. The inset in the bottom figure shows the separated relaxations at 337 K. Note the difference in the scale for dielectric loss in (a) and (b).(Kremer aand Hartmann. 2001: Dielectric News Letter)

Recent results of Alegaonkar et. al.[89] provide quantitave results for polyimide which is irradiated by an electron beam of 1 Mev energy. The measured loss factor is shown in fig. 5.41 for various flux of the electron beam at room temperature and two peaks are seen , one at a frequency of $\sim 10^4$ Hz and the other, a much larger one at ~1MHz. The higher frequency relaxation is broader suggesting a distribution of relaxation times. The higher loss with increasing flux is ascribed to conversion of crystalline regions to amorphous ones, increasing the disorder in the material. Further support to this reasoning is obtained by the observation that the density of the polyimide is lowered with increasing flux. This effect also explains the lowering of the dielectric constant (fig. 5.42). The lowering of the dielectric constant with cobalt-60 irradiation occurs due to the fact that the gamma rays generates defects in the polyimide into which Boron or flourine atoms diffuse and the polarization is lowered.

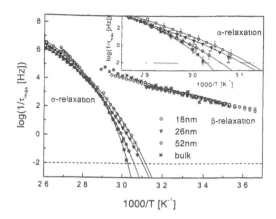

Fig. 5.40 Logarithm of the reciprocal of relaxation time corresponding to the maximum frequency as a function of 1/T in isotactic PMMA. The thickness of the film is the parameter (Kremer and Hartmann, 2001: Dielectric News Letter).

Polymeric insulation is extensively used in power equipments such as motors, transformers, cables and capacitors. The insulation is subjected to high electrical stress of the order of 20 MV/m, though this films have a much higher dielectric strength. The degradation that occurs in the material is a major cause of worry to design and operating engineers. The phenomenon of electro-luminescence, that is the generation of light due to electrical stress is the precursor of progressive deterioration and inception of partial discharges. Bamji and colleagues[90], [91] have studied this phenomenon in polyethylene.

Semi-crystalline polymers exhibit both α-and β-relaxations both of which are well defined. At temperatures $T \sim T_G$ the α-relaxation is due to the micro-Brownian motion of main chains and at $T < T_G$ the β-relaxation occurs. The ε''-log ω plots for β-relaxation curves are quite broad when compared with those for the α-relaxation.

The β-relaxation is usually attributed to the motion of side groups. In linear chain polymers with no branches segmental motion occurs. The segmental motion is frozen far below glass transition temperature, but considerable local motion such as rotation or rotational oscillation may exist at lower temperatures. Each mode is associated with a characteristic temperature and frequency.

In highly crystalline polymers, as in high density polyethylene, the α-relaxation is due to reorientation of the chain within the lamella or extended chain crystal. For short chains the reorientation occurs as a rigid rod but as chain length increases rotation-translation assisted by twisting of the chain is the process for the primary relaxation. The β-relaxation in polyethylene is due to large scale micro-Brownian motion in the disordered regions between the crystallites.

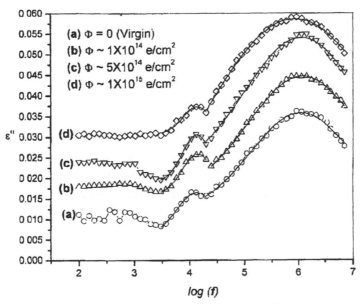

Fig. 5.41 Variation in the loss factor with frequency for polyimide samples irradiated with 1 MeV electrons at various electron flux (Algaonkar et. al. 2002, permission of A I P).

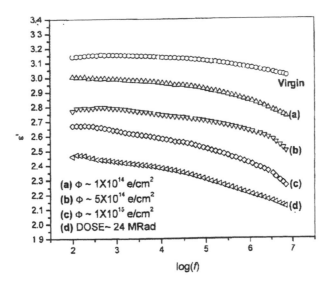

Fig. 5.42 Variation in dielectric constant with frequency for polyimide samples irradiated with 1 MeV electrons at different flux, φ. Plots a, b, c are for electron irradiated samples, plot d for sample irradiated with gamma ray and immersed in BF_3 solution. (Algaonkar et. al. 2002, permission of Amer. Inst. Phys.)

The origin of the γ-relaxation in polyethylene is not quite clear and attributed to impurities. The number of dipoles exhibiting this type of relaxation increases with decreasing crystallinity suggesting that the process occurs in the amorphous regions. In PCTFE the γ-relaxation has two components, designated as γ_c and γ_a one in the wholly crystalline region and the other in the wholly amorphous region. In polyethylene and PCTFE the γ-relaxation makes the dielectric loss curves very broad at low temperatures and the curves become narrower as the temperature is increased.

The studies on a large number of polymers have provided insight into the molecular motion in amorphous and crystalline solids. The methods of correlating the results of dielectric studies with the structure of solids continue to be a fruitful area of research. A quick-glance pictorial of relaxation in some of the polymers discussed is provided in Fig. 5.43.

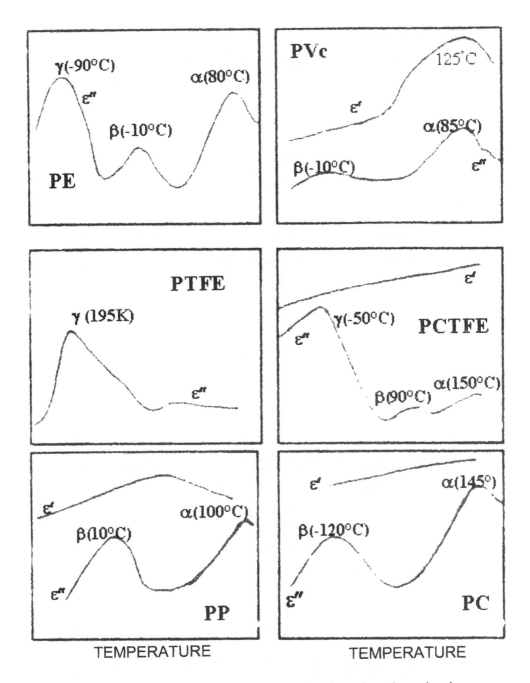

Fig. 5.43 Quick reference for relaxation mechanisms in selected polymers.

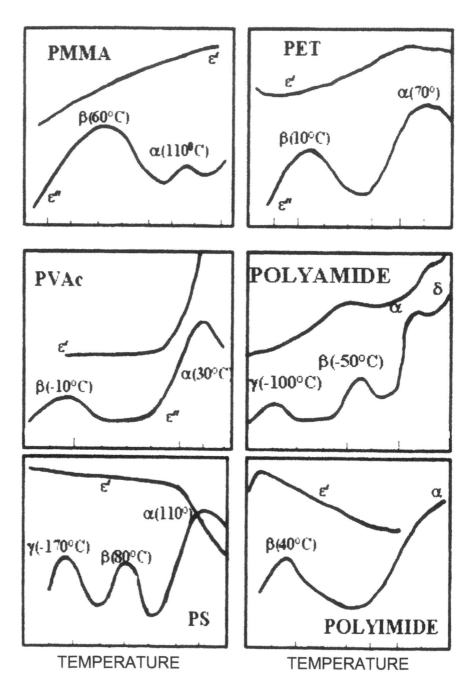

Fig. 5.43 (contd.). Quick reference of relaxation mechanisms in selected polymers.

5.7 REFERENCES

[1] A. K. Jonscher, "Dielectric Relaxation in Solids", Chelsea Dielectric Press, London, 1983

[2] K. H. Illinger, in Progress in Dielectrics, Vol. 4, pp. 37-101, Academic Press, London.

[3] R. Singh, R. P. Tandon, V. S. Panwar, and S. Chandra, J. Chem. Phys., **95 (1)**, (1991) 722.

[4] P. A. M. Steeman and F. H. J. Maurer, **33**, No. 20 (1992) 4230-4241.

[5] W. Reddish, J. Poly. Sci., Part C., (1966) 123-137.

[6] M. S. Dionisio, J. Moura-Ramos and G. Williams, Polymer, **34** (1993) 4105-4110. See p. 4106.

[7] Y. Ishida and K. Yamafuji, Kolloid Z., **177** (1961) 97.

[8] S. Havriliak and S. Negami, Polymer, London, **8** (1967) 161.

[9] M. L. Williams and J. D. Ferry, J. Polym. Sci., **11** (1953) p. 169.

[10] D. W. Kim and K. Yoshino, J. Phys. D: Appl. Phys., **33** (2000) 464-471.

[11] C. Hall, "Polymer Materials", The Macmillan Press Ltd., London, 1981; F. Bueche, "Physical Properties of Polymers", Interscience Publishers, New York, 1962.

[12] H. St.-Onge, IEEE Trans. Elec. Insul., vol. 15, pp. 359-361, 1980.

[13] F. Garwe, A. Schönhals, M. Beiner, K. Schröter and E. Donth, J. Phys.: Condens. Matter **6** (1994), 6941-6945.

[14] J. van Turnhout in "Thermally stimulated Discharges in Electrets", ch. 3 in "Electrets", Ed: G. M. Sessler, Springer Verlaag, Berlin, 1980.

[15] J. D. Hoffman, G. Williams and E. Passaglia, J. Poly. Sci., Part C (1966) pp. 173, 1966.

[16] L. A. Dissado and J. C. Fothergill, "Electrical Degradation and Breakdown in Polymers", Peter Peregrinus (IEE), London, 1992.

[17] G. L. Link, " Dielectric Properties of Polymers", Polymer Science, Vol. 2, Ed: A. D. Jenkins, North Holland Publishing Co., Amsterdam, 1972, p. 1281. See fig. 18.2

[18] I. T. Barrie, K. A. Buckingham, W. Reddish, Proc. IEE, 113 (1966) 1849.

[19] K. A. Buckingham and W. Reddish, Proc. IEE, 114 (1967) 1810.

[20] W. Reddish, Trans. Farad. Soc., **46** (1950) 459.

[21] A. J. Bur, Polymer, **26** (1984) 963.

[22] Z. Xia, R. Gerhard-Multhaupt, W. Künstler, A. Wedel and R. Danz, J. Phys. D: Appl. Phys., **32** (1999) L83 - L85.

[23] F. Krum and F. H. Muller, Kolloid Z., 164 (1959) p.81.

[24] P. Avakian, H. W. Starkweather,Jr., J. J. Fontanella and M. C. Wintergill, in "Dielectric Spectroscopy of Polymeric Materials, Ed: J. P. Runt and J. J. Fitzgerald, American Chem. Soc., Washingtom, 1997, pp. 379-393

[25] H. W. Starkweather, P. Avakian, J. J. Fontanella, M. C. Wintersgill, Macromolecules, **25** (1992) 7145.

[26] H. W. Starkweather, P. Avakian, R. R. Matheson, J. J. Fontanella, M. C. Wintersgill, Macromolecules, **25** (1992) 1475.

[27] H. Kramer and K. E. Heif, Kolloid zeitschrifr, 180 (1962) 114.

[28] R. N. Work, R. D. McCammon and R. G. Saba, J. Chem. Phys., **41** (1964) 2950.

[29] H. M. Banford, R. A. Fouracre, A. Faucitano, A. Buttafava and F. Martinitti, IEEE Trans. Diel. Elec. Insul., **3** (1996) 594.

[30] Y. Ishida, Kolloid Z., **168**, no. 1, (1960) 29.

[31] K. Deutsch, E. A. W. Hoff and W. Reddish, J. Poly. Sci., 13 (1954) 565.

[32] W. Reddish: J. Poly. Sci., Part C. pp. 123-137, 1966.

[33] A. H. Scott, D. J. Scheiber, A. J. Curtis, J. I. Lauritzen, Jr., and J. F. Hoffman, J. Res. Natl. Bur. Std., **66A** (1962) 269.

[34] D.W. Davidson and R.H. Cole, J. Chem. Phys., **19** (1951) 1484.

[35] M. L. Williams, R. F. Landel and J. D. Ferry, J. Am. Chem. Soc., **77** (1955) 3701.

[36] S. Matsuoka and Y. Ishida, J. Poly. Sci., Part C, No. 14, (1966) pp. 247-259.

[37] J. P. Runt and J. J. Fitzgerald, Editors, "Dielectric Spectroscopy of Polymeric Materials", A. Chem. Soc., 1997

[38] S. Matsuoka and Y. Ishida, J. Poly. Sci., Part C, No. 14 (1966) 247-259.

[39] Y. Ishida, M. Matsuo and K. Yamafuji, Kolloid Z., vol. **180**, p. 108, 1962. See page 110.

[40] Y. Ishida, K. Yamafuji, H. Ito and M. Takayanagi, Kolloid-Z., vol. 184, p. 97, 1962.

[41] G. J. Pratt and M. J. A. Smith, Br. Polym. J., **18** (1986), 103; also Polymer, **30** (1989) 1113-1116.

[42] Y. Ishida, Kolloid Z. **168** (1960) 29

[43] M. Matsuo, Y. Ishida, K. Yamafuji, M. Takayanagi, and J. Irie, Kolloid Z., **201** (1965) 89

[44] Y. Ishida, M. Matsuo and K. Yamafuji, Kolloid Z,. **180, No. 2** (1962) 108-118.

[45] Y. Ishida, Kolloid Z., **174** (1961) p. 124.

[46] Y. Ishida, K. Yamafuji, H. Ito and M. Takayanagi, Kolloid-Z, **184** (1962) 97

[47] R. Bergman, F. Alvarez, A. Alegria and J. Colmenero, J. Chem. Phys., **109** (1998) 7546.

[48] Y. Ishida and K. Yamafuji, Kolloid Zeitschrift, **177** (1961) 97-116

[49] G. J. Pratt and M. J. A. Smith, Polymer, **27** (1986) 1483-1488.

[50] S. Mashimo, R. Nozaki, S. Yagihara and S. Takeishi, J. Chem. Phys., **77** (1982) 6259-6262

[51] R. Nozaki and S. Mashimo, J. Chem. Phys., **87** (1987) 2271-2277.

[52] M. S. Dionisio and J. J. Moura-Ramos, Polymer, **34** (1993) 4105-4110.

[53] D. J. Mead and R. M. Fuoss, J. Am. Chem. Soc. **64** (1942) 2389

[54] F. Bröns and F. H. Müller, Kolloid Z. **119** (1950) 45

[55] K. Deutsch, E. A. W. Hoff and W. J. Reddish, J. Poly. Sci., **13** (1954) 565

[56] J. Heijboer, Kolloid z., **148** (1956) 36; Makromol. Chem., **35A** (1960) 86.

[57] L. Debrouckere and G. Offergeld, J. Poly. Sci., **30** (1958) 106

[58] G. P. Mikhailov, J. Poly. Sci., **30** (1958) 605.

[59] W. Reddish, Pure Appl. Chem. **5** (1962) 723.

[60] A. F. Lewis, J. Poly. Sci. B, Poly. Lett. Edn., **1** (1963) 649.

[61] N. G. McCrum and E. L. Morris, Proc. Roy. Soc., London, **A281** (1964) 258.

[62] N. G. McCrum, B. E. Read and G. Williams, "Anelastic and Dielectric Effects in Polymeric Solids", Wiley, New York, 1967.

[63] E. V. Thompson, J. Poly. Sci., A-2, **6** (1968) 433

[64] H. Sasabe and S. Saito, J. Poly. Sci., A-2, **6** (1968) 1401.

[65] Y. Kawamura, Y. Nagai, S. Hirose and Y. Wada, J. Poly. Sci., **A-2, 7** (1969) 1559.

[66] C. A. Solunov, and C. S. J. Ponevsky, J. Poly. Sci., Poly. Phys. Edn., **15** (1977) 969.

[67] P. Hedvig, "Dielectric Spectroscopy of Polymers", Adam Hilger, Bristol, 1977, p. 399

[68] A. S. Gilbert, R. A. Pethrick and D. W. Phillips, J. Appl. Poly. Sci., **21** (1977) 319.

[69] J. VanderschuerenA. Linkens, J. Electrostat., **3** (1977) 155.

[70] S. Mashimo, S. Yagihara and Y. Iwasa, J. Poly. Sci. Poly. Phy. Edu. **16** (1978) 1761.

[71] O. Yano and Y. Wada, J. Poly. Sci., Part A-2, **vol. 9**, (1971) pp. 669-686.

[72] J. C. Coburn and R. H. Boyd, Macromolecules, **19** (1986) 2238-45.

[73] K. Miyairi, J. Phys. D: Appl. Phys., **19** (1986) 1973-80.

[74] V. Adamec and J. H. Calderwood, J. Phys.D: Appl. Phys., **24** (1991) 969-74

[75] T. Tatsumi, E. Ito, R. Hayakawa, J. Poly. Sci., Phys. Edn., **30** (1992) 701.

[76] E. Dargent, J. Santais, J. Salter, J. M. Bayard and J. Grenet, J. Non-Crystalline Solids, **172-174** (1994) 1062-65.

[77] S. Osaki. Polymer, **35** (1994) 47-49.

[78] E. Neagu, P. Pissis, L. Apekis and J. L. G. Ribelles, J. Phys.D: Appl. Phys. **30** (1997) 1551-1560.

[79] G. Floudas and T. Reisinger, J. Chem. Phys., **111**(1999) p. 5201.

[80] G. Floudas, C. Gravalides, T. Reisinger and G. Wegner, J. Chem. Phys., vol. **111**, no. 21 (1999) 9847.

[81] S. Corezzi, E. Campani, P. A. Rolla, S. Capaccioli and D. Fioretto, J. Chem. Phys., **111** (1999) 9343.

[82] R. Srinivasan, R. R. Hall and D. C. Allbee, Appl. Phys. Lett., **63** (1993) 3382.

[83] L. E. Amborski, Ind. Eng. Chem., **2** (1963) p. 189.

[84] W. J. Wrasidlo, J. Poly. Sci., Poly. Phys. Edu., **11** (1973) 2143, also J. Macromol. Sci.-Phys., B, **3** (1972) 559.

[85] J. Melcher, Y. Daben and G. Arlt, IEEE Trans. Elec. Insul., **24**, No. 1 (1989) 31-38.

[86] P. K. Nixon, Lei Wu, S. R. Nagel, B. D. Williams and J. P. Carini, Phys. Rev. Lett., **65,** No. 9, (1990) 1108.

[87] G. Williams, IEEE Trans. Elec. Insul., EI-**17**, No. 6 (1982) 469-477.

[88] F. Kremer and L. Hartmann, Dielectric News letter, September 2001.

[89] P. S. Alegaonkar, V. N. Bhoraskar, P. Bolaya and P. S. Goyal, Appl. Phys. Letters, 80 (2002) p. 640.

[90] S. S. Bamji, Electrical Insulation Magazine, May/June, 1999, p. 9

[91] S. S. Bamji, Transactions on Diel. Elec. Insul., **6** (1993) 288.

A new scientific truth does not triumph by convincing its opponents and making them see the light, but rather because its opponents eventually die, and a new generation grows up that is familiar with it.
- Max Planck.

6

ABSORPTION AND DESORPTION CURRENTS

The response of a linear system to a frequency dependent excitation can be transformed into a time dependent response and vice-versa. This fundamental principle covers a wide range of physical phenomena and in the context of the present discussion we focus on the dielectric properties ε' and ε''. Their frequency dependence has been discussed in the previous chapters, and when one adopts the time domain measurements the response that is measured is the current as a function of time. In this chapter we discuss methods for transforming the time dependent current into frequency dependent ε' and ε''. Experimental data are also included and where possible the transformed parameters in the frequency domain are compared with the experimentally obtained data using variable frequency instruments.

The frequency domain measurements of ε' and ε'' in the range of 10^{-2} Hz-10 GHz require different techniques over specific windows of frequency spectrum though it is possible to acquire a 'single' instrument which covers the entire range. In the past the necessity of using several instruments for different frequency ranges has been an incentive to apply and develop the time domain techniques. It is also argued that the supposed advantages of the time domain measurements is somewhat exaggerated because of the commercial availability of equipments covering the range stated above[1]. The frequency variable instruments use bridge techniques and at any selected frequency the measurements are carried out over many cycles centered around this selected frequency. These methods have the advantage that the signal to noise ratio is considerably improved when compared with the wide band measurements. Hence very low loss angles of ~10 μ rad. can be measured with sufficient accuracy (Jonscher, 1983).

The time domain measurements, by their very nature, fall into the category of wide band measurements and lose the advantage of accuracy. However the same considerations of accuracy apply to frequencies lower than 0.1 Hz which is the lower limit of ac bridge techniques and in this range of low frequencies, $10^{-6} < f < 0.1$ Hz, time domain measurements have an advantage. Use of time-domain techniques imply that the system is linear and any unexpected non-linearity introduces complications in the transformation techniques to be adopted. Moreover a consideration often overlooked is the fact that the charging time of the dielectric should be large, approximately ten times (Jonscher, 1983), compared with the discharging time. The frequency domain and time domain measurements should be viewed as complementary techniques; neither scheme has exclusive advantage over the other.

6.1 ABSORPTION CURRENT IN A DIELECTRIC

A fundamental concept that applies to linear dielectrics is the superposition principle. Discovered nearly a hundred years ago, the superposition principle states that each change in voltage impressed upon a dielectric produces a change in current as if it were acting alone. Von Schweidler[2, 3] formulated the mathematical expression for the superposition and applied it to alternating voltages where the change of voltage is continuous and not step wise, as changes in the dc voltage dictate.

Consider a capacitor with a capacitance of C and a step voltage of V applied to it. The current is some function of time and we can express it as

$$i(t) = CV\phi(t) \tag{6.1}$$

If the voltage changes by ΔV_t at an instant T previous to t, the current changes according to the superposition principle,

$$\Delta i = C \Delta V_t \phi(t - T) \tag{6.2}$$

If a series of change in voltage occurs at times T_1, T_2,...etc. then the change in current is given by

$$i = C \sum_N \Delta V_N \phi(t - T_N) \tag{6.3}$$

If the voltage changes continuously, instead of in discrete steps, the summation can be replaced by an integral,

$$i = C \int_{-\infty}^{t} \frac{dV(T)}{dT} \phi(t - T) dT \tag{6.4}$$

The integration may be carried out by a change of variable. Let $p = (t - T)$. Then $dp = -dT$ and equation (6.4) becomes

$$i = -C \int_0^\infty \frac{dV(t-p)}{dp} \varphi(p)dp \tag{6.5}$$

The physical meaning attached to the variable p is that it represents a previous event of change in voltage. This equation is in a convenient form for application to alternating voltages:

$$v = V_{max} \exp[j(\omega t + \delta)] \tag{6.6}$$

where δ is an arbitrary phase angle with reference to a chosen phasor, not to be confused with the dissipation angle. The voltage applied to the dielectric at the previous instant p is

$$v = V_{max} \exp j[\omega(t - p) + \delta] \tag{6.7}$$

Differentiation of equation (6.7) with respect to p gives

$$\frac{dv}{dp} = -V_{max} j\omega \exp j[\omega(t - p) + \delta] \tag{6.8}$$

Substituting equation (6.8) into (6.5) we get

$$i = j\omega CV_{max} \int_0^\infty \exp j[\omega(t - p) + \delta] \varphi(p)dp \tag{6.9}$$

The exponential term may be split up, to separate the part that does not contain the variable as:

$$\exp[j\omega(t - p) + \delta] = \exp[j(\omega t + \delta)] \times \exp(-j\omega p) \tag{6.10}$$

Equation (6.9) may now be expressed as:

$$i = \omega CV_{max} \exp[j(\omega t + \delta)] \tag{6.11}$$

Substituting the identity

$$\exp(-j\omega p) = \cos(\omega p) - j\sin(\omega p) \tag{6.12}$$

we get the expression for current as:

$$i = \omega C V_{max} \exp[j(\omega t + \delta)]\{\int_0^\infty j\cos(\omega p)\varphi(p)dp + \int_0^\infty \sin(\omega p)\varphi(p)dp\} \tag{6.13}$$

The current, called the **absorption current**, consists of two components: The first term is in quadrature to the applied voltage and contributes to the real part of the complex dielectric constant. The second term is in phase and contributes to the dielectric loss.

An alternating voltage applied to a capacitor with a dielectric in between the electrodes produces a total current consisting of three components: (1) the capacitive current I_c which is in quadrature to the voltage. The quantity ε_∞ determines the magnitude of this current. (2) The absorption current I_a given by equation (6.13), (3) the ohmic conduction current I_c which is in phase with the voltage. It contributes only to the dielectric loss factor ε''. The absorption current given by equation (6.13) may also be expressed as

$$i_a = j\omega C_0 v \varepsilon_a^* = j\omega C_0 v(\varepsilon_a' - j\varepsilon_a'') \tag{6.14}$$

Equating the real and imaginary parts of eqs. (6.13) and (6.14) gives:

$$\varepsilon_a' = \varepsilon' - \varepsilon_\infty = \int_0^\infty \cos(\omega t)\varphi(t)dt$$

$$= \frac{1}{C_0 V_0}\int_0^\infty i(t)\cos(\omega t)dt \tag{6.15}$$

where C_0 is the vacuum capacitance of the capacitor and V_0 the applied voltage. Note that we have replaced p by the variable t without loss of generality.

$$\varepsilon_a'' = \varepsilon'' = \int_0^\infty \sin(\omega t)\varphi(t)dt$$

$$\tag{6.16}$$

The standard notation in the published literature for ε_a' is $\varepsilon' - \varepsilon_\infty$ as shown in equation (6.15). Equations (6.15) and (6.16) are considered to be fundamental equations of dielectric theory. They relate the absorption current as a function of time to the dielectric constant and loss factor at constant voltage.

To show the generality of equations (6.15) and (6.16) we consider the exponential decay function

$$\phi(t) = Ae^{-t/\tau} \tag{6.17}$$

where τ is a constant independent of t and A is a constant that also includes the applied voltage V. Substituting this equation in equations (6.15) and (6.16) we have

$$(\varepsilon' - \varepsilon_\infty) = A\int_0^\infty e^{-t/\tau} \cos\omega t\, dt \tag{6.18}$$

$$\varepsilon'' = A\int_0^\infty e^{-t/\tau} \sin\omega t\, dt \tag{6.19}$$

We use the standard integrals:

$$\int_0^\infty e^{-px} \sin qx = \frac{q}{p^2 + q^2}$$

$$\int_0^\infty e^{-px} \cos qx = \frac{p}{p^2 + q^2}$$

Equations (6.18) and (6.19) then simplify to

$$\varepsilon' - \varepsilon_\infty = \frac{A\tau}{1 + \omega^2\tau^2} \tag{6.20}$$

$$\varepsilon'' = \frac{A\omega\tau^2}{1 + \omega^2\tau^2} \tag{6.21}$$

The factor A is a constant with the dimension of s^{-1} and if we equate it to

$$A = \frac{\varepsilon_s - \varepsilon_\infty}{\tau} \tag{6.22}$$

Equations (6.20) and (6.21) become

$$\varepsilon' = \varepsilon_\infty + \frac{(\varepsilon_s - \varepsilon_\infty)}{1 + \omega^2\tau^2} \tag{6.23}$$

$$\varepsilon'' = \frac{(\varepsilon_s - \varepsilon_\infty)\omega\tau}{1 + \omega^2\tau^2} \tag{6.24}$$

These are Debye equations (3.28) and (3.29) which we have analyzed earlier in considerable detail. Recovering Debye equations this way implies that the absorption currents in a material exhibiting a single relaxation time decay exponentially, in accordance with equation (6.17). As seen in chapter 5, there are very few materials which exhibit a pure Debye relaxation.

The transformation from the time domain to frequency domain using relationships (6.18) and (6.19) also proves that the inverse process of transformation from the frequency domain to time domain is legitimate. This latter transformation is carried out using equations

$$\phi(t) = \frac{2}{\pi}\int_0^\infty (\varepsilon' - \varepsilon_\infty)\cos\omega t\, d\omega \tag{6.25}$$

$$\phi(t) = \frac{2}{\pi}\int_0^\infty \varepsilon''(\omega)\sin\omega t\, d\omega \tag{6.26}$$

Substituting equations (6.20) and (6.21) in these and using the standard integrals

$$\int_0^\infty \frac{\cos mx}{x^2 + a^2}\, dx = \frac{\pi}{2a}e^{-ma}$$

$$\int_0^\infty \frac{x\sin mx}{x^2 + a^2}\, dx = \frac{\pi}{2}e^{-ma}$$

equation (6.17) is recovered.

A large number of dielectrics exhibit absorption currents that follow a power law according to

$$I(t) = Kt^{-n} \tag{6.27}$$

where K is a constant to be determined from experiments. Carrying out the transformation according to equations (6.15) and (6.16) we get

$$\varepsilon' - \varepsilon_\infty = \frac{K}{2}\omega^{n-1}\Gamma(1-n)\cos[(1-n)\pi/2]; \quad 0 < n < 1 \tag{6.28}$$

$$\varepsilon'' = K\omega^{n-1}\Gamma(1-n)\sin[(1-n)\pi/2]|; \quad 0 < n < 2 \tag{6.29}$$

where the symbol Γ denotes the **Gamma function**. The left side of equation (6.28) is, of course, the dielectric decrement. For the ranges shown the integrals converge. The author has calculated[4] ε' and tan δ and fig. 6.1 shows the calculated values of the dielectric decrement and the loss factor versus frequency for various values of n, assuming $K = 1$. The power law (6.21) yields ε' that decreases with increasing frequency in accordance with dispersion behavior. However the loss factor decreases monotonically whereas a peak is expected.

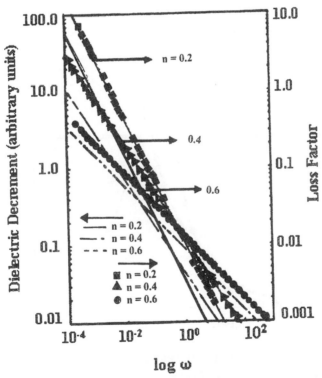

Fig. 6.1 The dielectric decrement and loss factor calculated by the author according to equation (6.28) at various values of the index n. The loss factor decreases with increasing n at the same value of ω (rad/s). The value of K in equations (6.28) and (6.29) is arbitrary.

Fig. 6.2 shows the shape of the loss factor versus log ω for various values of n in the range of $0.5 \le n \le 2$.

In this range we have to use a different version of the solution of eq. (6.14) and (6.15)[5] :

$$\int_0^\infty x^{p-1} \sin ax\, dx = \frac{\pi \sec \dfrac{p\pi}{2}}{2a^p \Gamma(1-p)}$$

$$\int_0^\infty x^{p-1} \cos ax\, dx = \frac{\pi \cosec \dfrac{p\pi}{2}}{2a^p \Gamma(1-p)}$$

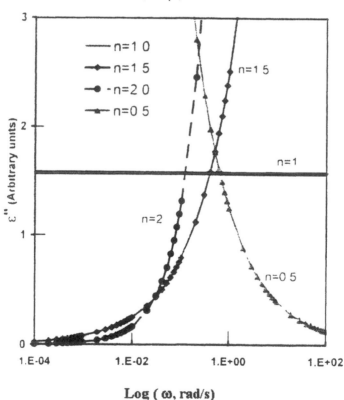

Fig. 6.2 Loss factor as a function of frequency at various values of $0.5 \le n \le 2.0$. ε'' is constant at n = 1. There is also a change of slope from negative to positive at n >1.0.

The slope of the loss factor curve depends on the value of *n* in the range $1 < n < 2$ is positive, in contrast with the range $0 < n < 1$. The loss factor decreases, remains constant or increases according as *n* is lower, equal to or greater than one, respectively. The calculated values do not show a peak in contrast with the measurements in a majority of polar dielectrics and one of the reasons is that the theory expects the current to be infinite at the instant of application of the voltage. Further the current cannot decrease with time according to the power law because if it does, it implies that the charge is infinite. The loss factor should be expressed as a combination of at least two power laws. Jonscher (1983) suggests an alternative to the power law, according to

$$i(t) \propto \frac{1}{(\omega_{max} t)^n + (\omega_{max} t)^{1+m}} \tag{6.30}$$

According to this equation a plot of t versus log I yields two straight lines. The larger slope at shorter times is $-n$ and the smaller slope at longer times is $-1-m$. The change over from one index to the other in the time domain occurs corresponding to the loss peak in the frequency domain. An experimental observation of such behavior is given by Sussi and Raju. The change of slope is probably associated with different processes of relaxation in contrast with the exponential decay, equation (6.17) for the Debye relaxation.

Jonscher[1] suggests that the absorption currents should be measured for an extended duration till the change of slope in the time domain is observed. This requirement is thought to neutralize the advantage of the time domain technique.

Combining equations (6.15) and (6.16) we can express the complex permittivity as

$$\varepsilon^* - \varepsilon_\infty = \frac{1}{C_0 V} \ell \, \mathrm{I}(t) \tag{6.31}$$

where C_0 is the vacuum capacitance and V is the height of the voltage pulse. ℓ is the symbol for Laplace transform defined as

$$\ell[F(t)] = \int_0^\infty \exp(-j\omega t) F(t) dt$$

6.2 HAMON'S APPROXIMATION

Let us consider equation (6.27) and its transform given by (6.29). If we add the component of the loss factor due to conductivity then the latter equation becomes

$$\varepsilon'' = \frac{\sigma_{dc}}{\omega \varepsilon_0} + K\omega^{n-1}\{\Gamma(1-n)\sin[(1-n)\pi/2]\} \tag{6.32}$$

Hamon[7] suggested the substitution

$$\omega t_1 = \{\Gamma(1-n)\sin[(1-n)\pi/2]\}^{-1/n} \tag{6.33}$$

noting that the right side of this equation is almost independent of n in the range $0.3 < n < 1.2$. This leads to the expression

$$\varepsilon'' \cong \frac{i(0.63/\omega)}{\omega C_0 V} \tag{6.34}$$

The equation is accurate to within $\pm$ 5% for the stated range of n, but it also has the advantage that there is no need to measure σ_{dc}.

6.3 DISTRIBUTION OF RELAXATION TIME AND DIELECTRIC FUNCTION

It is useful to recapitulate from chapter 3 the brief discussion of the distribution of relaxation times in materials that exhibit a relaxation phenomenon which is much broader than the Debye relaxation. Analytical expressions are available for the calculation of the distribution of the relaxation functions $G(\tau)$ considered there. To provide continuity we summarize the equations, recalling that α, β and γ are the fitting parameters.

6.3.1 COLE-COLE FUNCTION

$$G(\tau) = \frac{\sin \alpha \pi}{2\pi \cosh[(1-\alpha)\ln(\tau/\tau_0)] - \cos \alpha \pi} \tag{3.94}$$

6.3.2. DAVIDSON-COLE FUNCTION

$$G(\tau) = \frac{\sin \beta \pi}{\pi}(\frac{\tau}{\tau_0 - \tau})^\beta, \quad \tau < \tau_0 \tag{3.95}$$

$$= 0 \qquad \qquad \tau > \tau_0 \tag{3.96}$$

6.3.3 FUOSS-KIRKWOOD FUNCTION

The distribution function for this relaxation is given as:

$$G(\tau) = \frac{\delta}{\pi} \frac{\cosh(\frac{\delta\pi}{2})\cosh(\delta s)}{\cos^2(\delta\pi/2) + \sinh^2(\delta s)}; s = \log(\omega/\omega_p) \qquad (3.97)$$

6.3.4 HAVRILIAK-NEGAMI FUNCTION[8]

$$G(\tau) = \frac{1}{\pi} \frac{(\tau/\tau_H)^{\alpha\beta}\sin(\beta\theta)}{[(\tau/\tau_H)^{2(1-\alpha)} + 2(\tau/\tau_H)^{1-\alpha}\cos\pi(1-\alpha) + 1]^{\beta/2}}; \qquad (6.35)$$

$$\theta = arc\tan\left[\frac{(\omega\tau_H)^{1-\alpha}\cos(\alpha\pi/2)}{1 + (\omega\tau_H)^{1-\alpha}\sin(\alpha\pi/2)}\right] \qquad (6.36)$$

It is necessary to visualize these functions before we attempt to correlate the results that are obtained by applying them to specific materials. We use the variable $\log(\tau/\tau_0)$. Fig. (6.3) shows calculated distributions for the range of parameters as shown. For the Cole-Cole distribution the relaxation times are symmetrically distributed on either side of $\log(\tau/\tau_0) = 1$ while the Davidson-Cole distribution (fig. 6.4) is not only asymmetrical but $G(\tau) = 0$ at $\log(\tau/\tau_0) = 1$.

The Fuoss-Kirkwood distribution (fig. 6.5) is again symmetrical but it should be noted that the ordinate here is Log (ω/ω_p) and hence the height of the distribution increases in thr reverse order of the parameter β. The Havriliak-Negami distribution involves two parameters, α and β, and to discern the trend in change of distribution function we compute the relaxation times at various values of α with $\beta = 0.5$, and at various values of β with $\alpha = 0.5$. Fig. 6.6 shows the results which may be summarized as follows: (1) As α is increased the distribution become more narrow and the function $G(\tau)$ will attain a higher value at $\log(\tau/\tau_H) = 1$ though there appears to be a change of characteristics as α changes from 0.2 to 0.4. (2) As β increases the peak height increases and the distributions become broader.

Fig. 6.6 shows the application of equation of (6.35) to poly(vinyl acetate) PVAc which has been studied by a number of authors. We recall that PVAc is an amorphous polar polymer with $T_G \sim 30°C$. Nozaki and Mashimo[9] have measured the absorption currents and evaluated the Havriliak parameters through the numerical Laplace transformation technique. The present author has used their data to calculate $G(\tau)$[4]. The abrupt change of distribution function at T_G is evident. Whether this is true for other polymers needs to be examined.

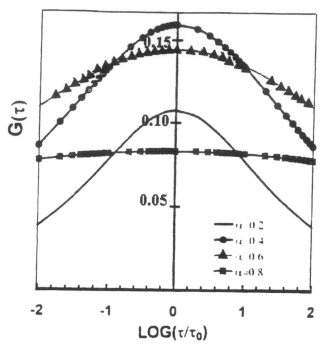

Fig. 6.3 Distribution of relaxation times according to Cole-Cole function (3.93). The distribution becomes broader with increasing value of α.

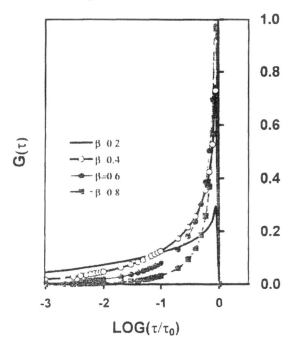

Fig. 6.4 Distribution of relaxation times according to Davidson-Cole function. $G(t) = 0$ for $t / t_0 > 1$.

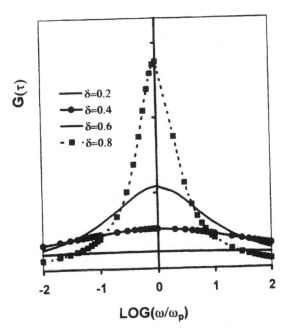

Fig. 6.5 Distribution of relaxation times according to Fuoss-Kirkwood relaxation, eq. (3.96). The distribution is symmetrical about $\omega = \omega_p$.

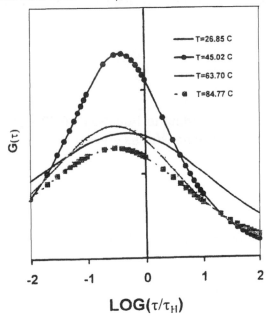

Fig. 6.6 Relaxation time in PVAc calculated by using the experimental results of Nozaki and Mashimo and values of α and β (1987). At T_g there is an abrupt change of $G(\tau)$.

6.4 THE WILLIAMS-WATTS FUNCTION

The detour to the distribution function of relaxation times was necessary due to the fact that it is an intermediate step in the numerical methods of transforming the time domain absorption currents to frequency domain dielectric loss factor. We shall, however, pursue the analytical methods a little further to indicate the potential and limitations of making aproximations to facilitate the transformation. We have already seen that an exponential decay function for absorption current results in a Debye relaxation. A power law exhibits a peak, not necessarily Debye relaxation, in the dielectric loss factor. Williams and Watt suggested a decay function[10], usually called the stretched exponential,

$$\phi(t) = \exp[-(\frac{t}{\tau_{w-w}})^\gamma]$$ (6.37)

where $0 < \beta \le 1$ and τ is an effective relaxation time. Sometimes this equation is called Kohlraush-Williams-Watts (KWW) equation because Kohlrausch used the same expression in 1863 to express the mechanical creep in glassy fibres (Alvarez et. al., 1991). For the time being we can drop the subscript for the relaxation time as confusion is not likely to occur. The complex dielectric constant is related to the decay function according to eq. (6.31)

$$\frac{\varepsilon^* - \varepsilon_\varsigma}{\varepsilon_\varsigma - \varepsilon_\infty} = \ell \left[-\frac{d\phi(t)}{dt} \right]$$ (6.38)

If the decay function is exponential according to Equation (6.37), or $\gamma = 1$ in equation (6.37), a reference to the Table of Laplace transforms gives

$$\frac{\varepsilon^* - \varepsilon_\infty}{\varepsilon_\varsigma - \varepsilon_\infty} = \frac{1}{1 + j\omega\tau}$$ (3.31)

which is the familiar Debye equation. Substitution of equation (6.37) in (6.38) yields

$$\frac{\varepsilon^* - \varepsilon_\infty}{\varepsilon_\varsigma - \varepsilon_\infty} = \ell \left[\frac{\gamma}{\tau_0} \left(\frac{t}{\tau_0} \right)^{-(1-\gamma)} \exp - \left(\frac{t}{\tau_0} \right)^\gamma \right]$$ (6.39)

The analytical expression for the Laplace transform of the right side is quite complicated. So we first substitute $\gamma = 0.5$ for which an analytical evaluation of the transform can be obtained as follows. For this special case equation (6.39) becomes

$$\frac{\varepsilon^* - \varepsilon_\infty}{\varepsilon_s - \varepsilon_\infty} = \frac{1}{2} \left(\frac{\pi}{\tau_0} \right)^{1/2} \ell \left[\frac{1}{\sqrt{\pi t}} \exp - 2k(t)^{1/2} \right] \tag{6.40}$$

where $k = (\tau_0)^{-1/2}$. This has the standard form found in tabulated Laplace transforms[11] giving

$$\frac{\varepsilon^* - \varepsilon_\infty}{\varepsilon_s - \varepsilon_\infty} = \frac{1}{2} \left(\frac{\pi}{\tau_0} \right)^{1/2} \frac{1}{(j\omega)^{1/2}} \exp(\frac{k^2}{j\omega}) erfc \frac{k}{(j\omega)^{1/2}} \tag{6.41}$$

Substituting

$$\sqrt{j} = \frac{1}{\sqrt{2}} + \frac{j}{\sqrt{2}}$$

the transform (6.41) becomes

$$\frac{\varepsilon^* - \varepsilon_\infty}{\varepsilon_s - \varepsilon_\infty} = \sqrt{\pi} \frac{1-j}{\sqrt{8\omega\tau_0}} w(z);$$

$$z = \frac{1+j}{\sqrt{8\omega\tau_0}}; \; w(z) = e^{-z^2} erfc(-jz) \tag{6.42}$$

The error function for complex arguments, w(z), has been tabulated in reference[12]. A sample calculation is provided here. Let $(\omega\tau_0) = 0.1$, $(\varepsilon_s - \varepsilon_\infty) = 2.0$, $\varepsilon_\infty = 2.25$. Then

$$z = 1.12 + j1.12; \quad w(z) = 0.1957 \, (Abramowitz \; and \; Stegun, 1972);$$
$$\frac{\varepsilon^* - \varepsilon_\infty}{\varepsilon_s - \varepsilon_\infty} = \sqrt{\pi} \times 0.1957 \times \frac{1-j}{0.894} = 0.39 - j0.39$$
$$\varepsilon^* = \varepsilon' - j\varepsilon'' = 3.03 - j1.47; \quad \varepsilon' = 3.03, \varepsilon'' = 1.47$$

The function w(z) may also be calculated for values close to 0.5 by interpolation because values for $\gamma = 1$ may also be calculated using the Debye expression. William-Watts have computed equation (6.42) for several polymers including PVAc and obtained reasonable agreement between the measured and calculated values of $\varepsilon''/\varepsilon''_{max}$ and $(\varepsilon' - \varepsilon_\infty) / (\varepsilon_s - \varepsilon_\infty)$ versus log (ω /ω_{max}).

The Laplace transform of equation (6.40) for other values of $0 < \gamma \le 1$ has been given as a summation by Williams et. al.[13]

$$\frac{\varepsilon^* - \varepsilon_\infty}{\varepsilon_1 - \varepsilon_\infty} = \sum_{n=1}^{\infty} (-1)^{n-1} \frac{1}{(\omega\tau_0)^{n\gamma}} \frac{\Gamma(n\gamma+1)}{\Gamma(n+1)} [\cos(n\gamma\pi/2) - j\sin(n\gamma\pi/2)] \tag{6.43}$$

The range of parameters for which convergence is obtained is also worked out as

$$0.25 \le \gamma \le 1.0; \qquad -1 \le \log\omega\tau_0 \le +4$$

Outside this range

$$0 \le \gamma \le 0.25; \qquad -4 \le \log\omega\tau_0 \le +4 \quad and$$

$$0.25 \le \gamma \le 1.0; \qquad -4 \le \log\omega\tau_0 \le -1$$

equation (6.43) was employed to calculate the real and imaginary parts separately. We then obtain the values of the following expressions:

$$\frac{\varepsilon' - \varepsilon_\infty}{\varepsilon_s - \varepsilon_\infty} \qquad \text{(Dispersion Ratio)}$$

$$\frac{\varepsilon''}{\varepsilon_s - \varepsilon_\infty} \qquad \text{(Absorption Ratio)}$$

We shall denote these ratios as the real part and imaginary part of the dielectric decrement ratio.

Watts et. al. employed a computer program to evaluate the equations (6.42) and (6.43). Their results are shown fig. 6.7.

The **dispersion ratios** plotted as a function of log ($\omega\tau_0$) changes approximately linearly for $\gamma = 0.1$ and the curves become steeper in the dispersion region as γ increases. The **absorption ratios** plotted as a function of $\log(\omega\tau_0)$ (fig. 6.8) exhibit the peak as the experimental results demand. In all cases $\log(\omega_{max}\tau_0) < 0$. Though the absorption curves appear to be similar to Davidson-Cole functions the latter are found to be much broader.

The curves become narrower for higher values of γ, the Debye relaxation being obtained for $\gamma = 1$, satisfying equation (3.31).

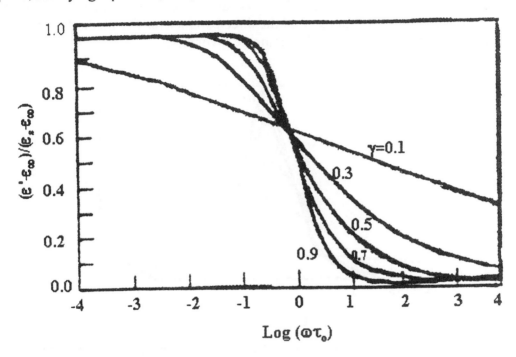

Fig. 6.7 Transformation from time domain to frequency domain using Laplace transform for odd values of the exponent, γ. The dispersion ratio, defined by the quantity shown on the ordinate (y-axis) is almost linear for small value of $\gamma = 0.1$. For higher values the familiar inverted S shape is generated (Williams et. al. 1971. with permission of the Faraday Soc.)

The half width of the absorption ratio curve may be used to determine the dielectric decrement ($\varepsilon_s - \varepsilon_\infty$) using the relation

$$\left(\varepsilon_s - \varepsilon_\infty\right) = k(\gamma)\Delta w \varepsilon''_{max} \tag{6.44}$$

where Δw is the half width of the absorption ratio curve and $k(\gamma)$ is a constant. For $\gamma = 1$ $k(\gamma) = 1.75$, and this corresponds to Fuoss-Kirkwood distribution.

The three parameters that appear in the Williams-Watt equation (6.31) are γ, τ and ($\varepsilon_s - \varepsilon_\infty$). Numerical techniques are required to find the Laplace transform or evaluate the integrals. As an example, the function $\phi(t)$ is approximated to a series of exponentials[14] and the Fourier transform of the resulting series is evaluated. The parameters γ and τ are fitted in terms of the peak and half-width of the dielectric loss function. The approximations for each value of γ require determination of a large number of parameters (about 30) by curve fitting techniques. A simpler method of calculating these parameters one by one has been proposed by Weiss et. al.[15] in contrast with other methods that involve simultaneous evaluations of all the three parameters. The method is demonstrated to apply in PVAc yielding results that compare favorably.

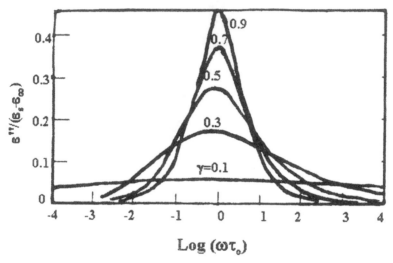

Fig 6.8 Transformation from time domain to frequency domain using Laplace transform for odd values of the exponent, γ. The absorption ratio, defined by the quantity shown on the ordinate (y-axis) is almost flat for small value of γ = 0.1. For higher values the familiar peak is observed (Williams et. al. 1971, with permission of Faraday Soc).

A summary for the function within square brackets in equation (6.38) is appropriate at this juncture from the point of view of using Laplace transforms for evaluating the dispersion ratios and absorption ratios. Let

$$\Psi(t) = -\frac{d\phi(t)}{dt} \qquad (6.45)$$

Then the relaxation formulas yield the following time dependent functions:

A. DEBYE FUNCTION

$$\Psi(t) = \frac{1}{\tau}\exp(-\frac{t}{\tau}) \qquad (6.46)$$

B. COLE-COLE FUNCTION

$$\Psi(t) = \frac{1}{\tau\,\Gamma(1-\alpha)}\left(\frac{t}{\tau}\right)^{-\alpha}\,; \quad for\,(\frac{t}{\tau}) \ll 1 \tag{6.47}$$

$$\Psi(t) = \frac{(1-\alpha)}{\tau\,\Gamma\alpha}\left(\frac{t}{\tau}\right)^{\alpha-2}\,; \quad for\,(\frac{t}{\tau}) \gg 1 \tag{6.48}$$

C. DAVIDSON-COLE FUNCTION

$$\Psi(t) = \frac{1}{\tau\,\Gamma\beta}\left(\frac{t}{\tau}\right)^{\beta-1}\exp\left(-\frac{t}{\tau}\right) \tag{6.49}$$

D. WILLIAMS AND WATTS FUNCTION

$$\Psi(t) = \frac{1}{\tau\,\Gamma\gamma}\left(\frac{t}{\tau}\right)^{\gamma-1}\exp\left(-\frac{t}{\tau}\right)^{\gamma} \tag{6.50}$$

E. GENERAL FUNCTION

$$\Psi(t) = \int_{-\infty}^{\infty} G(\tau)\exp(-\frac{t}{\tau})\,dt \tag{6.51}$$

This expression assumes a superposition of Debye processes for an incremental relaxation time $G(\tau)\,d(\tau)$ and follows directly from equation (6.46). The physical significance of superposition in the time domain and the frequency domain is explained by Fig. 6.9.

Fig. 6.10 shows the function $\Psi(t)$ as a function of (t/τ) for a constant value of the parameters α, β, $\gamma = 0.5$. For the purpose of comparison the exponential decay and the power law decay are also shown. For short times ($t/\tau \ll 1$) which correspond to high frequencies, the three functions, namely Cole-Cole, Debye-Cole and William-Watts, have the same dependence on time for chosen parameters. However for large values of t/τ the Davidson-Cole function drops off rapidly due to the exponential term while the decay according to William-Watt function is less sharp.

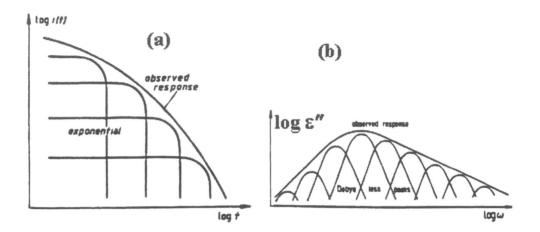

Fig. 6.9 Schematic illustration of superposition of (a) currents in the time domain, (b) loss factor in the frequency domain (Jonscher, 1983, with permission of Chelsea Dielectric Press).

6.5 THE G(τ) FUNCTION FOR WILLIAM-WATT CURRENT DECAY

Current measurements in the time domain give a single curve for a particular set of experimental conditions such as temperature and electric field. From these measurements the fitting parameters and τ are obtained using equations (6.46)-(6.50). From these parameters $G(\tau)$ is evaluated with the help of analytical equations (eqs. 3.94- 3.101). Using these values of τ, ε′ and ε″ may be determined with the help of equations (3.89) and (3.90). However, it is not possible to express the $G(\tau)$ function analytically for the William-Watt decay function except for the particular value of γ = 0.5. In this case the expression for $G(\tau)$ is (Alvarez, 1991)

$$G(\tau) = \frac{1}{2}\pi^{-\frac{1}{2}}x^{-\frac{3}{2}}e^{-\frac{1}{4}x}$$

(6.52)

where $x = \tau / \tau_{ww}$. For other values of γ the expression for $G(\tau)$ is (Alvarez, 1991)

$$G(\tau) = -\frac{1}{\pi}\sum_{k=0}^{\infty}\frac{(-1)^k}{k!}\sin(\pi\gamma k)\frac{\Gamma(\gamma k+1)}{x^{(\beta k+1)}}$$

(6.53)

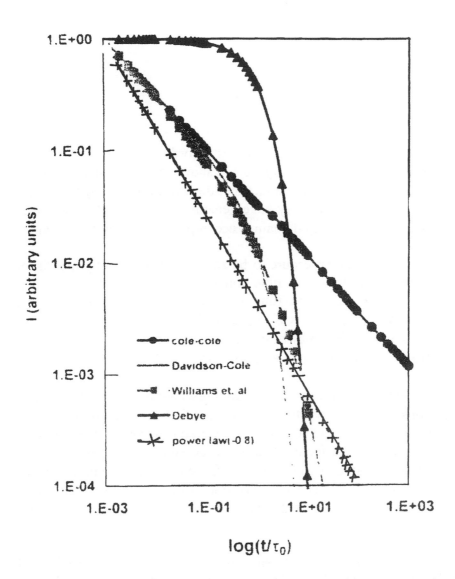

Fig. 6.10 Calculated current versus time for several relaxations, according to equations (6.46) – (6.50). 1-Cole-Cole, 2-Davidson-Cole, 3-Williams et. al., , 4-Debye, 5-t$^{-0.8}$.

Though expression (6.53) is analytical its evaluation is beset with difficulties such as wide ranging alternating terms and large trignometric functions. Hence, numerical techniques involving inversion of Laplace transform have been developed[16, 17].

Alvarez et. al. (1991) have also developed another important equation relating the William-Watt time domain parameter γ to the frequency domain parameters of Havriliak-Negami parameters, according to

$$\alpha\beta = \gamma^{1\,23} \tag{6.54}$$

$$\ln\left[\frac{\tau_H}{\tau_{ww}}\right] = 2.6(1-\gamma)^{0\,5}\exp(-3\gamma) \tag{6.55}$$

where τ_H and τ_{ww} are the relaxation times according to Havriliak-Negami and William-Watt functions, respectively. Fig. 6.11 shows these calculated relationships. The usefulness of Fig. 6-11 lies in the fact that the values of (α, β) for a given value of γ is not material specific. Hence they have general validity and the transformation from the time domain to the frequency domain is accomplished as long as the value of γ is known.

We now summarize the procedure to be adopted for evaluating the dispersion ratio and the absorption ratio from the time-domain current measurements. For an arbitrary dependence of the current on time we use equation (6.38) and numerical Laplace transform yields the quantities $\varepsilon'- \varepsilon_\infty$ and ε''. If the transient current is an analytical function of time, then $G(\tau)$ may be evaluated (section 6.3) and hence the quantity ($\varepsilon*- \varepsilon_\infty$)/($\varepsilon_s -\varepsilon_\infty$) from equations given in section (3.16).

In chapter 5 enough experimental data are presented to bring out the fact that the α- and β- relaxations are not always distinct processes that occur in specific temperature ranges. As the temperature of some polymers is lowered towards T_G there is overlap of the two relaxations, (see Fig. 5.8 for three different possible ways of merging). For example in PMMA the α-relaxation merges with the β- relaxation at some temperature which is a few degrees above T_G[18]. The β- relaxation occurs at higher frequencies than the α-relaxation, and it also has a weaker temperature dependency. It is present on either side of T_G.

In the merging region, two different ansatzes have been proposed for the analysis of dielectric relaxation. One is the simple superposition principle according to

$$\phi(t) = \phi_\alpha(t) + \phi_\beta(t) \tag{6.56}$$

where ϕ_α and ϕ_β are the normalized decay functions for the α- and β- relaxations respectively.

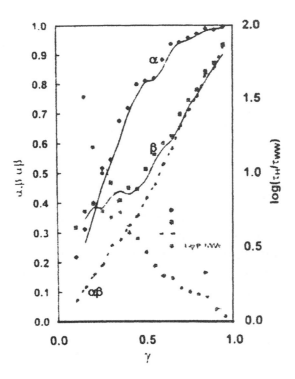

Fig. 6.11 Calculated relation between time domain parameters (γ) and H-N parameters (α, β), (Alvarez et. al., 1991, with permission of A. I. P.)

The normalized decay function is defined as

$$at\ t = 0 \quad \phi(t) = 1; \quad at\ t = \infty \quad \phi(t) = 0 \tag{6.57}$$

The second ansatz is called the **Williams ansatz** which is expressed as

$$\phi(t) = A\phi_\alpha t + (1 - A)\phi_\alpha(t)\,\phi_\beta(t) \tag{6.58}$$

where the individual decay functions are normalized according to equation (6.57). It is necessary to determine ϕ_α and ϕ_β at temperatures where the superposition from the other relaxation does not occur.

The relative merits of the superposition and the Williams ansatz in the merging region region of PMMA has been examined by Bergman et. al. (1998). The procedure adopted for transforming the frequency domain data into time domain current involves two steps:
(1) Determine $G(\tau)$ from ε''- ω data by inverse Laplace transformation, eq. (3.90).
(2) Determine $\phi_\alpha(t)$ and $\phi_\beta(t)$ from $G(\tau)$ according to equation (6.51).

(3) Determine $\phi(t)$ from equation (6.58).

Fig. 6.12 shows the result of step (1) above in PMMA, using broad band dielectric spectroscopy in the frequency range of $10^{-2} - 10^9$ Hz and temperature range of 220 - 490K. The loss factors from these measurements are shown in fig. (5-26). Fig. (6-13) shows the time domain currents. The current amplitude is seen to decrease with faster decay at higher temperatures.

A few comments with regard to numerical Laplace transformation is in order. Mopsik[19] has adapted a cubic spline to the original data and uses the spline to define integration. The method is claimed to be computationally stable and more accurate. For an error of 10^{-4} or less, only ten points per decade are required for all frequencies that correspond to the measurement window. Provencher (1982) has developed a program called CONTIN for numerical inverse Laplace transformation, which is required to derive $G(\tau)$ from the KWW function. Imanishi et. al.[20] propose an algorithm for the determination of $G(\tau)$ from $\varepsilon''- \omega$ data.

6.6 EXPERIMENTAL MEASUREMENTS

The experimental arrangement for measurement of absorption currents is relatively simple and a typical setup is shown in fig. 6.14[21]. The transformation from the time domain to the frequency domain involves the assumption that the current is measured in the interval 0 to infinity, which is not attained in practice. The necessity to truncate the integral to a finite time t_{max} involves the assumption that the contribution of the integrand at $t > t_{max}$ ceases to contribute to ε' and ε''[22]. The lowest frequency at which ε' and ε'' are evaluated depends on the longest duration of the measurement and the current magnitude at that instant.

In the data acquisition system the measured current is converted from analog to digital and Shannon's **sampling theorem** states that a band limited function can be completely specified by equi-spaced data with two or more points per cycle of the highest frequency. The high frequency limit f_n is given by the time interval between successive readings or the sampling rate according to $2f_n = 1/\Delta t$. For example a sampling rate of 2 s^{-1} results in a high frequency cut off at 1 Hz. At high sampling rates a block averaging technique is required to obtain a smooth variation of current with time.

For low frequency data the current should be measured for extended periods and for a higher f_n the time interval should be smaller. These requirements result in voluminous data but the problem is somewhat simplified by the fact that the currents are greater at the beginning of the measurement.

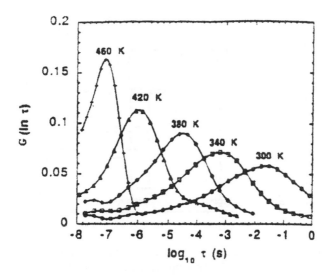

Fig. 6.12 The distribution of relaxation times at selected temperatures in PMMA derived from loss factor measurements. H-N function is used to calculate inversion from frequency domain data. The relaxation times decrease and the distributions become narrower as the temperature is increased (Bergman et. al., 1998, with permission of A. Inst. Phys.)

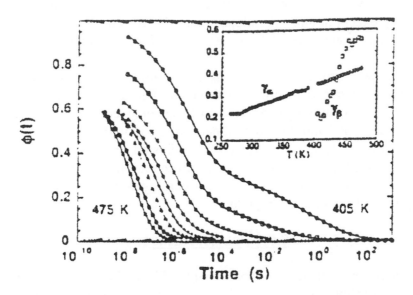

Fig. 6.13 Time domain current functions calculated using data shown in fig. 6.12. The temperature increases in steps of 10K. The inset shows the temperature dependence of the KWW stretching parameters for the α- and β- relaxations in PMMA (Bergman et. al. 1998, with permission of A. I. P.)

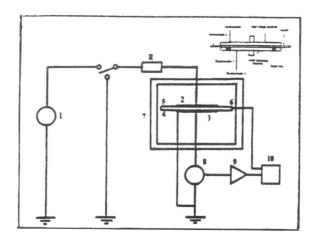

Fig 6.14 Schematic diagram for measuring absorption and desorption current. 1-variable DC supply, 2-High voltage electrode, 3-Measuring electrode, 4-Guard Electrode, 5-Sample, 6-Thermocouple, 7-Environmental chamber, 8-Electrometer, 9-Amplifier, 10-Recorder/computer. Inset shows details at the electrode (Sussi and Raju, 1994, with permission of Sampe journal).

The current may be obtained at greater intervals at later periods and the intermediate values obtained by interpolation. Quadratic schemes of interpolation have been found to be satisfactory except at short durations where the current decay is the steepest. In this range the log I(t) versus log(t) data are usually used for quadratic interpolation.

The sample under study is charged using a DC power source for the required time, a sensitive electrometer and recording equipment. The latter consists of a variable gain amplifier and A/D converter connected to a personal computer. The data acquisition system is capable of handling 16 channels of input with individual control for each channel. The frequency of sampling can be adjusted up to 4 kHz though after a few minutes of starting the experiment a much lower sampling rate is adequate to limit the volume of data. A block averaging scheme is employed with the number of readings in a block being related to the sampling rate.

6.6.1 POLY(VINYL ACETATE)

As already mentioned in the previous chapter PVAc is one of the most extensively investigated amorphous polymers, both in the frequency domain and the time domain providing an opportunity to compare the relative merits of the two methods over overlapping ranges of measurements. Table 6.1 lists selected publications.

Table 6.1

Dielectric Constant and Dielectric Loss Data in PVAc.

Authors & ref. No.	Year	Method	Temp. range °C	Frequency range (Hz)
Mead and Fuoss[23]	1941	F-D	50-100	60-10^4
Broens and Muller[24]	1955	F-D	20-65	6-10^7
Thurn and Wolf[25]	1956	F-D	-100 to 150	2×10^6
Ishida et. al[26]	1962	F-D	- 61 to 150	5 - 10^6
Saito[27]	1963	F-D and T-D	38.1-78.7	0.1-10^6
Block, et. al.	1972	T-D	-	10^{-4}-1
Mashimo, et. al.[28]	1982	T-D and F-D	23.2-28.25	10^{-6} –10^6
Weiss, et. al.	1985	T	66.7°C	502-7194
Nozaki and Mashimo[29]	1986	T-D & F-D	28-93	10^{-3} to 10^7
Nozaki and Mashimo[30]	1987	F-D	26-85	10^{-6} –10^6
Shioya and Mashimo[31]	1987	T	26-85	6×10^{-5}-10^6
Rendell et. al[32]	1987	T	26-85	10^{-7} to 10^3
Murthy[33]	1990	T	26-85	10^{-6} –10^6
Dionisio et. al.[34]	1993	F-D	45-70	16 - 10^5

F - D = Frequency Domain, T - D = Time Domain, T = Theory

Other studies are due to Veselovski and Slutsker[35], Hikichi and Furuichi[36], Patterson[37], Sasabe and Moynihan[38], Funt and Sutherland[39].

Fig. 6.15 shows the complex dielectric constant in PVAc due to Nozaki and Mashimo (1986). The Havriliak-Negami distribution function is found to apply and the evaluated parameters α and β are shown in Table 6.2. In a subsequent publication Shioya and Mashimo (1987) have found that Cole-Cole distribution also applies well for the data.

The measured data have also been analyzed according to William-Watson function; while the agreement between the H-N parameters and the W-W parameter is good at lower frequencies, some differences prevail at the higher frequencies. This is attributed to a secondary relaxation process at higher frequencies, in addition to the main relaxation process. The secondary relaxation is thought to be due to side group rotation, the local mode relaxation of the main chain or co-operative relaxation of both motions.

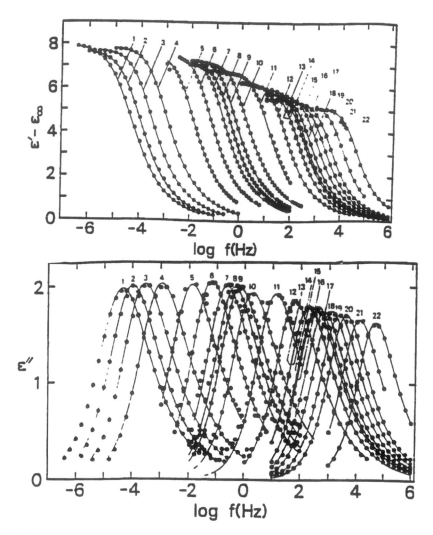

Fig. 6.15 Dispersion (a) and absorption (b) curves in PVAc at various temperatures: 1-26.8, 2-28.01, 3-29.41, 4-30.98, 5-34.95, 6-37.38, 7-40.20, 8-41.34, 9-42.3, 10-45.02, 11-50.20, 12-54.85, 13-58.80, 14-58.98, 15-61.23, 16-61.7, 17-63.70, 18-66.07, 19-68.91, 20-71.53, 21-77.62, 22-84.77 (°C).Solid equation (Nozaki and Mashimo, 1987). (with permission of the A. Inst. Phy.)

We recall that the WLF equation holds true for many polymers at $T > T_G$. For PVAc this equation is found to hold true above 45°C (Nozaki and Mashimo, 1987), and the constants are given by

$$\log f_m = \frac{900.5}{T - 245.3} \tag{6.59}$$

At lower temperatures the WLF equation fails and the constants in this region are

$$\log f_m = -3.01 \times 10^4 \frac{1}{T} + 96.1 \tag{6.60}$$

This deviation from WLF equation is not observed if one evaluates the frequency at maximum loss according to W-W function (Mashimo et. al., 1982). The reasons are not quite clear if we suppose that the W-W function is a mathematical tool to specify the shape of the absorption currents. Equation (6.60) is Arrhenius law with an activation energy of 6 eV. At T_G the distribution of relaxation time shows an abrupt change as shown in Fig. 6.6.

Table 6.2
Dielectric Relaxation Parameters in PVAc [Nozaki and Mashimo, 1986].

T (K)	τ_0 (s)	$\Delta\varepsilon$	ε_∞	α	β	μ (D)
301.25	5.70×10^3	7.922	3.139	0.390	0.821	2.458
301.71	2.94×10^3	8.017	3.138	0.388	0.813	2.490
302.90	9.25×10^2	8.053	3.135	0.402	0.819	2.515
303.60	4.90×10^2	7.960	3.133	0.414	0.822	2.498
305.80	5.64×10	8.040	3.126	0.421	0.823	2.549
309.16	1.86×10	7.279	3.117	0.450	0.860	2.367
311.15	3.26	7.444	3.111	0.445	0.844	2.440
315.65	1.20	6.800	3.098	0.467	0.864	2.298
326.64	7.16×10^{-3}	6.220	3.068	0.501	0.880	2.231
326.25	1.38×10^{-2}	6.004	3.068	0.513	0.856	2.161
330.47	4.15×10^{-2}	5.909	3.056	0.500	0.859	2.174
335.05	1.01×10^{-3}	5.582	3.043	0.526	0.868	2.110
339.30	3.26×10^{-4}	5.289	3.031	0.530	0.872	2.050
343.75	1.03×10^{-4}	5.197	3.018	0.551	0.869	2.060
349.80	3.02×10^{-5}	4.800	3.001	0.560	0.870	1.972
365.95	3.23×10^{-6}	4.170	2.955	0.590	0.873	1.865

$\Delta\varepsilon = \varepsilon_s - \varepsilon_\infty$; (With permission of J. Chem. Phys.)

The last column in Table 6.2 is the dipole moment of the monomer according to Onsager's equation and the method of calculating it has been demonstrated in section 2.8.

Murthy (1990) has discussed the dielectric relaxation in PVAc and suggest that the deviation from the WLF equation may be removed if a power law is assumed instead of the WLF equation (5.19),

$$\log f_m = A[\frac{T - T_k}{T}]^s; \quad T > T_k \tag{6.61}$$

The large value of s (12.1 for PVAc) combined with the introduction of another arbitrary temperature T_k and lack of physical basis limits the usefulness of this equation.

6.7 COMMERCIAL DIELECTRICS

Absorption currents are measured, as described, during the application of a step voltage lasting from a few minutes to several hours. The recorded current is the sum of the polarization currents and conduction currents due to the electric field impressed. The absorption current is reversible; if the applied electric field is removed and the two electrodes are short circuited the current flows in the opposite direction and its magnitude is exactly equal to the absorption current. The conduction current, however, is not reversible, since the electric field is removed. To remove the effect of applied field on the absorption current the discharging current is preferably measured using the same setup shown in fig. 6.14. The discharging current is also called **desorption current,** though we use the common term "absorption current" to denote both charging and discharging conditions.

There are a number of reasons for the generation of absorption currents. These have been variously identified as[40]:

> (1) electrode polarization,
> (2) dipolar orientation,
> (3) charge injection leading to space charge effects,
> (4) hopping of charge carriers between trapping sites, and
> (5) tunneling of charge carriers from the electrodes into the traps.

An attempt to identify the mechanism involves a study of the absorption currents by varying several parameters such as the electric field applied during charging, the temperature, sample thickness and electrode materials. Wintle[41] has reviewed these processes (except the hopping process) and discussed their application to polymers. Table 6.3 summarizes the characteristics responsible for the transient currents.

The mechanisms of generation of absorption current in specific materials is still an active area of study and there is only agreement in the broad generality. Measurements to discriminate the various processes should include a systematic study of the influence of parameters such as field strength, electrode material, sample thickness, water content, temperature, and thermal history[42]. Indicators, which are looked for in experimental results, may be summarized as below:

Table 6.3
Summary of the characteristics of the mechanism responsible for transient currents. L is the thickness and F is the electric strength.

Process	field dependence of isochronal current	thickness dependence of isochronal current at constant field	electrode material dependence	Temperature dependence	time dependence $I \propto t^{-n}$	relation between charging and discharge transients
Electrode polarization	$\propto F^p$ where $p=1$	not specified	strongly dependent through blocking parameter	thermally activated	initially $n = 0.5$ followed by $n > 1$	mirror images
Dipole orientation, uniformly distributed in bulk	$\propto F^p$ where $p=1$	independent	independent	thermally activated	$0 \leqslant n \leqslant 2$	mirror images
Charge injection forming trapped space charge	related to mechanism controlling charge injection	independent	related to mechanism controlling charge injection	related to mechanism controlling charge injection	$0 \leqslant n \leqslant 1$	dissimilar
Tunneling	$\propto F^p$ where $p=1$	L^{-1}	strongly dependent	independent	$0 \leqslant n \leqslant 2$	mirror images
Hopping	$\propto F^p$ where $p=1$	independent	independent	thermally activated	$0 \leqslant n \leqslant 2$	mirror images

(Das-Gupta, 1997, with permission of IEEE).

1. Identical absorption and desorption currents, except in sign, implies that charge carriers are generated in the bulk and not injected from the electrodes. If dipolar processes are present, polymers as a rule exhibit a distribution of relaxation times with several overlapping processes. This is reflected as a change of slope of the power law index over short segments of time.

2. Oxidation of a polymer introduces additional dipoles of a different kind and these are likely to reside closer to the surface. The transient currents will then be inversely proportional to the thickness.

3. Charging current independent of the electrode material provides additional support for the absence of charge injection from the electrodes. Two different electrode materials with a large difference in work function may be employed in four different configurations to check the injection of charge carriers from the electrodes.

4. Independence of current on electrode material indicates lack of electrode polarization and the power law index for this process is close to 0.5.

5. Currents that are independent of thickness and temperature suggest that tunneling is at least not the principal mechanism for the transient current.

6. Hopping of ionic or electronic charge generally shows a transient decay of the discharging current divided in two successive domains having n~1 and n~0, respectively.

7. Presence of moisture generally reduces the activation energy for dipolar motions shifting the α-peak to lower temperatures. This is reflected in the absorption currents increasing at room temperature.

Lowell[43] has provided an analysis of the absorption and conduction currents by developing a model of the behavior of charges trapped on a polymer network. The theory is claimed to agree with experimentally observed current-time relationship in polymers.

6.7.1 ARAMID PAPER

Aramid Paper which is an aromatic polyamide is a high temperature insulating material and finds increasing applications in electrical equipment such as motors, generators, transformers and other high energy devices. There are a growing number of uses for this material in nuclear and space applications due to its excellent capability of withstanding intense levels of radiation such as beta, gamma and X-rays. The dielectric properties of this material are of obvious interest, and Sussi and Raju (1992) have measured the absorption currents covering a wide range of parameters.

To ensure repeatable results the samples were conditioned by heating for several hours (~ 5 hours) at high temperatures (~260°C). Two different electrode materials, aluminum and silver of 50 nm thickness, were vacuum deposited on the samples at reduced pressure of 0.1 mPa. A guard ring which was also vapor deposited was employed to reduce effects due to stray fields. Before each measurement a blank run was performed in order to free the sample from extraneous charges. This run consisted of raising the temperature of the sample at a constant rate to 200°C and short circuiting the electrodes for twelve hours. The influences of several factors on the absorption currents are described below.

A. TIME DEPENDENCE

Fig. 6.16 shows typical charging currents in a 127 μm thick sample at a temperature of 200°C and various poling electric fields for a time duration of 10^4 s. The charging currents decay very slowly and no polarity dependence is observed. There is no appreciable dependence of the current on the sample thickness or electrode material.

A similar slow time dependence of charging current has been reported for other insulating materials such as SiO_2 and SiO with aluminum electrodes[44], solid n-octadecane[45] and polycarbonate above its glass transition temperature of $150°C$[46]. Lindmeyer suggests that a slow filling of traps by charge carriers in conjunction with a space charge effect is responsible for the slow decay of current with respect to time. It has also been suggested that that a high electric field can induce traps in a polymer with the trap creation rate being proportional to the applied field. This theory, also, possibly explains the reason for not being able to observe a steady current under certain experimental conditions, even after as long as 10^5s.

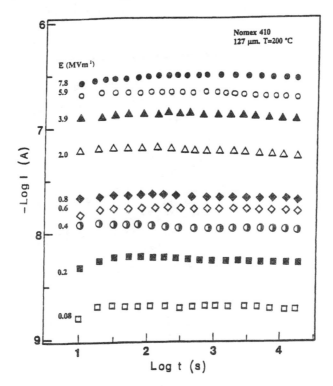

Fig. 6.16 Charging currents at various electric fields at 200°C in aromatic polyamide with silver electrodes. The current decays very slowly with time. This is just one of several types of decay of current in polymers (Sussi and Raju, 1992, with permission of Sampe journal).

Fig. 6.17 shows the influence of temperature on the absorption currents at a constant electric field. The decay of current is more rapid in the lower temperature range of 70 - 170°C. Such behavior has also been observed in linear polyamides in which the conduction currents in the range of 25 - 80°C and 120 - 155°C decay more rapidly than in the intermediate range of 80 - 110°C. Obviously there is a change of conduction mechanism above 150°C in aramid paper.

As stated earlier a comparison of charging and discharging currents permits reliable conclusions to be drawn with regard to the origin of charging current. Fig. 6.17 shows the discharging currents at 200°C after poling at various fields in the range of 0.08 – 8 MVm⁻¹. The current decays fast for a duration of about 10^3 s where the power law holds with a value of n ~1.0. For times longer than 10^3 s the current decay is much less pronounced and the decay constant in this range of time is between 0.1- 0.4.

The data in Figs. 6.17 and 6.18 clearly show that the charging and the discharging currents are dissimilar for times up to 10^4s. Such dissimilarities have been observed in other polymers as well and there is no agreement on the processes responsible for them.

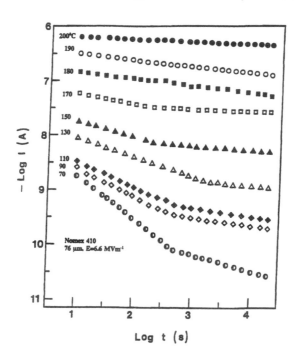

Fig. 6.17 Charging currents in aromatic polyamide at various temperatures and constant electric field of 6.6 MVm-1 with silver electrodes. A change of mechanism is clearly seen above 150°C. (Sussi and Raju, with permission of Sampe Journal).

Lindmeyer (1965) discusses these dissimilarities with the aid of energy diagrams and attributed the initial transient during charging to empty traps. The current can be as large as allowed by the injecting barrier. As the traps become filled, the current reduces to the space charge limited current with traps. In the discharging period, on the other hand, the trapped carriers will be emptied towards both electrodes, the external circuit registering a smaller current.

Das Gupta and Brockley[47], on the other hand, concluded that the dissimilarity between the charging and discharging currents is due to the onset of quasi steady state conduction current which is superimposed on the transient current. If this reasoning holds good then the dissimilarity should increase with electric field or temperature.

It is relatively easy to interpret the results of fig. 6.18 in terms of the universal relaxation theory of Jonscher (chapter 4) who suggests that the time domain response of the current has two slopes or two values for the power law index , equation (6.30). An increase in the value of index at longer periods is attributed to a dipolar mechanism whereas a decrease at longer periods is attributed to interfacial polarization. On this basis it is suggested that the relaxation changes from dipolar to interfacial as the temperature is increased from 90°C to 200°C.

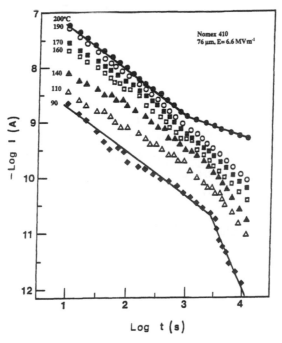

Fig. 6.18 Discharging currents at various temperatures and constant electric field of 6.6 MVm-1 with silver electrodes. Charging and discharging currents are dissimilar indicating that dipolar processes are not dominant (Sussi and Raju, 1992, with permission of Sampe Journal).

Aramid paper is manufactured out of aromatic polyamide in the form of fibers which are amorphous and flocs (like corn flakes) which are crystalline, the finished product having a crystallinity of approximately 50%. The material abounds with boundaries between the crystalline and amorphous regions. Both the number of distributed trap levels and the mobility of charge carriers are likely to be higher resulting in a higher conductivity in amorphous regions in comparison with the crystalline parts. This difference in conductivity is possibly the origin of the observed interfacial polarization.

B. TEMPERATURE DEPENDENCE

From Figs. 6.17 and 6.18 it is inferred that the isochronal discharging currents increase with temperature at the same electric field in the temperature range of 90 - 200°C. The currents show no significant peaks and from the slopes an approximate activation energy of 0.21 ± 0.05 eV may be evaluated.

C. FIELD DEPENDENCE

The isochronal charging current shows a linear relationship with the electric field at each temperature according to

$$I(t) = k(t)E^p \qquad (6.62)$$

where k is a constant independent of E and p is approximately equal to one. Das-Gupta and Joyner[48] and Das-Gupta et. al.[49] have observed similar relationship in poly(ethylene terephthalate) (PET) at 173 K. The magnitude of isochronal current is observed to depend upon the time lapsed in contrast with aramid paper which shows only a marginal dependence on time. This difference is attributed to the fact that the charging currents vary slowly with time (Fig. 6.16).

D. EFFECT OF ELECTRODE MATERIAL

The charging currents are found to be dependent on the electrode material, silver yielding higher currents than aluminum. The power law index n is also higher for silver. However the discharging currents do not show any significant dependence on the electrode material.

E. LOW FREQUENCY DIELECTRICLOSS FACTOR (ε'')

The theoretical basis for transforming the time domain data into ε''-log ω has been dealt with in considerable detail in earlier sections of this chapter. The Hamon approximation given by equation

$$\varepsilon'' \cong \frac{i(0.63/\omega)}{\omega C_0 V} \qquad (6.34)$$

is easy to apply to the measured data. Fig. 6.19 shows the calculated values of ε'' at low frequencies in the temperature range of 100-190°C. Since dc conductivity contributes to the loss factor appropriate correction has been applied. Two peaks are observed at each temperature and they correspond to relaxation times of 1.6×10^3s and 2×10^4s. These values are comparable to the relaxation time of 1×10^4s at 150°C obtained by Govinda Raju[50] from studies of thermally stimulated discharge currents in aramid paper. This study has shown that dipolar mechanism is responsible for the TSD currents in the temperature range cited and the observed peak at the higher frequency in fig. 6.19 (the first peak) is attributed to the dipolar mechanism. We also note that the low frequency peak (second peak) is more dominant at higher temperatures.

Sussi and Raju (1992) also examine whether the low frequency peak is possibly due to interfacial polarization according to the Maxwell Wagner equation, considering that the amorphous and crystalline regions have different conductivity. The parameters evaluated are shown in Table 6.3. Note the large increase of the loss factor at 200°C which is possibly due to an intensified Maxwell-Wagner effect. Higher temperatures increase the conductivity in amorphous regions more rapidly than in crystalline region. It is interesting to compare Fig. 6.19 with Fig. 5.17 which also demonstrates a large increase in ε'' as the temperature is increased.

6.7.2 COMPOSITE POLYAMIDE

The demand for electrically insulating polymers has increased over the years due to their high levels of electrical, chemical and mechanical integrity, making them ideally suited for several applications. Some organic polymers are capable of withstanding temperatures higher than 180°C and also have capability for withstanding radiation. The mechanical strength of aramid paper can be increased substantially for industrial applications by using it in conjunction with other polymers and such materials are referred to here as composites. For example an industrially available material is comprised of aramid-polyester-aramid (A-P-A) with three layers of equal thickness.

The isochronal currents measured by Sussi and Raju[51] in A-P-A are shown in fig. 6.20 with regard to the influence of temperature on discharging currents. For each curve the current increases with temperature and shows a well defined peak in the range of 80-100°C. The position of this peak tends to shift to a higher temperature with decreasing times, suggesting the presence of thermal activation processes (Sussi and Raju, 1992). A comparison with fig. 6.16 shows that the observed peaks are due to the intermediate layer of polyster which has T_G in the vicinity of the observed peak. The dipolar contribution is suggested as a possible mechansm. Above 130°C the isochronal currents increase almost monotonically with time and the increase is attributed to interfacial polarization as explained in the previous section.

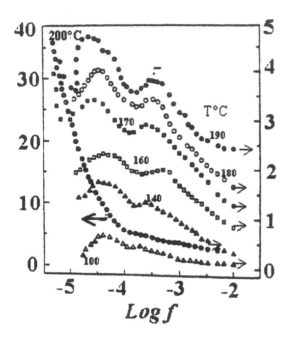

Fig. 6.19 Frequency dependence of dielectric loss factor ε'' at various temperatures. Pre-applied field of 6.6 MV m-1; charging time 5h; electrodes made of silver (Sussi and Raju, 1992). (with permission of Sampe Journal.)

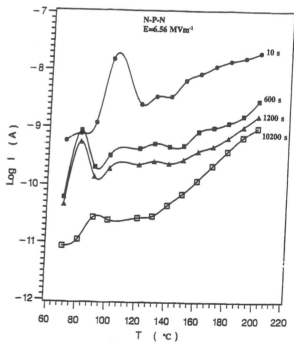

Fig. 6.20 Temperature dependence of the isochronal discharge currents in composite aramid-polyester-aramid polymer at prescribed times. The electric field is 6.56 MVm^{-1}, 154 μm sample with aluminum electrodes. (Sussi and Raju, 1994: With permission of Sampe Journal)

6.7.3 POLY(ETHYLENE TEREPHTHALATE)

The absorption currents in PET have been measured by a number of authors. Das-Gupta and Joyner (1976) distinguish two temperature regions, 83-273K and 273-430K. In both regions the isochronal currents (I-T plots) are proportional to the electric field. No thickness dependence or electrode material dependence is observed. The charging and discharging transients are mirror images and two thermally activated energies of 0.65 and 1.6 eV are measured in the regions mentioned above respectively. The lower energy is identified with the β-relaxation and the higher energy with the α-relaxation.

Table 6.4
Loss factor in aramid paper (Sussi and Raju, 1992)

T (°C)	ε''	$\sigma \, (\Omega \, m)^{-1}$	τ (s)
\multicolumn{4}{c}{(a) High frequency peak}			
110	0.76	1.75×10^{-15}	1.6×10^3
130	1.14	5.48×10^{-15}	1.1×10^3
140	1.31	1.20×10^{-14}	6.3×10^2
150	2.14	2.62×10^{-14}	5.5×10^2
160	2.03	4.38×10^{-14}	3.1×10^2
\multicolumn{4}{c}{(b) Low frequency peak}			
110	0.89	7.0×10^{-16}	5.6×10^3
140	1.70	2.3×10^{-15}	5.0×10^3
160	2.27	4.0×10^{-15}	4.0×10^3

(with permission from SAMPE journal)

More recent results are due to Thielen et.al[52] who discovered that the traditional use of three electrodes for measurement of current is likely to damage the ultra-thin films (<12-15 μm). This leads to spurious and noisy current, the preferred arrangement of electrodes being the simple two electrode system with lateral contacts. Fig. 6.21[53] shows the absorption current measurements and the power law index according to t^{-n}. While there is a slight but significant reduction in the current with increase of thickness, the index remains the same for all electric fields, ~ 0. 75 ± 0.03, which is close to the value for thicker films.

Isochronal currents for electric fields up to 50 MV/m showed that the current is ohmic in nature for short durations of 2-5 minutes whereas at longer durations (~210 min) the current increases exponentially. The electrode materials do not significantly influence the intensity and slope of the current. The transient current is attributed to the dipoles whereas Schottky mechanism is found to hold true for the steady state current. Water content increases the steady state current more than it increases the transient current.

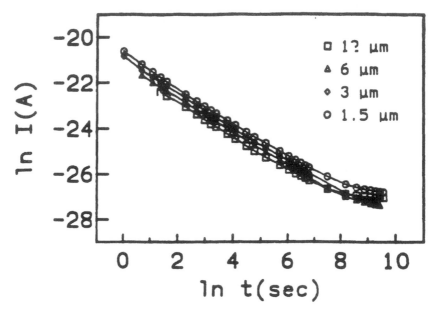

Fig. 6.21 Transient conductivity observed in Mylar ™ films of various thicknesses (12. 6, 3 and 1.5 µm) for times ranging from 1 to 10^4 s T=23°C, RH=52%, $E_p=4×10^7$ V/m (Thielen et. al., 1994, with permission of American Inst. Phys).

There are two relaxation peaks observed in PET. The high temperature measurements covered by the Neagu et. al.[54] falls into the range 20-110°C where both α- and β-relaxations are detected. In fact the β- relaxation is centered about −110°C[55] and attributed to local motions of glycol methylene and carboxylic groups. It is characterized by a broad distribution. The dipoles corresponding to this relaxation are oriented in a short time < 1s and not measured by Thielen et. al (1994). The α-process is due to the motion of main chain and the distribution of relaxation times extends down to room temperature.

The measured current is probably due to this relaxation and the relaxation time in the range of 10^3-10^4 s agrees with the expected value for this relaxation.

6.7.4 FLUOROPOLYMER

The fluoropolymer, commercially known as Tefzel™ (E. I. Dupont De Nemours & Co. Inc) is a copolymer of tetrafluoro-ethylene and ethylene considered as a high temperature insulating material. As far as the author is aware there is only one study carried out on the dielectric properties (ε' and ε'') and absorption currents in this high temperature dielectric[56] and these data are presented below.

A. DIELECTRIC CONSTANT AND LOSS FACTOR

Fig. 6.22 shows ε' and ε'' as a function of temperature and frequency. The dielectric constant decreases with increasing temperature at each frequency. Further the loss factor shows a well defined peak in the region 100-120°C, the loss peak shifting to higher frequencies as the temperature is increased. These are characteristics of dipolar relaxation and plots of the molar polarizability versus 1/T (eq. 2.53) do not yield a straight line. It is evidence for the non-applicability of Debye theory for dipolar relaxation on the basis of a single relaxation time. This result is not surprising since most polymers are characterized by a distribution of relaxation times.

B. ABSORPTION CURRENTS

The absorption currents are measured at $5 \leq E \leq 40$ MV m^{-1} and $50° \leq T \leq 200°C$ by Wu and Raju (1965). Fig. 6-23 shows the time dependence of isothermal charging currents at 20 MVm^{-1} poling field and various temperatures. Results at other temperatures and electric fields shows a similar trend, suggesting that the mechanism for absorption current remains the same over these ranges of parameters. The material exhibits absorption currents that increase with time during the initial 1 – 100 s and then decrease slowly. Similar behavior is found in PE in which the conduction mechanism is identified mainly as space charge enhanced Schottky injection[57].

The absorption current also depends significantly on the electrode material, aluminum electrodes yielding higher currents than silver electrodes. This dependence suggests possible injection of charge carriers from the cathode and the accepted mechanism is field assisted thermionic emission. The current dependence on the electrode material can then be explained by the work function differences. Others have also reported similar results in PET and polyimide.

Injection of electrons from the cathode due to field assisted thermionic emission generally leads to a space charge layer within the thickness of the material and accumulation of charge carriers can also occur in the metal/polymer interface. The space charge, if sufficiently intense, will reduce the electric field at the cathode resulting in a decreased electron injection. The observed increase in current with time can only be explained on the basis of accumulation of positive charges within the polymer and close to the cathode. The ions originate within the bulk of the material or are injected from the anode and drift towards the cathode, thereby increasing the cathode field.

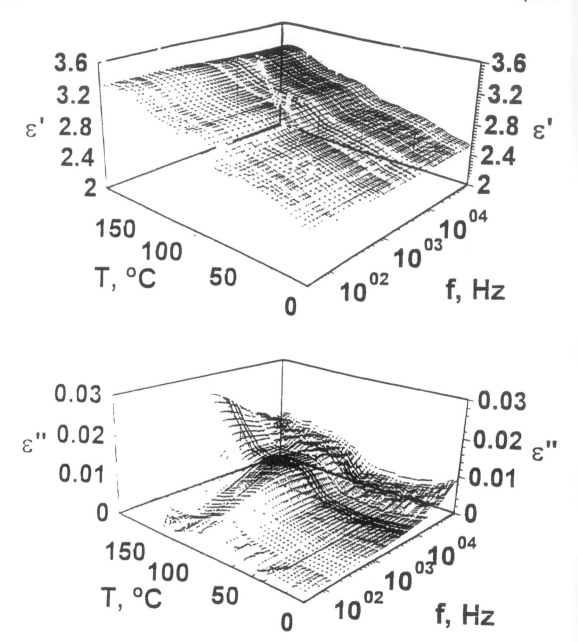

Fig. 6.22 Real and imaginary part of the complex dielectric constant in Tefzel (author's measurements).

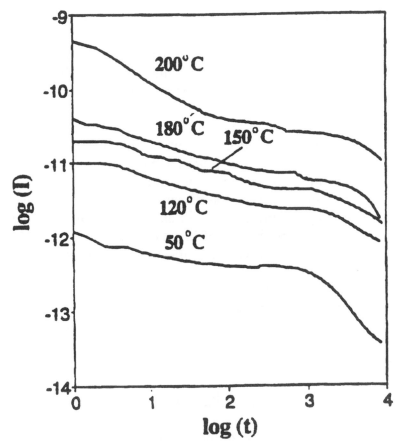

Fig. 6.23 Isothermal charging currents at 20 MV/m at various temperatures (electrodes silver) in fluoropolymer (Wu and Raju, 1995, with permission of IEEE).

There are several sources for ion generation. One of them is impurity ions, the current due to which tends to decrease if successive measurements are repeated on the same sample within a short interval of time. A second source of ions is due to autoionization of one of the atoms in the monomer, as demonstrated by Raju (1971) in aromatic polyamide. There is no experimental evidence in the literature to support this mechanism in this copolymer, and it is shown in chapter 7 that ionic conduction is not a feasible mechanism in this material.

Poole-Frenkel mechanism (Ch. 7) suggests that the electric field lowers the barrier for transition of charge carriers in the localized states. The mechanism is not specific to charge carriers of a specific type and is similar for donors (creating electrons) or acceptors (generating holes). Holes generated within the bulk drift towards the cathode and contribute to the intensification of the field.

An additional mechanism, which has been suggested to account for an increase in current with time, is the increase of the number of traps in the bulk due to the influence of the electric field. The new traps that are created permit more injection resulting in an increase of the current. Such an increase in the number of traps is likely to saturate after a certain time of application of voltage and the surface potential of the polymer is expected to vary according to fig. (6.23).

At longer time duration the current decreases with time and this is the more common behavior for many polymers. Lindmeyer [1965] suggests that gradual filling of traps by charge carriers leads to this type of behavior. Further the polymer-metal interface is sensitive to structural features such as the density of localized molecular states in the polymer and tunneling range of the delocalized electrons in the metal electrode. Unless the charge is moved away from the site of injection, further injection is reduced. This effect added to the bulk trapping causes decay of charging currents at longer periods.

Isothermal absorption currents are also measured by interchanging the polarity of the electrodes. The currents were approximately the same as the current under positive polarity fields (fig. 6.23). This is considered to be reasonable because both electrodes are vacuum deposited with the same type of metal under identical conditions and the change in surface conditions is not significant enough to cause a change in emission.

6.7.5 POLYIMIDE

Chohan et. al.[58] have measured the absorption and desorption currents in Kapton and the two currents were dissimilar in selected temperatures in the range 39°-204°C. Fig. 6-24 shows the low frequency loss factors versus the frequency, which exhibit peaks that shift to lower frequencies as the temperature decreases. This observation is attributed to the interfacial polarization either at the electrode-dielectric interface or in boundaries between crystalline and amorphous regions. There is need for systematic study of transient currents in this important polymer over a wider range of temperature and electric field.

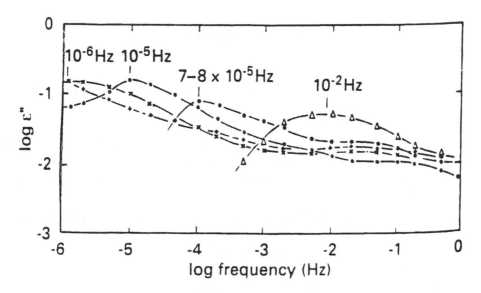

Fig. 6.24 Hamon transform results from desorption currents. Temperature (°C) as shown. The peak at 39°C is outside the frequency window (Chohan et. al. 1995). (with permission of Chapman and Hall).

6.8 REFERENCES

[1] A. K. Jonscher, "Dielectric Relaxation in Solids", Chelsea Dielectrics Press, London 1983

[2] E. Von Schweidler, Ann. Der. Physik, **24** (1907) 711.

[3] M. F. Manning and M. E. Bell, "Electrical conduction and related phenomena in solid dielectrics", Reviews of Modern Physics, **12** (1940) 215-256,

[4] G. R. Govinda Raju, Annual Report, Conference on Electrical Insulation and Discharge Phenomena, 2001.

[5] I. S. Gradshteyn, I. M. Ryzhik, Yu. V. Geronimus and M. Yu. Tseytlin, "Tables of integrals, series, and products", Academic Press, New York, 1965.

[6] M A. Sussi and G. R. Govinda Raju, SAMPE Journal, **28** (1992) 29-36.

[7] B. V. Hamon, Proc. IEE, Proc. IEE (London), **99** (1952) 151.

[8] F. Alvarez, A. Alegría and J. Colmenero, , Phys. Rev., B, Vol. **44**, No. 14 (1991) 7306-7312.

[9] R. Nozaki and S. Mashimo, J. Chem. Phys. **87** (1987) 2271.

[10] G. Williams and D. C. Watts, Trans. Farad. Soc., Vol. **66** (1970) 80.

[11] M. Abramowitz and L. Stegun, Handbook of Mathematical Functions, N.B.S., formula 118.

[12] M. Abramowitz and L. Stegun, Handbook of Mathematical Functions, N.B.S., page 325.

[13] G. Williams, D.C. Watts, S. B. Dev and A. M. North, Trans. Farad. Soc., **67** (1972) pp. 1323-1335.

[14] C. T. Moynihan, L. P. Boesch and N. L. Laberge, Phys. Chem. Glasses, **14** (1973) 122.

[15] G. H. Weiss, J. T. Bendler and M. Dishon, J. Chem. Phys., **83** (1985) 1424-1427.

[16] S. W. Provencher, Comput. Phys. Commun., **27** (1982) 213.

[17] J. –U. Hagenah, G. Meier, G. Fytas and E. W. Fischer, Poly. J. **19** (1987) p. 441.

[18] R. Bergman, F. Alvarez, A. Alegria and J. Colmenero, J. Chem. Phys., **109** (1998) 7546-7555.

[19] F. I. Mopsik, IEEE Trans. Elec. Insul., **EI-20** (1985) 957-963.

[20] Y. Imanishi, K. Adachi and T. Kotaka, J. Chem. Phys., **89** (1988) 7593.

[21] M. A. Sussi and G. R. Govinda Raju, Sampe Journal, **28**, (1992) 29.

[22] H. Block, R. Groves, P. W. Lord and S. M. Walker, J. Chem. Soc., Farad. Trans. II, **68** (1972) 1890.

[23] D. J. Mead and R. M. Fuoss, J. Am. Chem. Soc., vol. **63** (1941) 2832.

[24] O. Broens and F. M. Muller, Kolloid Z., Vol. **141** (1955) 20.

[25] H. Thurn and K. Wolf, Kolloid Z., Vol. **148** (1956) 16.

[26] Y. Ishida, M. Matsuo and K. Yamafuji, Kolloid Z., vol. **180** (1962) 108.

[27] S. Saito, Kolloid Z., vol. 189 (1963) 116.

[28] S. Mashimo, R. Nozaki, S. Yagihara, and S. Takeishi, J. Chem. Phys., 77 (1982) 6259.

[29] R. Nozaki and S. Mashimo, J. Chem. Phys., 84 (1986) 3575.

[30] R. Nozaki and S. Mashimo, J. Chem. Phys., 87 (1987) 2271.

[31] Y. Shioya and S. Mashimo, J. Chem. Phys., 87, (1987) 3173.

[32] R. W. Rendell, K. L. Ngai, and S. Mashimo, J. Chem. Phys., vol. 87 (1987) 2359.

[33] S. S. N. Murthy, J. Chem. Phys., 92 (1990) 2684.

[34] M. S. Dionisio, J. J. Moura-Ramos and G. Williams, Polymer, 34 (1993) 4106.

[35] P. F. Veselovskii and A. Slutsker, A. I., Zh. Tech., 25 (1955) 989, 1204.

[36] K. Hikichi and J. Furuichi, Rept. Prog. Poly. Phys. (Japan), 4 (1961) 69.

[37] G. D. Patterson, J. Poly. Sci, Poly. Phys., 15 (1977) 455.

[38] H. Sasabe and C. T. Moynihan, J. Poly. Sci., Poly. Phys. Edu., 16 (1978) 1447.

[39] B. L. Funt and T. H. Sutherland, Can. J. Chem., 30 (1952) 940.

[40] D. K. Das Gupta IEEE Trans. Diel. Elec. Insu., 4 (1997) 149-156.

[41] H. J. Wintle, J. Non-cryst. Solids, 15 (1974) 471.

[42] A. Thielen, J. Niezette, G. Feyder and J. Vanderschuren, J. Appl. Phys., 76 (1994) 4689.

[43] J. Lowell, J. Phys. D: Appl. Phys., 23 (1990) 205.

[44] J. Lindmeyer, J. Appl. Phys., 36 (1965) 196.

[45] I. E. Noble and D. M. Taylor, J. Phys. D: Appl. Phys., 13 (1980) 2115.

[46] J. Vanderschueren and J. Linkens, J. Appl. Phys., 49 (1978) 4195.

[47] D. K. Das Gupta and R. S. Brockley, J. Phys. D., Appl. Phys., 11 (1978) 955.

[48] D. K. Das Gupta and K. Joyner, J. Phys. D: Appl. Phys., 9 (1976) 824.

[49] D. K. Das Gupta, K. Doughty and R. S. Brockley, J. Phys. D: Appl. Phys., 13 (1980) 2101.

[50] G. R. Govinda Raju, IEEE Trans. Elec. Insul., EI-27 (1992) 162.

[51] M. A. Sussi and G. R. Govinda Raju, SAMPE journal, 30 (1994) 41; , 30 (1994) 50.

[52] A. Thielen, J. Niezette, J. Vanderschuren and G. Feyder, J. Appl. Poly. Sci., 49 (1993) 2137, 1993.

[53] A. Thielen, J. Niezette, G. Feyder and J. Vanderschuren, J. Appl. Phys., 76 (1994) 4689.

[54] E. Neagu, P. Pissis, L. Apekie and J. L. G. Ribelles, J. Phys. D: Appl. Phys., 30 (1997) 1551.

[55] C. D. Armeniades and E. Baer, J. Poly. Sci. A2 (1971) 1345.

[56] Z. L. Wu and G. R. Govinda Raju, IEEE Trans. Elec. Insul., 2 (1995) 475.

[57] Y. Suzoki, E. Muto, T. Mizutani, and M. Ieda, J. Phys. D. Appl. Phys., 18 (1985) 2293.

[58] M. H. Chohan, H. Mahmood and Farhana shaw, Journal of Materials Science Letters, 14 (1995) 552.

Mathematics may be defined as the subject in which we never know what we are talking about, nor whether what we are saying is true.
- Bertrand Russell (Mysticism and Logic, 1918)

7

FIELD ENHANCED CONDUCTION

The dielectric properties which we have discussed so far mainly consider the influence of temperature and frequency on ε' and ε'' and relate the observed variation to the structure and morphology by invoking the concept of dielectric relaxation. The magnitude of the macroscopic electric field which we considered was necessarily low since the voltage applied for measuring the dielectric constant and loss factor are in the range of a few volts.

We shift our orientation to high electric fields, which implies that the frequency under discussion is the power frequency which is 50 Hz or 60 Hz, as the case may be. Since the conduction processes are independent of frequency only direct fields are considered except where the discussion demands reference to higher frequencies. Conduction current experiments under high electric fields are usually carried out on thin films because the voltages required are low and structurally more uniform samples are easily obtained. In this chapter we describe the various conduction mechanisms and refer to experimental data where the theories are applied. To limit the scope of consideration photoelectric conduction is not included.

7.1 SOME GENERAL COMMENTS

Application of a reasonably high voltage ~500-1000V to a dielectric generates a current and let's define the macroscopic conductivity, for limited purposes, using Ohm's law. The dc conductivity is given by the simple expression

$$\sigma = \frac{I}{AE} \tag{7.1}$$

where σ is the conductivity expressed in $(\Omega\ m)^{-1}$, A the area in m^2, and E the electric field in $V\ m^{-1}$. The relationship of the conductivity to the dielectric constant has not been theoretically derived though this relationship has been noted for a long time. Fig. 7.1 shows a collection of data[1] for a range of materials from gases to metals with the dielectric constant varying over four orders of magnitude, and the temperature from 15K to 3000K. Note the change in resistivity which ranges from 10^{26} to 10^{-14} Ω m. Three linear relationships are noticed in barest conformity. For good conductors the relationship is given as

$$\log \rho + 3\log \varepsilon' = 7.7 \tag{7.2}$$

For poor conductors, semi-conductors and insulators the relationship is

$$\log \rho + 12.5\log \varepsilon' = 24 \tag{7.3}$$

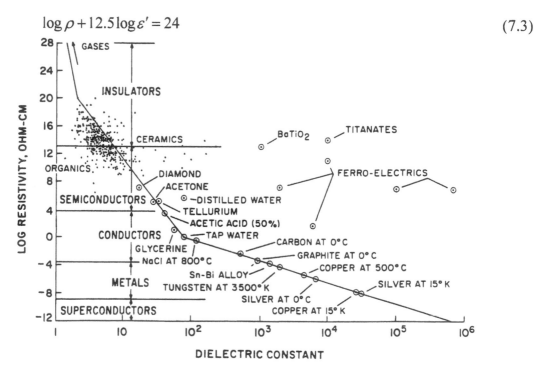

Fig. 7.1 Relationship between resistivity and dielectric constant (Saums and Pendleton, 1978, with permission of Haydon Book Co.)

Ferro-electrics fall outside the range by a wide margin. The region separating the insulators and semi-conductors is said to show "shot-gun" effect. Ceramics have a higher dielectric constant than that given by equation (7.3) while organic insulators have lower dielectric constant. Gases are asymptotic to the y-axis with very large resistivity and ε' is close to one. Ionized gases have resistivity in the semi-conductor region.

From the definition of complex dielectric constant (ch. 3), we recall the following relationships (Table 7.1):

Table 7.1
Summary of definitions for current in alternating voltage

Quantity	Formula	Units
Charging current, I_c	$\omega C_0 \varepsilon' V$	amperes
Loss current, I_L	$\omega C_0 \varepsilon'' V$	amperes
Total current, I	$\omega C_0 V(\varepsilon'^2 + \varepsilon''^2)^{1/2}$	amperes
Dissipation factor, tanδ	$\varepsilon'' / \varepsilon'$	none
Power loss, P	$\omega C_0 \varepsilon' V^2 \tan\delta$	Watts

7.2 MOTION OF CHARGE CARRIERS IN DIELECTRICS

Mobility of charge carriers in solids is quite small, in contrast to that in gases, because of the frequent collision with the atoms of the lattice. The frequent exchange of energy does not permit the charges to acquire energy rapidly, unlike in gases. The electrons are trapped and then released from localized centers reducing the drift velocity. Since the mobility is defined by $W_e = \mu_e E$ where W_e is the drift velocity, μ_e the mobility and E the electric field, the mobility is also reduced due to trapping. If the mobility is less than $\sim 5 \times 10^{-4}$ m^2 / Vs the effective mean free path is shorter than the mean distance between atoms in the lattice, which is not possible in principle. In this situation the concept of the mean free path cannot hold.

Electrons can be injected into a solid by a number of different mechanisms and the drift of these charges constitutes a current. In trap free solids the Ohmic conduction arises as a result of conduction electrons moving in the lattice of conductors and semi-conductors. In the absence of electric field the conduction electrons are scattered freely in a solid due to their thermal energy. Collision occurs with lattice atoms, crystal imperfections and impurity atoms, the average velocity of electrons is zero and there is no current. The mean kinetic energy of the electrons will, however, depend on the temperature of the lattice, and the rms speed of the electrons is given by $(3kT/m)^{1/2}$.

If an electric field, E, is applied the force on the electron is $-eE$ and it is accelerated in direction opposite to the electric field due to its negative charge. There is a net drift velocity and the current density is given by

$$J = N_e\,e\,\mu\,E = N_e\,e\,\mu\frac{V}{d} \tag{7.4}$$

where N_e is the number of electrons, μ the electron mobility, V the voltage and d the thickness.

We first consider Ohmic conduction in an insulator that is trap free. The concept of collision time, τ_c, is useful in visualizing the motion of electrons in the solid. It is defined as the time interval between two successive collisions which is obviously related to the mobility according to

$$\mu = e\tau_c\,m* \tag{7.5}$$

where m* is the effective mass of the electron which is approximately equal to the free electron mass at room temperature.

The charge carrier gains energy from the field and loses energy by collision with lattice atoms and molecules. Interaction with other charges, impurities and defects also results in loss of energy. The acceleration of charges is given by the relationship, a = F/m* = e E/m* where the effective mass is related to the bandwidth W_b. To understand the significance of the band width we have to divert our attention briefly to the so-called **Debye characteristic temperature**[2].

In the early experiments of the ninteenth century, Dulong and Petit observed that the specific heat, C_v, was approximately the same for all materials at room temperature, 25 J/mole-K. In other words the amount of heat energy required per molecule to raise the temperature of a solid is the same regardless of the chemical nature. As an example consider the specific heat of aluminum which is 0.9 J/gm-K. The atomic weight of aluminum is 26.98 g/mole giving C_v = 0.9 × 26.98 = 24 J/mole-K. The specifc heat of iron is 0.44 J/gm-K and an atomic weight of 55.85 giving C_v = 0.44 × 55.85 = 25 J/mole-K. On the basis of the classical statistical ideas, it was shown that C_v = 3 R where R is the universal gas constant (= 8.4 J/g-K). This law is known as **Dulong-Petit law** (1819).

Subsequent experiments showed that the specific heat varies as the temperature is lowered, ranging all the way from zero to 25 J/mole-K, and near absolute zero the specific heat varies as T^3. Debye successfully developed a theory that explains the increase of C_v as T is increased, by taking into account the coupling that exists between individual atoms in a solid instead of assuming that each atom is a independent vibrator, as the earlier approaches had done. His theory defined a characteristic temperature for each material, Θ, at which the specific heat is the same. The new relationship is $C_v(\Theta) = 2.856R$. Θ is called the Debye characteristic temperature. For aluminum $\Theta = 395$ K, for iron $\Theta = 465$ K and for silver $\Theta = 210$ K. Debye theory for specific heat employs the Boltzmann equation and is considered to be a classical example of the applicability of Boltzmann distribution to quantum systems.

Returning now to the bandwidth of the solids, W_b may be smaller or greater than $k\Theta$. Wide bandwidth is defined as $W_b > k\Theta$ in contrast with narrow bandwidth where $W_b < k\Theta$. In materials with narrow bandwidth the effective mass is high and the electric field produces a relatively slow response. The mobility is correspondingly lower.

The band theory of solids is valid for crystalline structure in which there is long range order with atoms arranged in a regular lattice. In order that we may apply the conventional band theory a number of conditions should be satisfied (Seanor, 1972).

1. According to the band theory the mobility is given by

$$\mu = \frac{4e\lambda}{3 \times 10^2 (2\pi m * kT)^{1/2}} \quad m^2 V^{-1} s^{-1} \tag{7.6}$$

where λ is the mean free path of charge carriers which must be greater than the lattice spacing for a collision to occur. This may be expressed as

$$\lambda \approx 10^{-6} \mu \left(\frac{m*}{m_e}\right)^{1/2} > a \tag{7.7}$$

where m_e is the mass of the electron (9.1×10^{-31} kg) and a the lattice spacing.

2. The mean free path should be greater than electron wavelength (1 eV $= 2.42 \times 10^{14}$ Hz $= 1.3$ μm). This condition translates into the condition that the relaxation time τ should be greater than $(h/2\pi kT)$, 2.5×10^{-14} s at room temperature. τ is related to μ according to equation (7.5).

3. Application of the uncertainty principle yields the condition that $\mu > (e\,a^2\,W_b/2\pi hkT)$. For a lattice spacing of 50 nm we get $\mu > 3.8\,W_b/kT$.

If these conditions are not satisfied then the conventional band theory for the mobility can not be applied. The charge carrier then spends more time in localized states than in motion and we have to invoke the mechanism of hopping or tunneling between localized states. Charge carriers in many molecular crystals show a mobility greater than 5×10^{-4}

m^2 V s^{-1} and varies as $T^{-1.5}$. This large value of mobility is considered to mean that the band theory of solids is applicable to the ordered crystal and that traps exist within the bulk.

The Einstein equation $D/\mu = kT/eV$ where D is the diffusion co-efficient may often be used to obtain an approximate value of the mobility of charge carriers. For most polymers a typical value is $D = 1 \times 10^{-12}$ m^2 s^{-1} and substituting $k = 1.38 \times 10^{-23}$ J/K, e = 1.6×10^{-19}C and T = 300 K we get $\mu = 4 \times 10^{-11}$ m^2 $V^{-1}s^{-1}$ which is in the range of values given in Table 7.2.

Table 7.2

Mobility of charge carrier in polymers [Seanor, 1972]

polymer	Mobility ($\times 10^{-8}$ m^2 V s^{-1})	Activation energy (eV)
Poly(vinyl chloride)	7	
Acrylonitrile vinylpyridine copolymer	3	
Poly-N-inyl carbazole	$10^{-3} - 10^{-2}$	$0.4 - 0.52$
Polyethylene	10^{-3}	0.24 (Tanaka, 1973)
Poly(ethylene terephthalate)	1×10^{-2}	0.24 (Tanaka, 1973)
Poly(methyl methacrylate)	2.5×10^{-7}	0.52 ± 0.09
Commercial PMMA	3.6×10^{-7}	0.48 ± 0.09
Poly-n- butyl-methacrylate	2.5×10^{-6}	0.65 ± 0.09
Lucite	3.5×10^{-9}	0.52 ± 0.09
Polystyrene	1.4×10^{-7}	0.69 ± 0.09
Butvar	4.85×10^{-7}	0.74 ± 0.09
Vitel	4.0×10^{-7}	1.08 ± 0.13
polyisoprene	2.0×10^{-8}	1.08 ± 0.13
Silicone	3.0×10^{-10}	1.73 ± 0.17
Poly(vinyl acetate)		
Below T_G	2.2×10^{-8}	0.48 ± 0.09
Above T_G		1.21 ± 0.09

(with permission from North Holland Publishing Co.)

This brief discussion of mobility may be summarized as follows. If the mobility of charge carriers is greater than 5×10^{-4} m^2 V s^{-1} and varies as T^{-n} the band theory may be applied. Otherwise we have to invoke the hopping model or tunneling between localized states as the charge spends more time in localized states than in motion. The temperature dependence of mobility is according to exp $(-E_\mu / kT)$. If the charge carrier spends more time at a lattice site than the vibration frequency the lattice will have time to relax and within the vicinity of the charge there will be polarization. The charge is called a **polaron** and the hopping charge to another site is called the hopping model of conduction. Methods of obtaining mobilities and their limitations have been commented upom by Ku and Lepins (1987), and, Hilczer and Małecki (1986). Table 7.3 shows the wide range of mobility reported in polyethylene.

Table 7.3
Selected Mobility in polyethylene[3]

Mobility ($\times 10^{-8}$ m^2 V s^{-1})	Author
1.0×10^{-7} (20°C)	Wintle (1972)[1]
1.6×10^{-5} (70°C)	Davies (1972)[2]
2.2×10^{-4} (90°C)	Davies (1972)[3]
500	Tanaka (1973)[4]
10 to 1×10^{-5}	Tanaka and Calderwood (1974)[5]
1.0×10^{-7}	Pélissou et. al. (1988)[6]
1.0×10^{-8} (20°C)	Nath et. al. (1990)[7]
4.2×10^{-7} (50°C)	Lee et. al. (1997)[8]
2.3×10^{-6} (70°C)	Lee et. al. (1997)

Glarum has described trapping of charge carriers in a non-polar polymer[4]. The charge moves in the conduction band along a long chain as far as it experiences the electric field. At a bend or kink if there is no component of the electric field along the chain, the charge is trapped as it cannot be accelerated in the new direction. The trapping site is effectively a localized state and the charge stops there, spending a considerable amount of time. Greater energy, which may be available due to thermal fluctuations, is required to release the charge out of its potential well into the conduction band again. In the trapped state there is polarization and therefore some correspondence is expected between conductivity and the dielectric constant as shown in Fig. 7.1.

---

[1] H. J. Wintle, J. Appl. Phys., **43** (1972) 2927).
[2] Quoted in Tanaka and Calderwood (1974).
[3] Quoted in Tanaka and Calderwood (1974).
[4] T. Tanaka, J. Appl. Phys., **44** (1973) 2430.
[5] T. Tanaka and J. H. Calderwood, **7** (1974) 1295
[6] S. Pélissou, H. St-Onge and M. R. Wertheimer, IEEE Trans. Elec. Insu. 23 (1988) 325.
[7] R. Nath, T. Kaura, M. M. Perlman, IEEE Trans. Elec. Insu. **25** (1990) 419
[8] S. H. Lee, J. Park, C. R. Lee and K. S. Luh, IEEE Trans. Diel. Elec. Insul., **4** (1997) 425

A brief comment is appropriate here with regard to the low values of mobility shown in Table 7.2. The various energy levels in a dielectric with traps are shown in Fig. 7.2. For simplicity only the traps below the conduction band are shown. The conduction band and the valence band have energy levels E_c and E_v respectively. The Fermi level, E_F, lies in the energy gap somewhere in between the conduction band and valence band. Generally speaking the Fermi level is shifted towards the valence band so that $E_F < \frac{1}{2}$ $(E_c - E_v)$. We have already seen that the Fermi level in a metal lies in the middle of the two bands, so that the relation $E_F = \frac{1}{2}$ $(E_c - E_v)$ holds. The trap level assumed to be the same for all traps is shown by E_t and the width of trap levels is $\Delta E_t = E_c - E_t$.

Using Fermi-Dirac statistics the ratio of the number of free carriers in the conduction band, n_c, and in the traps, n_t is obtained as [Dissado and Fothergill, 1992]

$$\frac{n_c}{n_t} = \frac{N_{eff}}{N_t} \exp\left(-\frac{E_c - E_t}{kT}\right)$$

(7.8)

where N_{eff} and N_t is the effective number density of states in the conduction band, and the number density of states in the trap level, respectively. The ratio

$$\frac{n_c}{n_t} \approx \frac{n_c}{n_c + n_t}; \qquad n_t \gg n_c$$

(7.9)

is the fraction of charge carriers that determines the current density. Obviously the current will be higher without traps as the ratio will be unity. This ratio will be referred to in the subsection (7.4.6) on space charge limited current in insulators with traps.

Equation (7.8) determines the conductivity in a solid with traps present in the bulk. The change in conductivity due to a change in temperature, T, may be attributed to a change in mobility by invoking a thermally activated mobility according to

$$\mu = \mu_0 \exp\left(-\frac{E_c - E_t}{kT}\right)$$

(7.10)

In an insulator it is obvious that the number of carriers in the conduction band, n_c, is much lower than those in the traps, n_t, and the ratio on the left side of equation (7.8) is in the range of 10^{-6} to 10^{-10}. The mobility is 'unfairly' blamed for the resulting reduction in the current and the mobility is called **trap limited**. We will see later, during the discussion of space charge currents that this blame is balanced by crediting mobility for an increase in current by calling it **field dependent.**

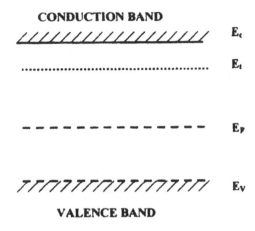

CONDUCTION BAND

E_c

E_t

E_p

E_v

VALENCE BAND

Fig. 7.2 A simplified diagram of energy levels with trap energy being closer to conduction level.

There is a minimum value for the mobility for conduction to occur according to the band theory of solids. Ritsko[5] has shown that this minimum mobility is given by

$$\mu_{min} = \frac{2\pi e a^2}{8h} \tag{7.11}$$

where a is the lattice spacing. For a spacing of the order of 1 nm the minimum mobility, according to this expression, is $\sim 10^{-4}$ m^2 Vs^{-1} which is about 6-10 orders of magnitude higher than the mobilities shown in Table 7.1. Dissado and Fothergill (1992) attribute this to the fact that transport occurs within interchain of the molecule rather than within intrachain.

The mobilitiy of electrons in polymers is $\sim 10^{-10}$ m^2 V^{-1}s^{-1} and at electric fields of 100 MV/m the drift velocity is 10^{-2} m/s. This is several orders of magnitude lower than the r.m.s. speed which is of the order of 10^5 m/s.

7.3 IONIC CONDUCTION

While the above simple picture describes the electronic current in dielectrics, traps and defects should be taken into account. For example in ionic crystals such as alkali halides the crystal lattice is never perfect and there are sites from which an ion is missing. At sufficiently high electric fields or high temperatures the vibrational motion of the neighboring ions is sufficiently vigorous to permit an ion to jump to the adjacent site. This mechanism constitutes an ionic current.

Between adjacent lattice sites in a crystal a potential exists, and let ϕ be the barrier height, usually expressed in electron volts. Even in the absence of an external field there will be a certain number of jumps per second of the ion, from one site to the next, due to thermal excitation. The average frequency of jumps v_{av} is given by

$$v_{av} = \frac{1}{av^2}\left(\frac{kT}{\hbar}\right)^3 \exp\left(-\frac{\phi}{kT}\right) = v_0 \exp\left(-\frac{\phi}{kT}\right) \tag{7.12}$$

where v is the vibrational frequency in a direction perpendicular to the jump, a the number of possible directions of the jump and the other symbols have their usual meaning. v is approximately 10^{12} Hz and substituting the other constants the pre-exponential factor comes to $\sim 10^{16}$ Hz. An activation energy of $\phi = 0.2$ eV gives the average jump frequency of $\sim 10^{11}$ Hz.

In the absence of an external electric field, equal number of jumps occur in every direction and therefore there will be no current flow. If an external field is applied along x-direction then there will be a shift in the barrier height. The height is lowered in one direction by an amount $eE\lambda$ where λ is the distance between the adjacent sites and increased by the same amount in the opposite direction (Fig. 7.3). The frequency of jump in the +E and −E direction is not equal due to the fact that the barrier potential in one direction is different from that in the opposite direction. The jump frequency in the direction of the electric field is:

$$v_{av \to E} = v_0 \exp\left(-\frac{\phi}{kT}\right)\exp\left(\frac{+E}{2kT}e\lambda\right) \tag{7.13}$$

In the opposite direction it is

$$v_{av \leftarrow E} = v_0 \exp\left(-\frac{\phi}{kT}\right)\exp\left(\frac{-E}{2kT}e\lambda\right) \tag{7.14}$$

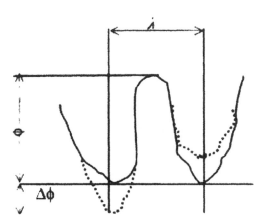

Fig. 7.3 Schematic illustration of the lowering of the barrier height due to the electric field.

The net jump frequency is

$$V_{net} = V_{av \to E} - V_{av \leftarrow E} \tag{7.15}$$

Substituting equations (7.13) and (7.14) into this equation we get

$$V_{net} = 2V_0 \exp\left(-\frac{\phi}{kT}\right) \sinh\frac{eE\lambda}{2kT} \tag{7.16}$$

The drift velocity of the carrier is $V_{net} \times \lambda$ and expressed as

$$W_d = 2V_0\lambda \exp\left(-\frac{\phi}{kT}\right) \sinh\frac{eE\lambda}{2kT} \tag{7.17}$$

This equation may be expressed as

$$E\mu(E) = 2V_0 \lambda \exp(-\frac{\phi}{kT}) \sinh(\frac{eE\lambda}{kT}) \tag{7.18}$$

leading to

$$\mu(E) = \frac{2v_0 \lambda}{E} \exp(-\frac{\phi}{kT}) \sinh(\frac{eE\lambda}{kT})$$

(7.19)

$$= \frac{2\mu_0}{E} \sinh(\frac{eE\lambda}{kT})$$

This expression is referred to as **field dependent mobility** though μ_0 is independent of the electric field. If the electric field satisfies the condition $eE\lambda << kT$ ($E < 10$ MV/m) then we can substitute in equation (7.17) the approximation

$$Sinh\frac{eE\lambda}{kT} \approx \frac{eE\lambda}{kT}$$

(7.20)

and equation (7.17) may be rewritten as

$$W_d = \frac{e\,E\lambda^2 v_0}{kT} \exp\left(-\frac{\phi}{kT}\right)$$

(7.21)

The current density is equal to

$$J = NeW_d = Nv_0 \frac{Ee^2\lambda^2}{kT} \exp\left(-\frac{\phi}{kT}\right)$$

(7.22)

where N is the number of charges per unit volume. The current is proportional to the electric field, or in other words, Ohm's law applies.

If the electric field is high then the approximation $eE\lambda << kT$ is not justified and in strong fields the forward jumps are much greater in number than the backward jumps and we can neglect the latter. The current then becomes[6, 7]

$$J = Nev_{\to F}\lambda = j_0 \exp\left(-\frac{\phi}{kT} + \frac{Ee\lambda}{2kT}\right)$$

(7.23)

where the pre-exponential is given by

$$J_0 = \frac{e^2 \lambda^2 N (kT)^2}{h^3 v^2}$$ (7.24)

In practice it is sufficiently accurate to express the current density using equation (7.17) as

$$J = 2Nev_0 \lambda \exp\left(-\frac{\phi}{kT}\right) \sinh\left(\frac{eE\lambda}{kT}\right)$$ (7.25)

To demonstrate the relative values of the two terms in brackets in the right side of eq. (7.25) we substitute typical values; $\phi = 1.0$ eV, T = 300 K, E = 100 MV/m, $\lambda = 10$ nm, $k = 8.617 \times 10^{-5}$ eV/K $= 1.381 \times 10^{-23}$ J/K. We get $\phi/kT = 38.68$ and $Ee\lambda/2kT = 19.33$. Therefore both terms have to be considered in equation (7.23) for calculating the current and neglecting the first term, as found in some publications, is not justified.

Dissado and Fothergill (1992) point out the characteristics of low field conduction as follows:
1. At room temperature, 293 K, very low mobilities are encountered. Typical values are $\lambda = 0.2$ nm, $v = 2 \times 10^{13}$ s^{-1}, $\phi = 0.2$ eV, the mobility is 4.9×10^{-9} m^2 V^{-1}s^{-1}. If ϕ increases to 1.0 eV the mobility is 1.8×10^{-24} m^2 V^{-1}s^{-1}.
2. If ionic conduction occurs there will be a build up of charges at the electrodes and initially the current will decay for an applied step voltage.

Experimental evidence reveals that the distance between sites changes in very strong fields. The difficulty in verifying ionic conduction is that the current may also be due to electrons. Transfer of charge and mass characterizes ionic current. Further the space charge created will induce polarization and uneven distribution of the potential. The current decreases with time under space charge limited conditions. The activation energy for ion transport is several eV higher than that for electron current which is usually less than one eV.

Not withstanding these differences it is quite difficult to distinguish the electron current from ion current due to the fact that space charge and polarization effects are similar. Activation energies and mobilities are also likely to be in the same range for both mechanisms. Fig. 7.4[8] shows a summary of the range of temperature where the ionic conduction mechanism occurs at low electric fields.

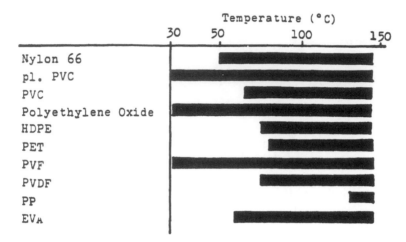

Fig. 7.4 Range of temperature for observing ionic conduction in polymers (Mizutani and Ida, 1988, with permission of IEEE).

7.4 CHARGE INJECTION INTO DIELECTRICS

Several mechanisms are possible for the injection of charges into a dielectric. If the material is very thin (few nm) electrons may tunnel from the negative electrode into unoccupied levels of the dielectric even though the electric field is not too high. Field emission and field assisted thermionic emission also inject electrons into the dielectric. We first consider the tunneling phenomenon.

7.4.1 THE TUNNELING PHENOMENON

In the absence of an electric field there is a certain probability that electron tunneling takes place in either direction. If a small voltage is applied across a rectangular barrier there will be a current which may be expressed as[9]

$$J = \frac{e^2 (2m\phi)^{1/2}}{h^2 s} V \exp\left[-\frac{4\pi s}{h}(2m\phi)^{1/2} \right] \tag{7.26}$$

where ϕ is the barrier height above the Fermi level, m the mass of an electron, s, the width of the barrier.

For high electric fields, $E = V/s$ and the current is given as

$$J = \frac{e^2 E^2}{8\pi\, h\phi}\exp\left[-\frac{8\pi}{3heE}(2m)^{1/2}\phi^{3/2}\right] \tag{7.27}$$

The current is independent of the barrier height at constant electric field. The theory has been further refined to include barriers of arbitrary shapes as rectangular shape is a simplified assumption. An equivalent barrier height is defined as

$$\phi = \frac{1}{\Delta s}\int_{s_1}^{s_2}\phi(x)\,dx$$

where $\Delta s = s_2 - s_1$. The current density is given by

$$j = \frac{6.2\times10^{14}}{(\Delta s)^2}\left\{\begin{array}{l}\phi\exp(-0.1\Delta s\,\phi^{1/2})\\[4pt]\qquad-(\phi+V)\exp\left[-0.1\Delta s(\phi+V)^{1/2}\right]\end{array}\right\} \tag{7.28}$$

where the barrier width Δs is expressed in nm, the voltage V in Volts, ϕ in eV and the current density in A/m^2. Equation (7.28) is independent of temperature and a small correction is applied by using the expression (Dissado and Fothergill, 1992)

$$\frac{J(V,T)}{J(V,0)} = 1 + \frac{3\times10^{-11}(\Delta s\,T)^2}{\phi} \tag{7.29}$$

where the left side is the ratio of the current density at temperature T to that at absolute zero. At T = 300 K the second term on the right side is

$$\frac{3\times10^{-11}(1.75\times300)^2}{1.5} = 5.5\times10^{-6}$$

which is negligible with respect to one as far as measurement accuracy is concerned. Marginal dependence of the conduction current on the temperature is often considered as evidence for tunneling mechanism.

A few comments with regard to the measurement of current through thin dielectric films are appropriate here. It is generally assumed that equation (7.27) is applicable and the measured currents are used to derive the thickness of the film. The thickness can also be derived from the capacitance of the sample. If the thickness is obtained this way and substituted in equation (7.27) the calculated currents are lower than measured ones by several orders of magnitude. One of the reasons for the discrepancy is that the thickness of the film may not be uniform. The current depends on this quantity to a considerable extent and in reality the current flows through only the least thick part and therefore the current density may be larger.

7.4.2 SCHOTTKY EMISSION

Thermionic field emission has been referred to in Ch. 1 and an applied electric field assists the high temperature in electron emission by lowering the potential barrier for thermionic emission. The carrier injection from a metal electrode into the bulk dielectric is governed by the Schottky theory

$$J = AT^2 \exp\left[-\frac{\phi - \beta E^{1/2}}{kT}\right] \tag{7.30}$$

where A, a constant independent of E and β, is given by

$$\beta = \left[\frac{e^3}{4\pi\varepsilon_0\varepsilon_\infty}\right]^{1/2} \tag{7.31}$$

and, ε_0 and ε_∞ are the permittivity of free space and relative permittivity at high frequencies. Theory [Dissado and Fothergill, 1992] shows that constant A is given by 4π $e\, m\, k^2/h^3$ and substitution of the constants gives a value of $\sim 1.2 \times 10^6$ A/m^2 K. The expression for current (7.28) consists of the product of two exponentials; the first exponential is the same as in Richardson equation for thermionic emission. The second exponential represents the decrease of the work function due to the applied field, and a lowering of the potential barrier.

$$A = \frac{4\pi\, e\, m\, k^2}{h^3} \tag{7.32}$$

According to equation (7.30) a plot of ln (J/T^2) versus $E^{1/2}$ gives a straight line. The intercept is the pre-exponential and the slope is the term in brackets. In practice the straight line will only be approximate due to space charge and surface irregularities of the cathode, particularly at low electric fields. Lewis[10] found that the pre-exponential term is six or seven orders of magnitude lower than the theoretical values, possibly due to formation of a metal oxide layer of 2 nm thick. The probability of crossing the resulting barrier explains the discrepancy.

Miyoshi and Chino[11] have measured the conduction current in thin polyethylene films of 50-120 nm thickness and their experimental data are shown in fig. 7.5. The measured resistivity is compared with the calculated currents, both according to Schottky theory, equation (7.30), and the tunneling mechanism, equation (7.27). Better agreement is obtained with Schottky theory, the tunneling mechanism giving higher currents. Lily and McDowell[12] have reported Schottky emission in Mylar.

7.4.3 HOPPING MECHANISM

Hopping can occur from one trapping site to the other; the mechanism visualized is that the electron acquires some energy, but not sufficient, to move over a barrier to the next higher energy. Tunneling then takes place, and from the probabilities of tunneling and thermal excitation the conductivity is obtained as:

$$\sigma = A \exp\left[-\frac{B}{T^n}\right] \tag{7.33}$$

where A is a constant of proportionality and $\frac{1}{4} \leq n \leq \frac{1}{2}$.

7.4.4 POOLE-FRENKEL MECHANISM

The Poole-Frenkel mechanism[13] occurs within the bulk of the dielectric where the barrier between localized states (Fig. 7.6) can be lowered due to the influence of the high electric field. For the mechanism to occur the polymer must have a wide band gap and must have donors or acceptors. Because the localized states of the valence band do not overlap those in the conduction band, the donors or acceptors do not gain enough energy to move into the conduction band or valence band, unlike in normal semiconductors. In terms of the energy band the donor levels are several kT below the conduction band and the acceptor levels are several kT above the valence band. We essentially follow the treatment of Dissado and Fothergill (1992).

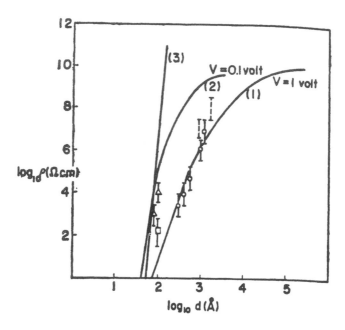

Fig. 7.5   Relationship of specific resistivity and thickness for thin polyethylene films. I=Theoretical Schottky curve (experimental values at 1V); II=Theoretical Schottky curve (experimental values at 0.1V); III-Theoretical tunneling curve. (Miyoshi and Chino, 1967, with permission of Jap. Jour. Appl. Phys.)

For the sake of simplicity we assume that the insulator has only donors as the treatment is similar if acceptors, or both donors and acceptors, are present. As mentioned above, we further assume that there are no thermally generated carriers in the conduction band and the only carriers that are present are those that have moved from donor states due to high electric field strength.
Let

N_D = Number of donor atoms or molecules per m^3.
N_0 = Number of non-donating atoms or molecules per m^3.
N_c = Number of electrons in the conduction band.

Then the relationship

$$N_c = N_D - N_0 \tag{7.34}$$

holds assuming that each atom or molecule donates one electron for conduction. As an electron is ionized it moves away from the parent atom but a Coloumb force of attraction exists between the electron and the parent ion given by

$$F = \frac{-e^2}{4\pi \varepsilon_0 \varepsilon_r r^2} \tag{7.35}$$

where ε_r is the dielectric constant and r the distance between charge centers.

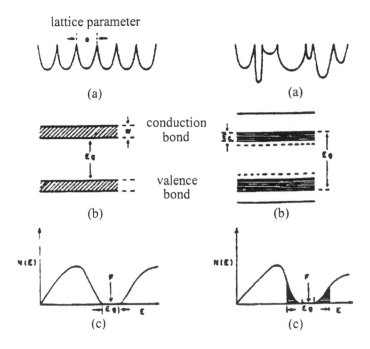

Fig. 7.6 Band formulation for (1) ordered and (2) disordered systems. (a) Potential wells; (b) band structure; (c) density of states; F = Fermi level; W = bandwidth; E_c = critical energy for band motion; E_g = forbidden energy gap. The shaded areas in 2(b) and 2(c) represent localized states. (Seanor, 1972, with permission of North Holland Publishing Co., Amsterdam.)

The potential energy associated with this force is

$$V(r) = \frac{-e^2}{4\pi \varepsilon_0 \varepsilon_r r} \tag{7.36}$$

Note that $V(r) \to 0$ as $r \to \infty$ and the dimension of the ionized donor assumes importance. The electric field changes the potential energy to

$$V(r) = \frac{-e^2}{4\pi\,\varepsilon_0\,\varepsilon_r\,r} - e\,E\,r \tag{7.37}$$

To find the condition for the potential energy to be maximum we differentiate equation (7.37) and equate it to zero

$$\frac{dV(r)}{dr} = \frac{e^2}{4\pi\varepsilon_0\,\varepsilon_r\,r^2} - e\,E = 0 \tag{7.38}$$

Solving for r and noting the maximum by r_m

$$r_m = \left(\frac{e}{4\pi\,\varepsilon_0\,\varepsilon_r\,E}\right)^{1/2} \tag{7.39}$$

The change in the maximum height of the barrier with and without E is

$$\Delta V_m = e\,E\,r_m = -\left(\frac{e^3 E}{4\pi\,\varepsilon_0\,\varepsilon_r}\right)^{1/2} \tag{7.40}$$

The interchange of electrons between the donor states and conduction band is in dynamical equilibrium and therefore N_c and N_o are constants. The rate of thermal excitation of electrons is, therefore, equal to the **rate of capture** of electrons by the donors.

The rate of thermal excitation of electrons to conduction band is given by

$$R_e(T) = N_0 v_0 \exp\left(-\frac{\phi_{eff}}{kT}\right)^{1/2} \tag{7.41}$$

where v_0 is the attempt to escape frequency and ϕ_{eff} is the reduced barrier height

$$\phi_{eff} = \varepsilon_D - \Delta V_m \tag{7.42}$$

where ε_D is the barrier height in the absence of the electric field.

The rate of capture of electrons by donors =

Number density of conduction electrons (N_c) × Number density of ionized donors ($N_D - N_0$) × drift velocity of electrons (W_e) × capture cross section (S). viz.,

$$R_c = N_c(N_D - N_0)W_e S = N_c^2 W_e S \tag{7.43}$$

where W_e is the thermal velocity of electrons in the conduction band. Under equilibrium conditions, we have the rate of excitation equal to the rate of recombination

$$N_c^2 W_e S = N_0 v_0 \exp\left(-\frac{\phi_{eff}}{kT}\right) \tag{7.44}$$

The jump frequency v_0 is approximately equal to

$$v_0 \approx N_{eff} W_e S \tag{7.45}$$

where N_{eff} is the effective density of charge carriers in the conduction band. Substituting equations (7.34) and (7.45) in (7.44) gives

$$N_c^2 W_e S = N_{eff} W_e S(N_D - N_c)\exp\left(-\frac{\phi_{eff}}{kT}\right) \tag{7.46}$$

In dielectrics the number density of conduction electrons is usually quite small when compared with the number density of donors, and we can make the approximation that $N_D \gg N_c$. Equation (7.46) then gives

$$N_c = \sqrt{N_{eff} N_D}\, \exp\left(-\frac{\phi_{eff}}{2kT}\right) \tag{7.47}$$

Substituting equation (7.42) and (7.40) in (7.47) gives

$$N_c = \left(N_{eff} N_D\right)^{1/2} \exp\left(-\frac{\varepsilon_d}{2kT}\right)\exp\left(\frac{e^{3/2}E^{1/2}}{(4\pi\,\varepsilon_0\,\varepsilon_r)^{1/2}\,kT}\right) \tag{7.48}$$

The conductivity $\sigma = N_c e \mu$ may now be expressed as

$$\sigma = \left(N_{eff} N_D\right)^{1/2} e\mu \exp\left(-\frac{\varepsilon_D}{2kT}\right)\exp\left(\frac{e^{3/2}E^{1/2}}{(4\pi\,\varepsilon_0\,\varepsilon_r)^{1/2}\,kT}\right) \tag{7.49}$$

This is the Poole-Frenkel expression from which the current may be calculated for a given electric field. A plot of logσ versus $E^{1/2}$ yields a straight line with slope M_{pf} and intercept C_{pf}:

$$M_{pf} = \frac{e^{3/2}}{kT(4\pi\varepsilon_0\varepsilon_r)^{1/2}}; \quad C_{pf} = e\mu \log(N_{eff}N_D)^{1/2} \tag{7.50}$$

The same plot of log J verses $E^{1/2}$ is also used as for determining whether the Schottky emission occurs and to resolve between the two mechanisms we have to apply additional tests as explained below.

The conduction mechanism generally does not remain the same as the temperature is raised and it is not unusual to find two mechanisms operating simultaneously, though one may predominate over the other depending on the experimental conditions. This fact has been demonstrated by the recent conduction current measurements in isotactic (iPP) and syndiotactic polypropylene (sPP)[14]. The current in pristine iPP is observed to be larger than in crystallized iPP possibly because impurities and uncrystallizable components collect at the boundaries of large spherulites. The relatively low breakdown field ($\sim$ 20 MV m^{-1}) of iPP compared with sPP is thought to support for this reasoning. In contrast sPP, which shows smaller spherulites, does not show appreciable dependence of conductivity on the electric field. The influence of electric field on current in sPP is shown in fig. 7.7. Ohmic conduction is observed at field strengths below 10 MV/m, and for higher fields the current increases faster.

Schottky injection mechanism which is cathode dependent may be distinguished from Poole-Frenkel mechanism which is bulk dependent in the following way. From the slope of log I versus $E^{1/2}$ the dielectric constant ε_r is calculated according to relationship (7.30) and compared with the dielectric constant measured with bridge techniques. The theoretical dielectric constant according to Poole-Frenkel mechanism is four times that due to Schottky emission. These considerations lead to the conclusion that at low temperatures, < 70°C, Schottky emission is applicable whereas at higher temperatures ion transport is the most likely mechanism.

The hopping distance may be calculated by application of eq. (7.25) and a hopping distance of approximately 3.3 *nm* is obtained in sPP. Hopping distances of 6.5 *nm* and 20 *nm* have been reported[15, 16] in bi-axially oriented and undrawn iPP respectively. The molecular distance of a repeating unit in PP is 0.65 nm [Foss, 1963] and therefore the ionic carriers jump an average distance of five repeating units. The barrier height obtained from eq. (7.25) is 0.82 eV for PP which means that the energy is not adequate to facilitate ionic motion below 70°C.

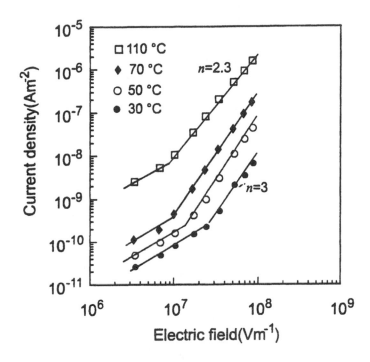

Fig. 7.7 Electric field dependence of current density at various temperatures in sPP (Kim and Yoshino, 2000, with permission of J. Phys. D: Appl. Phys.)

Another example of non-applicability of Poole-Frenkel mechanism in spite of linear relationship between $\log\sigma$ and $E^{1/2}$ in linear low density polyethylene [LLDPE] is shown in fig. (7.8)[17]. From the slopes of the plots the dielectric constant, ε_∞, is obtained as 12.8 which is much higher than the accepted value of 2.3 for PE. A three dimensional analysis of the Poole-Frenkel mechanism has been carried out by Ieda et. al.[18] who obtain a factor of two in the denominator of the second exponential of eq. (7.49). Application of their theory gives $\varepsilon_\infty = 3.2 \pm 0.3$ which is still considered to be unsatisfactory. Schottky theory gives $\varepsilon_\infty = 0.96$ which is obviously wrong. Nath et. al.[19] suggest an improvement to the space charge limited current which predicts a straight line relationship between plots of $\log(I/T^{2})$ versus $1/T$ at constant electric field. The parameters obtained for LLDPE are: activation energy 0.83 eV, trap separation distance = 2.8 nm, trap concentration = 4.5×10^{25} m^3.

The Poole-Frenkel expression may be simplified by neglecting the density of carriers that move against the electric field. The current density is then expressed as[20]

$$J = \frac{ev\lambda}{2}(N_cN_t)^{1/2}\exp\left[\left(-\frac{1}{2kT}\right)\left\{\phi - 2(1.25\alpha)^{1/2}\left(\frac{2.43\alpha}{\lambda^2} + eE\right)^{1/2}\right\}\right] \qquad (7.51)$$

where α is a constant equal to $e^2/(4\pi\,\varepsilon_0\,\varepsilon_r)$ and all other symbols have already been defined. Substituting the following values a current density of 3.4×10^{-9} A/m^2 is obtained:

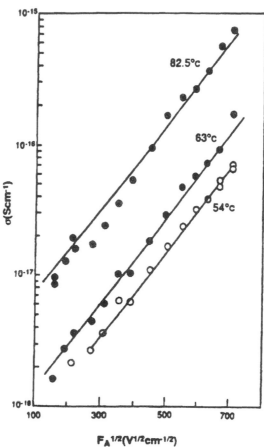

Fig. 7.8 Typical Poole-Frenkel plots (log σ vs. E$^{1/2}$) of the steady state conductivity of LLDPE (Nath etal 1990, with permission of IEEE).

$E = 4$ MV/m, $e = 1.6\times10^{-19}$ C, $\lambda = 2.5\times10^{-9}$ m, $v = 1\times10^{13}$ s^{-1}, $N_t = 10^{19}$ traps/m^3, $N_c = 10^{23}$ /m^3, $T = 300$K, $\phi = 1.9$ eV, $\varepsilon_r = 2.3$. This value is in reasonable agreement with measured current density.

7.4.5 SPACE CHARGE LIMITED CURRENT (TRAP FREE)

Charge injected from the electrode moves through the bulk and eventually reaches the opposite electrode. If the rate of injection is equal to the rate of motion charges do not accumulate in a region close to the interface and the electrodes are called Ohmic. Ohmic conductors are sources of unlimited number of charge carriers. If the mobility is low which is the normal situation with many polymers then the charges are likely to accumulate in the bulk and the electric field due to the accumulated charge influences the conduction current. A linear relationship between current and the electrical field does not apply anymore except at very low electric fields. At higher fields the current increases much faster than linearly and it may increase as the square or cube of the electric field. This mechanism is usually referred to as **space charge limited current (SCLC).**

A study of space charge currents yields considerable information about the charge carriers. In developing a theory for SCLC we assume that the charge is distributed within the polymer uniformly and there is only one type of charge carrier. In experiments it is possible to choose electrodes to inject a given type of charges and if both charges are injected from the electrodes recombination should be taken into account. With increased electric field the regions of space charge move towards each other within the bulk and coalesce. The number and type of charge carrier, and its mobility and thickness of space charge layer influence SCLC.

In gas discharges the mobility of positive ions is lower than that of electrons by 3-4 orders of magnitude. However in the solid state with traps this is not always the case. For example Kommandeur and Schneider[21] studied the effect of illumination on anthracene and found that the current was a function of illumination and also due to the differences in carrier mobility, though equal number of holes and electrons are generated. With illumination focused on the positive side the holes move rapidly to the negative electrode. When the negative side is illuminated electrons drift slowly creating space charge within the solid.

Mott and Gurney[22] first derived the current due to space charge limited current in crystals assuming that there were no traps. The following assumptions are made:
(1) The electrodes are ohmic and electrons are supplied at the rate of their removal. This means that there is no potential barrier at the electrode-insulator interface. This assumption yields the boundary condition

$$n(0) = \infty; \quad E(0) = 0 \tag{7.52}$$

(2) The current is a function of the number and drift velocity of electrons and not dependent on the position in the sample, z, measured from the cathode. Denoting the number of electrons and the electric field at z as n(z) and E(z) respectively, the current density is:

$$j = n(z)e\mu E(x) \neq f(z) \tag{7.53}$$

(3) Poisson's equation is applicable

$$\frac{d^2V}{dz^2} = \frac{en(z)}{\varepsilon_0 \varepsilon} \tag{7.54}$$

(4) There is no discontinuity of the electric field within the dielectric, that is

$$\int_0^d E(z)dz = V \tag{7.55}$$

The current through the dielectric is composed of three components: Drift, diffusion and displacement. Adding these contributions the current density is expressed as:

$$J = n(z)e\mu E(z) - eD\frac{dn}{dz} + \varepsilon_0 \varepsilon \frac{dE}{dt} \tag{7.56}$$

where D is the diffusion co-efficient and other symbols have already been defined. Under steady state conditions the last term is zero, and for injecting electrodes there is no accumulation of charges near the electrodes. Further, the space charge density within the bulk is assumed uniform so that the diffusion term on the right side of equation (7.56) may be neglected. Using equations (7.54) and (7.55) in (7.56) the current density is given by

$$J = n(z)e\mu E(z) = \varepsilon_0 \varepsilon \mu E(z)\frac{dE(z)}{dz} \tag{7.57}$$

This equation may be integrated and rewritten as

$$E = \left(\frac{2Jz}{\varepsilon_0 \varepsilon \mu}\right)^{1/2} + z_0 \tag{7.58}$$

where z_0 is an integration constant. At $z = 0$ the electric field is $E(0) = 0$ and therefore

$$E = \left(\frac{2Jz}{\varepsilon_0 \varepsilon \mu}\right)^{1/2}$$ (7.59)

Substituting this equation in (7.55) we get

$$V = \int_0^d (\frac{2J}{\varepsilon_0 \varepsilon \mu})^{1/2} z^{1/2} \, dz$$ (7.60)

where d is the thickness of the insulator. Integration gives the current density as

$$J = \frac{9\varepsilon_0 \varepsilon \mu}{8d^3} V^2$$ (7.61)

This equation is similar to Child's law for space charge due to thermionic emission in gases, though the exponent of V is 3/2 for gases. This equation is also referred to as **Mott and Gurney square law**. The current density is inversely dependent on thickness even though the electric field is maintained constant.

Differentiating equation (7.60) we get

$$\frac{dV}{dz} = \left(\frac{2J}{\varepsilon_0 \varepsilon \mu}\right)^{1/2} z^{1/2}$$ (7.62)

Substituting for J from equation (7.61) the electric field at any point within the dielectric is obtained as

$$E(z) = \frac{3V}{2d^{3/2}} z^{1/2}$$ (7.63)

The number density of charge carriers is given by

$$n(z) = \frac{3\varepsilon_0 \varepsilon V}{4ed^{3/2} z^{1/2}}$$ (7.64)

The electric field increases and the number of charge carriers decreases towards the positive electrode as shown by Fig. 7.9.

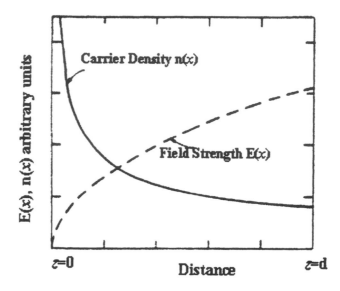

Fig. 7.9 Schematic variation of field strength, eq. (7.63) and carrier density, eq. (7.64) across the sample under trap free space charge limited current flow (Seanor 1972). (With permission of North Holland Co.)

At low voltages the current density is given by the Ohmic current

$$J = ne\mu E \tag{7.65}$$

At higher voltages the current is SCLC limited and given by equation (7.61). The slope of the plot of J versus V increases from unity to two though the transition is not as abrupt as the theory indicates. The voltage at which transition from Ohmic conduction to SCLC occurs is obtained by equating (7.65) and (7.61). The transition voltage is given as

$$V = \frac{8ed^2n}{9\varepsilon_0\varepsilon_r} \tag{7.66}$$

Experimentally the transition voltage is determined by the departure from linearity of the J versus V plot. The carrier density is then calculated using equation (7.64) and the mobility calculated from equation (7.65). Alternately μ may be obtained using equation (7.61).

7.4.6 SPACE CHARGE LIMITED CURRENT (WITH TRAPS)

We extend the concepts of the previous sub-section to an insulator in which the traps exist and investigate the nature of space charge limited current. The basic concept is that if traps are present the SCLC will be reduced by several orders of magnitude. Rose[23] proposed that neither the space charge density nor the field distribution is altered by trapping, but the current-voltage relationship should be modified by a trapping parameter θ, which is given by

$$\theta = \frac{n_c}{n_t + n_c} \tag{7.67}$$

and the current density due to SCLC is given by[24]

$$J = \theta \frac{9\varepsilon_0 \varepsilon \mu V^2}{8d^3} \tag{7.68}$$

for the trap free situation $n_t = 0$ and $\theta = 1$. With traps present the current is always lower than that without. The numbers n_c and n_t are given by

$$n_c = N_v \exp\left(-\frac{E_f - E_v}{kT}\right) \tag{7.69}$$

$$n_t = \frac{N_t}{1 + \exp[(E_f - E_t)/kT]} \tag{7.70}$$

where N_v is the density of states in the valence band and N_t the density of traps. The energy symbols have already been defined in fig. 7.2. For traps lying below E_f by several kT, that is, $(E_f - E_t)/kT > 1$ (shallow traps) equation (7.70) may be approximated to

$$n_t = N_t \exp\left(-\frac{E_f - E_t}{kT}\right) \tag{7.71}$$

The number density of occupied traps dominate and we can assume that $n_t > n_c$ and the ratio of n_c/n_t is given as

$$\theta = \frac{N_v}{N_t}\exp(-\frac{E_t - E_v}{kT})$$
(7.72)

Substituting for θ in equation (7.68) we get the current density as

$$J = \frac{9}{8}\varepsilon_0\,\varepsilon\,\mu\frac{N_v}{N_t}\frac{V^2}{d^3}\exp\left(-\frac{E_t - E_v}{kT}\right)$$
(7.73)

This is Child's law for an insulator with traps. Since μ is the mobility of free carriers an effective mobility may be defined as:

$$\mu_{eff} = \frac{N_v}{N_t}\exp\left(-\frac{E_t - E_v}{kT}\right)$$
(7.74)

Substituting this equation in (7.73) gives

$$J = \frac{9}{8}\varepsilon_0\varepsilon\mu_{eff}\frac{V^2}{d^3}$$
(7.75)

Note that equations (7.71) - (7.75) are expressed for the case of holes being the carriers and the trap levels are counted from the valence band.

Reverting back to the assumption that the free carriers are electrons we recall the assumption that $E_c - E_F \gg kT$, and this condition is called the shallow traps. As the voltage is increased the current increases rapidly and at sufficiently high voltages the number density of injected electrons is approximately equal to the density of traps. As the last traps are being filled θ approaches unity rapidly, the current increases (equation (7.68)), and the Fermi level will rise above the bottom of the trap energy level.

The number density of occupied trap sites approaches the number density of trap levels, $n_t \rightarrow N_t$ in equation (7.71), and the condition of all traps filled is called **trap filled voltage limit** (TFVL). We refer to this voltage as V_{TFL}. The rapid increase of current can easily be misunderstood as breakdown but this can be verified by lowering the voltage and short circuiting the electrodes for a considerable time, ~2-3 hrs, and in the case of TFVL the currents are reproduced with sufficient accuracy. Once this condition is reached a further increase in voltage exhibits a current dependence according to V^2 as predicted by Mott and Gurney, equation (7.61).

Under trap filled conditions the space charge is almost entirely due to the charges in the traps and the charge density is

$$\rho = e N_t \tag{7.76}$$

The voltage at which TFVL occurs is given by the Poisson's equation

$$\frac{d^2V}{dz^2} = \frac{e N_t}{\varepsilon_0 \varepsilon_r} \tag{7.77}$$

which has the simple solution

$$V_{TFL} = \frac{e N_t}{2\varepsilon_0 \varepsilon_r} d^2 \tag{7.78}$$

The trap density at the trap filled conditions is given by the above expression as

$$N_t = \frac{2\varepsilon_0 \varepsilon_r V_{TFL}}{e d^2} \tag{7.79}$$

A slightly different result is obtained if the trap density at V_{TFL} is calculated according to

$$N_t = \frac{\int_0^d n(z)\,dz}{\int_0^d dz} \tag{7.80}$$

Substituting for n(z) from equation (7.64) we get

$$N_t = \frac{3\varepsilon_0 \varepsilon V_{TFL}}{2 e d^2} \tag{7.81}$$

This equation differs from (7.79) only in the numerical constant.

It is not unusual to find that the distribution of traps is continuous with an exponential distribution (Ma, et al., 1999) instead of being discrete, and the number of traps may be expressed by invoking a characteristic temperature, T_t, according to

$$N_t(E) = \frac{N_t}{kT_t} \exp\left(-\frac{E}{kT_t}\right) \tag{7.82}$$

where $kT_t = E_t$ is the characteristic trap energy. The expression for current in the trap filled limited region is

$$J = N_c \, \mu_e \, e^{1-m} \left(\frac{2m+1}{m+1}\right)^{m+1} \left(\frac{\varepsilon_0 \varepsilon m}{N_t(m+1)}\right)^m \frac{V^{m+1}}{d^{2m+1}} \tag{7.83}$$

where μ_e is the electronic mobility. We can summarize the I –V characteristics in the condensed phase as follows. Four different characteristics are observed depending upon the mechanism that is operative (Fig. 7.10).

1. Region 1: Ohmic conduction due to thermally generated free carriers. Equation (7.4) applies. Slope of line is +1.
2. Region 2: Space charge limited current (SCLC) in trap free materials. Voltage is less than V_{TFL}. Equation (7.61) applies. Slope of the line is +2
3. Region 3: Trap filled limited current. Voltage is equal to V_{TFL}. Equation (7.73) applies. Slope of the line is >2.
4. Region 4: Trap filled space charge current. Voltage is greater than V_{TFL}. Equation (7.75) applies. Slope of the line is +2.

An example of current increasing very rapidly in the trap filled region is shown in Fig. 7.11, (Ma et. al., 1999) in a 136 nm electroluminiscent polymer with aluminum electrodes. The initial rise is linear due to Ohmic conduction followed by the trap free space charge current at higher voltages. Fig. 7.12 shows the conduction currents in thin PVAc films at voltages in the range of 4-300 V[25]. The four regions described above are clearly seen.

The parameters derived at T = 30°C are as follows: $\sigma = 6.82 \times 10^{-12} \ \Omega^{-1} \ m^{-1}$, $\theta = 2.8 \times 10^{-2}$, $\mu_0 = 2 \times 10^{-13} \ m^2 V^{-1} s^{-1}$, $\mu_e = \mu_0 \, \theta = 5.6 \times 10^{-15} \ m^2 V^{-1} s^{-1}$, $n_c = 1 \times 10^{20} \ m^3$, $n_t = 1 \times 10^{21} \ m^3$, $E_c - E_t = 0.17$ eV.

7.5 SPACE CHARGE PHENOMENON IN NON-UNIFORM FIELDS

The mobility of charge carriers is independent of the electric field in the ohmic region of conduction, and under space charge limited conditions, the mobility becomes field dependent as given by equation (7.19). A closer examination reveals that the drift velocity of charge carriers as a function of the electric field depends upon the electric field as shown in Fig. 7.13[26].

$$J = 6\pi \varepsilon \varepsilon_0 \mu n \frac{V_i^2}{(r_0^{0.5}[2 + \frac{r_i^3}{5r_0^3}] - 2.245 r_i^{0.5})^2}$$

(7.84)

Three regions are identified in the Log W_d versus Log E plot: (a) The low field Ohmic region, (b) the high field space charge region where the electric field is non-linear, and (c) the region of negative resistance. In a concentric spherical geometry with the inner and outer electrodes having radii of r_i and r_0 respectively it is assumed that the inner electrode is at V_1 and the outer electrode is at $V(r_0)=0$. Injection occurs from the inner sphere and the current density is given as (Zeller, 1987), for $r_0 >> r_i$, When r_0 is large the current is zero.

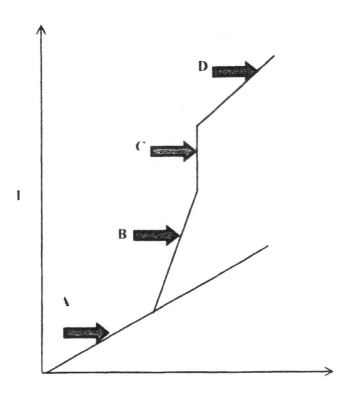

Fig. 7.10 I-V characteristics for space charge limited currents. A-Ohm's Law, B-Trap limited SCLC, C-Trap filled limit, D-Trap free SCLC. (Dissado and Fothergill, 1993.)

D Ma *et al*

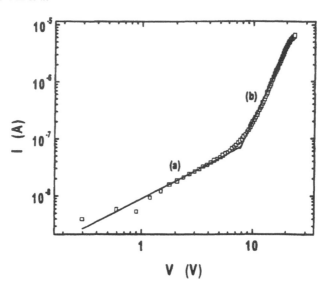

Fig. 7.11 Experimental (Open Square) and theoretical (full line) I-V characteristics of an Al/polymer/al device, (d=136 nm). The theoretical curve is obtained by using equation (7.4) and (7.83) at high voltages. The active area of the device is 4 mm2 (Ma et. al., 1999, with permission of J. Phys. D: Appl. Phys.)

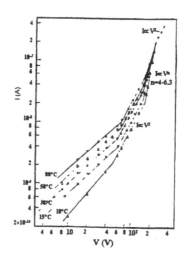

Fig. 7.12 I-V characteristics on a log-log scale for a typical film 1-2 mm at temperatures shown (Chutia and Barua, 1980, with permission of J. Phys. D: Appl. Phys.)

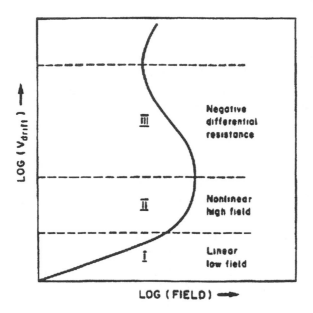

Fig. 7.13 The local field versus local drift velocity relation can be divided into three regions. The field, current and charge distributions for three regions are qualitatively different and discussed separately (Zeller, 1987, with permission of IEEE).

In region II the current increases rapidly with a field dependent mobility and the influence of space charge is present. In view of the non-uniformity of the electric field it is difficult to express the mobility as a function of field and only a qualitative description is possible.

In region III the I-V characteristics become unstable at a critical field, exhibiting a negative resistance. The current distribution becomes uneven in space leading to filamentary nature, and current pulses are observed in the external circuit. Zeller (1987) gives field relationships from which the spatial dependence of charge is calculated for homogeneous and filamentary space charge. The filamentary region is shown to extend far deeper in the inter-electrode gap and the volume of charge density is also higher for the filaments when compared with the homogeneous space charge.

7.6 CONDUCTION IN SELECTED POLYMERS

This section shall present experimental evidence for the applicability of the theories discussed above in some polymers. Several different schemes are available in general for determining the mechanism of conduction as shown in Fig. 7.13 and preferably each chosen scheme should be supplemented by a second scheme. This procedure also acts as a cross check on some conclusions that one is tempted to draw on a limited set of data. A brief description of Fig. 7.14 follows:

A. Current measurement involves application of a dc or a time varying voltage. Under dc the current is measured at constant temperature and this is known as **isothermal measurements**. Current measurement at the same time elapsed after application of voltage is called **isochronal measurements.** Either scheme may be employed depending upon the data acquisition technique. Absorption and desorption currents have been discussed in ch.6 and this chapter concerns transport currents. A study of thermally stimulated processes require measurement of **thermally stimulated discharge** (TSD) or **thermally stimulated polarization** (TSP) currents. These techniques and associated poling techniques such as thermal, corona, and X-ray poling, are discussed in ch. 10.

Conduction currents have been measured for several decades with conflicting results in the same material due to insufficient control of the experimental conditions or without realizing the importance of the morphology. Even experiments on single crystals often lead to different interpretations (Seanor, 1972). For example single crystals of polyethylene have demonstrated negative resistance and this observation has been reinterpreted as due to local heating. Thin films of poly (ethylene terephthalate) exhibit different charging characteristics depending upon the rate of crystallization.

Conductivity and activation energy of conductivity have been observed to decrease with increasing crystallinity in a number of polyesters. However, the more recent investigation of Tanaka et. al[27] shows that the increased crystallinity of low density polyethylene makes electron transport easier and the mean free path longer. The choice of electrode materials influences the parameters derived from the I-V characteristics. For example in light emitting polymers (Ma et. al, 1999) use of gold, palladium or nickel anode instead of copper exhibits the trap filled space charge limited currents.

In polypropylene the isotactic configuration has much higher conductivity when compared with the syndiotactic structure (Kim, 2000). The iPP has much larger spherulites and its higher conductivity is attributed to this factor. In contrast sPP has more minute spherulites improving its electrical characteristics. In polyamides conduction is electronic at low temperatures and protonic at higher temperatures. In polyethylene doped with carbonyl sites the electric field distribution is found to be independent of temperature in the temperature range of 49-82°C at constant applied field. The same characteristic is also observed in undoped polyethylene[28].

These brief comments are meant to stress the importance of specifying morphology, crystallinity, moisture content, heat treatment and electrode preparation in conductivity studies. They also indicate the reasons for the difficulty in obtaining repeatable results between different investigations, but also, often under apparently the same conditions in the same laboratory. A review of published literature on conduction mechanisms in polymers can be found in ref. [29, 30, 31] and we refer to these publications only as required.

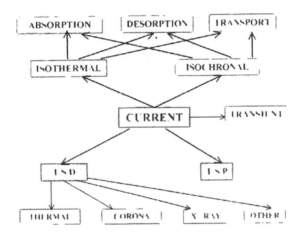

Fig. 7.14 Schemes for measuring current in polymers.

7.6.1 CONDUCTION IN POLYETHYLENE

Polyethylene is an extensively investigated material because of its simple structure and technological importance. Selected studies are shown, without claiming completeness, in Table 7.4. Fig. 7.15 shows the range of conductivity observed in PE[32] with some recent data added. The energy spectrum proposed by Tanaka[33] for LDPE is shown in Fig. (7.16). From photon absorption measurements the separation of conduction band from valence band is determined as 7.35 eV. Such large band gaps are characteristics of insulators and fairly extensive data are available in the literature for other polymers. The probability of an electron escaping to the conduction band by thermal excitation at room temperature is given by exp(- 7.35/ kT) ~ 10^{-124} and this never occurs.

The Fermi level is closer to the conduction band than to the valence band at 1.6 eV and donor levels exist 1.95 eV, 2.35 eV and 2.95 eV below the Fermi level. The probability of an electron moving to the conduction band due to thermal excitation from the 1.95 eV level at 80°C is ~ 10^{-28} which must be considered as negligible. Donors at such large energy levels are known as deep donors. Impurities, however, may have donors at closer levels increasing the probability. Electronic traps exist below the conduction band but above the Fermi level.

The conduction band has a tail with no sharp edge, but ending near about 1 eV below the conduction band. Electrons move by hopping from site to site in the tail of the conduction band to cause thermally activated mobility. The activation energy E_μ, not to be confused with donor level, is found to be 0.24 eV and μ_0 is found to have a value of 5×10^{-4} $m^2V^{-1}s^{-1}$ giving a mobility, at room temperature, that is too high by at least three order of magnitude. The difference cannot be ascribed to impurities as they are likely to increase the discrepancy. Tanaka, et. al.[34] have observed that increased crystallinity of LDPE makes electron transport easier, which may explain the discrepancy.

Another factor is that the pulse measurement gives a higher value of the mobility when compared with the steady state measurements [Tanaka 1973]. Kumar and Perlman (1992) point out the differences that should be taken into account while considering condction in LLDPE and HDPE. In LLDPE there is a lowering of the barrier for the Poole-Frenkel mechanism to be effective.

In HDPE a larger boundary concentration results in a greater number of trapped charges. The effective field is lower at the boundaries in the steady state, resulting in negligible field lowering. In LDPE mobility in amorphous regions is field independent. In HDPE a high degree of branching gives rise to high concentration of defects in the amorphous regions ~ 10^{26} m^{-3}. These defects become major hopping sites and the mobility is field dependent.

7.6.2 CONDUCTION IN FLUOROPOLYMERS

We consider poly(tetrafluoroethylene) and also a co-polymer of tetrafluoroethylene and ethylene, the latter having a trade name of Tefzel™ by Dupont Nemours Inc®. Both are high temperature materials with desirable mechanical and thermal properties.

Table 7.4
Electrical conduction in polyethylene
Units: $E = MV\ m^{-1}$, $\mu_0 = m^2\ V^{-1}s^{-1}$, $N_t = m^{-3}$, $\varepsilon = eV$. Unless otherwise stated, current – field measuring technique is employed.

Major Conclusions	Reference
Tunneling mechanism in single crystal polyethylene	Van Roggen[35] (1962)
Thermally activated ionic conduction, $\varepsilon_a = 0.84 – 1.06\ eV$, $N_t = 1.0 \times 10^{26}$	Lawson[36], (1965)
Space charge in bulk increases cathode-dielectric potential barrier, $E = 2.63- 13.2$, $\varepsilon_a = 1.11-1.45$	Taylor and Lewis[37] (1971)
Technique: Surface charge decay; hopping transport mechanism; Iodine doping increases μ significantly. $\mu_0 = 7 \times 10^{-4}- 5.8 \times 10^{2}$, $N_t = 1.2 \times 10^{19} –9 \times 10^{27.}$ $\varepsilon_a = 0.92 – 1.2$	Davies[38] (1972)
Poole-Frenkel mechanism in LDPE	Das-Gupta and Barbarez[39] (1973)
Deep donors at 1.95, 2.35 & 2.95 eV below conduction band, $\mu_0 = 5 \times 10^{-4}$, $E_\mu = 0.24$, Poole-Frenkel mechanism with hopping at $25°C \le T \le 80°C$	Tanaka[40] (1973)
Space charge limited current possibly with Poole-Frenkel effect	Tanaka & Calderwood[41], (1974)
Trapping of free charge carriers at $27°C \le T \le 90°C$, $\varepsilon_a = 1.0\ eV$	Adamec& Calderwood[42] (1978)
Technique: Absorption current; $E > 1\ MVm^{-1}$, Charge carrier hoping process in LDPE	Das-Gupta and Brockley[43] (1978)
Technique: Absorption and desorption currents, Space charge limited currents; distributed energy of traps	Pelissou et. al. (1988)[44]
Trapping of injected charge carriers, hopping due to field dependent mobility, $\lambda = 2.2\ nm$, $N_t = 9.4 \times 10^{25}$, $\varepsilon_a = 1.15\ eV$	Kumar and Perlman[45] (1992)
Semi-empirical theory in divergent fields	Boggs[46] (1995)
Ohmic conduction below 50°C and SCLC above, in LDPE. Removal of low molecular wt. Species increases σ	Lee et. al. (1997)[47]

(A) PTFE

Compared to polymers like polyethylene and PVAc the number of investigations published on PTFE are relatively few and a brief summary is in order. Lilly and

McDowell[48] measured conduction currents in 254 μm films at fields in the range of 7 – 12 MVm⁻¹, at just three temperatures of 162.5, 181.5 and 200°C using electrodes having an area of 46×10^{-4} m². The measured current is observed to vary with the field according to $\gamma E^{1/2}$ where γ is a geometric constant which is unity for absolutely parallel plane electrodes without any fringing effects (Table 7.5). They concluded that the mechanism was both ionic and electronic, each dominant at different electric field and each influenced by space charge.

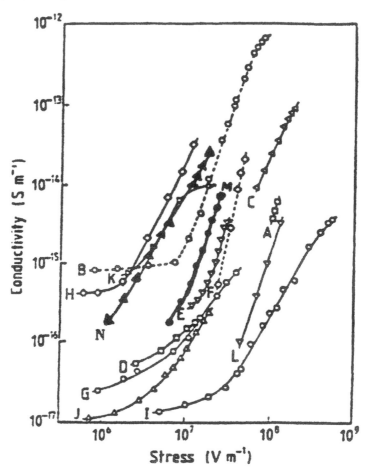

Fig. 7.15 Field strength dependence of conductivity in LDPE at about 50°C (dashed curves at room temperature according to: A, Lawson (1965); B, Stetter (1967),; C, Bradwell, et al (1971); D, Taylor and Lewis (1971); E, Das Gupta and Barbarez (1973); F, Cooper, et al (1973); G, Röhl and Fischer (1973); H, Beyer and Von Olshausen (1974); I, Garton and Parkmaan (1976); J, Johnsen and Weber (1978); K, Das Gupta and Brockley (1978); L, Adamec and Calderwood (1981); M, Nath and Perlman (1990), N, Lee, et al (1997). (Adamec and Calderwood, 1981; with permission of Inst. of Phys.), supplemented with the data of Nath and Lee.

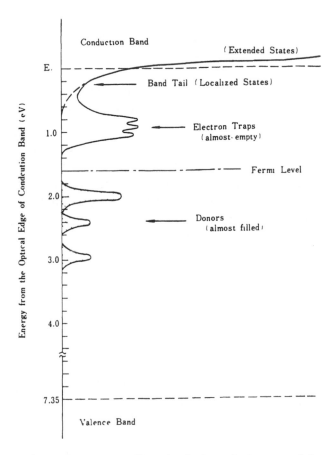

Fig. 7.16 Proposed energy spectrum for polyethylene. It does not claim to be quantitative for the density of states (Tanaka, 1973, with permission of J. Appl. Phys.)

Vollmann and Poll[49] investigated currents in radio frequency sputtered thin films 100 – 500 nm thick deposited by different methods on different electrode materials but the investigations were limited to room temperature. Over the whole range of electric fields and film thickness investigated, no dependence of current on the electrode material is observed. Method of polymerization also does not influence the currents. The results are interpreted in terms of a modified Poole–Frenkel effect for dielectrics with high impurity content.

Fig. 7.17 shows the currents measured by Raju and Sussi[50] in PTFE in the temperature range of 40 – 200°C. For $E \leq 4$ MVm^{-1} and $T \leq 100$°C Ohmic conduction is observed in agreement with Vollmann and Poll (1975) in the same temperature range. For higher electric field strengths ionic conduction, according to equation (7.25), shows reasonable fit and the hopping distance between ionic sites may be evaluated. Though $\log I - E^{1/2}$ plots yield straight lines with better fit, the Schottky injection or Poole-Frenkel mechanism is not plausible due to the fact that the dielectric constant evaluated according to equation (7.31) is too high.

Müller[51] and Lilly, et al. (1968) suggested that the space charge alters the electric field within the bulk which in turn influences the conduction mechanism. The electric field is expressed as γE where γ, called the **field modification factor**, is a geometric factor and E is the geometric electric field. Müller observed that γ could be as high as 139. Lilly and McDowell found values less than one in Mylar (PET) and values in the range of 13.4 – 25.6 in PTFE. The fit for ionic conduction could be improved by assuming a space charge modified field. Summary of results for films of two thickness are given in Table 7.5. The hopping distance appears to be a function of temperature and electric field which has also been observed in other polymers.

(B) TEFZEL®

As far as the author is aware there is only one study carried out on conduction mechanism in this high temperature dielectric[52]. We recall, from chapter 6, that the absorption current and desorption current in Tefzel® are not mirror images and the difference between them is defined as the transport current for which the equations in this chapter are developed. We also recall that evidence was produced for hetero space charge effects at the electrodes resulting in a field intensification.

Fig. 7.18 shows the $\log I - 1/T$ plots at various electric fields and the slope decreases at lower temperatures. This is considered to be due to increased accumulation of charges leading to greater field intensification. Tunneling mechanism is unlikely due to the fact that the current is temperature dependent. Ionic conduction, equation (7.25), is also unlikely because $\log I - E$ plots are not straight lines. Schottky theory, equation (7.30), appears to be valid, particularly if space charge effects are taken into account by substituting γE in place of E in the exponential term. Whenever Schottky injection is invoked the Poole-Frenkel mechanism should be verified as both mechanisms predict a $\log I - E^{1/2}$ linearity. The dielectric constant, ε_∞, calculated using both theories are shown in Table 7.6.

A comparison of the dielectric constant data shown in Fig. 6.22 shows that Schottky model applies below 100°C and for higher temperatures the Poole-Frenkel mechanism is a reasonable theory. The field modification factor, γ, is also in the same range as those obtained in polyethylene and polyimide.

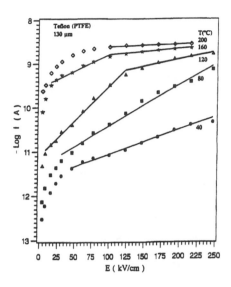

Fig. 7.17 Conduction currents in PTFE at various temperatures (Sussi and Raju, 1990).

7.6.3 AROMATIC POLYIMIDE

The aromatic polyimide is used in many electronic components such as integrated circuits and thin film substrates. Several grades such as device grade, corona resistant, Kapton®, and Kapton-H® film are available with resistivity varying in the range of $10^9 - 10^{17}$ Ωm over a temperature of 23° - 350°C.

Table 7.5

Hopping distance and field modification factor in PTFE

Thickness	50 µm		130 µm	
T (°C)	λ(nm)	γ	λ(nm)	γ
40	3.42	1.00	3.02	1.0
60	3.91	1.48		
80	4.55	1.17	5.70	2.55
100	4.09	1.03		
120	1.78	0.13	9.00, (2.69[a])	2.76
140	1.07	0.10		
160	0.68	0.02	5.57, (1.22[a])	0.62
180	0.56	0.01		
200	0.30	0.01	0.47[a]	

[a] Electric field > 10 MVm^{-1}

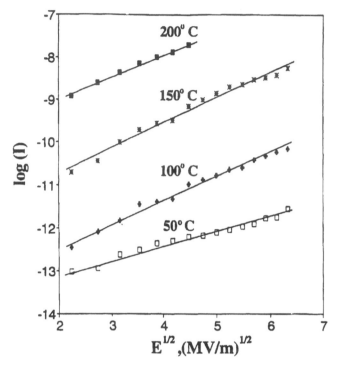

Fig. 7.18 Variation of transport currents in Tefzel with E½ at various constant temperatures (Wu and Raju, 1995, with permission of IEEE)

Table 7.6

Comparison of ε_∞ derived from Schottky and the Poole-Frenkel model (Wu and Raju, 1995).

T (°C)	ε_{Sch}	γ	ε_{P-F}
50	2.34	1.07	9.39
70	2.97	0.95	11.91
80	2.98	0.95	11.93
90	1.48	1.34	5.91
100	0.82	1.81	3.27
150	0.57	2.16	2.29
200	0.65	2.03	2.58

In device grade polyimide Smith, et. al.[53] observed that the current varied according to t^{-n}, with $0.6 < n < 0.8$ in the time range of $1 - 16,000$ s and temperature range of $23 - 125°C$. In contrast Hanscomb and Calderwood (1973) obtained $0.2 < n < 0.6$ at short times in the range $80\mu s < t < 2$ ms and temperature range of $150 - 175°C$.

The Debye theory of dipolar relaxation assumes that the dipoles are independent and rotate like a rigid body leading to the expression "rigid rotator". Smith, et. al. (1987) have developed a molecular dipole theory in which the single-phonon-assisted tunneling between traps occurs at low electric fields (E < 10 MVm^{-1}). A phonon is a fundamental concept of energy equal to the quantum of energy of lattice vibration in a crystal and is equal to hν_{phonon}. The phonon frequency is between 10^{12} – 10^{13} Hz.

The polarization current due to phonon assisted tunneling is derived by Smith et. al. (1987) as

$$J = \frac{Ae^2 E \lambda^2 N_t^{\ 2}}{3kT}(2\nu_{phonon})^{1-n} t^{-n}$$ (7.85)

where A is a constant with the dimension of (length)^3, N_t the number density of traps and n the index of current decay. For polyimide the index is found to be a function of temperature according to $n(T) = a - bT = 1.4 - 0.016$ T. By substituting typical values for polyimide, A = 1.7×10^{-23} m^3, λ = 10 nm, N_t = 4 $\times 10^{23}$ m^{-3}, the frequency is obtained as 0.016 Hz. Obviously this cannot be phonon frequency and Smith et. al. suggest that polarization can still occur due to a hopping charge carrier or a defect rather than phonon assisted, explaining the low frequency.

A comparison of results of several publications (Table 7.7) on the transport current covers a wide range of different polyimides, experimental conditions, method of sample preparation, etc. Some general conclusions may be drawn. Generally the low field transport current varies linearly with electric field, with a temperature activation energy in the range of 0.4 eV to about 1.4 eV. The linear regime leads to a dependence of Log I $\propto$ E or E$^{1/2}$ at high electric fields. The activation energy is also observed to increase at higher temperatures (fig. 7.19).

The decreasing slope at longer times and increasing slope at higher temperatures may be indicative of ionic transport and the question that arises is whether phonon assisted tunneling, which is invoked to explain absorption current, also explains the conduction currents. The current density in this case is given by

$$J = 2e\lambda kTN_t \nu_{phn} \exp\left(-\frac{\Delta}{kT} - \frac{e^2 E^2 \lambda^2}{16\Delta kT}\right) \sinh\left(\frac{eE\lambda}{2kT}\right)$$ (7.86)

where Δ is the energy difference between two trapping levels. Swamy, et. al[54] have shown that the measured current agrees with equation (7.86) confirming the phonon model both for polarization and transport.

7.6.4 AROMATIC POLYAMIDE

Aromatic polyamide has excellent dielectric properties and it is a candidate material for high voltage, high temperature applications. It is available as continuous filament yarn and as paper, which is made from a mixture of short length staple fibers and small fibrous particles called fibrids. However its poor mechanical strength has led to the development of films that have Young's modulous that is higher than many heat resisting polymers such as PET, PEEK and PI[55]. The dielectric constant and loss factor of the Polyamide films are shown in Fig. 7-20.

The conduction currents in aramid polyamide (paper) has been reported by Raju[56] as shown in Fig. 7-21. The mechanism of conduction has been shown to be ionic in nature with an activation energy in the range of 1.2 to 1.6 eV. Conduction occurs as a result of hopping between sites, the hopping distance being calculated as 4.4-8 nm. The mechanism of conduction in linear polyamides (nylons) has been shown to be due to protons with an activation energy of 1.29 eV[57].

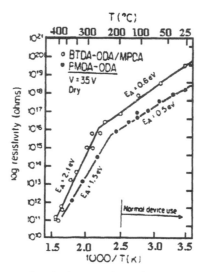

Fig. 7.19 Transport current in device grade polyimide: resistivity vs temperature (Ohmic behavior assumed for the high temperature data (Smith, et. al. 1987, with permission of Journal Electronic. Mat.)

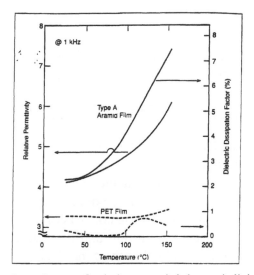

Fig. 7.20 Temperature dependence of relative permittivity and dielectric dissipation factor of aramid film specimens (Yasufuku, 1995, with permission of IEEE).

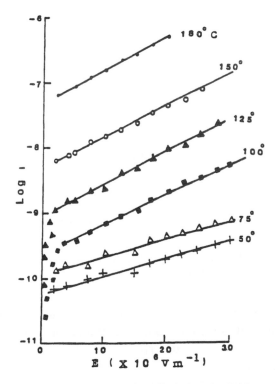

Fig. 7.21 Isochronal currents as a function of applied electric field at various constant temperature for 127 mm thick aramid paper (Raju, 1992, with permission of IEEE).

The energy of a hydrogen bond in polyamide is 0.22 eV and the energy required for self ionization is 0.75 eV. Assuming a mobility of 10^{-12} $m^2V^{-1}s^{-1}$ and using the measured conductivity, the concentration of ions is evaluated as $5 \times 10^{18}m^3$. Though this is an order of magnitude lower than that in other polyamides, a justification is provided on the basis that the aromatic content of aramid paper is high, with 85% of carbon replaced by the benzene ring.

7.7 NUMERICAL COMPUTATION

Advent of powerful personal computers has become a new tool to evaluate the I-V characteristics and the temporal variation of space charge within the thickness of the material. While great progress has been achieved in simulation methods in gaseous discharges (ch. 9) the literature on numerical computation in solid dielectrics is rather sparse[58, 59, 60]. The current through a dielectric consists of three components; the transport or conduction current, the polarization current and the displacement current. Assuming that the continuity equations are applicable the time dependent current density is given by

$$J(t) = \varepsilon_0 \varepsilon_r \frac{\partial}{\partial t} E(x,t) + j_p(x,t) + j_c(x,t) \tag{7.87}$$

$$j_c(x,t) = e\mu\, n(x,t) E(x,t) - e D \frac{\partial}{\partial x} n(x,t) \tag{7.88}$$

$$j_p(x,t) = \frac{\partial}{\partial t} P(x,t) = \frac{1}{\tau}\left[P_s(x,t) - P(x,t)\right] \tag{7.89}$$

$$P_s(x,t) = N\mu_0 \left\{ \coth\left[\mu_0 \frac{E(x,t)}{kT}\right] - \frac{kT}{\mu_0\, E(x,t)} \right\} \tag{7.90}$$

The symbol (x, t) in these equations means that the denoted parameter is a function of space and time co-ordinates, D the diffusion co-efficient, P the polarization, P_s the steady state polarization, n the number density of charge carriers, N the number density of dipoles and μ_0 the permanent dipole moment.

These equations are apparently intimidating though they are relatively easy to formulate and program for numerical solution. Some explanations of these equations follow. Equation (7.87) has three terms on the right side. The first term is the displacement current C dV/dt and decreases to a relatively small value at large t. The second term is the current due to polarization and the last term is the conduction current component. Equation (7.88) is the conduction current where the first term on the right side is the transport or drift term and the second term is the diffusion term. The mobility is always assigned a positive sign because the charge carriers, whether electrons (moving against E) or holes (moving along E) generate current in the same direction. The diffusion term is negative because diffusion occurs from a region of higher density gradient to lower gradient.

The polarization current, equation (7.89), is identical to eq. (3.11) and becomes zero in the steady state. Equation (7.90) is the Langevin function, identical to eq. (2.48). When the parameter $\mu E/kT \ll 1$ the Langevin function may be approximated to $1/3$ ($\mu E/kT$) though this is not required in numerical computations.

The electric field is computed by applying Poisson's equation at each co-ordinate

$$\frac{\partial}{\partial x}\left[\varepsilon_0\varepsilon_r E(x,t) + P(x,t)\right] = en(x,t) \tag{7.91}$$

The electrical field should satisfy the condition

$$\int_0^d E(x,t)\,dx = E_a d \tag{7.92}$$

where d is the thickness of the material and E_a the applied electric field. The Schottky emission, equation (7.30), is assumed and the boundary condition is $n(x, 0) = P(x, 0) = 0$.

The known (experimental) parameters are E_a, T and d . The material parameters are N, μ_0 and ε_r. The parameters determined are μ ($m^2 V^{-1}s^{-1}$), τ (s) and A in the Schottky equation.

The simulation is applied to poly(vinylidene fluoride) (PVDF) assuming the following parameters: $N = 2 \times 10^{28} \text{ m}^{-3}$, $\mu_0 = 7 \times 10^{-30}$ C m, $\varepsilon_r = 2.2$, T = 353 K, $E_a = 30$ MV/m and d = 15 μm. A relaxation time of 26 s (mechanical β-relaxation) and a mobility of 3×10^{-12} $m^2 V^{-1}s^{-1}$ is obtained. A comparison between experimental and theoretical currents is shown in fig. 7. 21. Dipole motion is found to affect E considerably, which in turn affects j_c and J except at low values of τ. The J – t profiles depend on the dipolar relaxation time and carrier mobility, and differ from those in non-polar dielectrics.

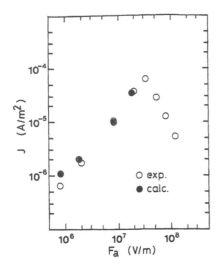

Fig. 7.22 The experimental (open circle) and the calculated J-E characteristics (closed circles) of PVDF after 480s charging at 353 K (Kaneko et. al., 1992, with permission of Jpn. J. Appl. Phys.)

Table 7.7
Summary of selected conduction studies in aromatic polyimide

T (K)	E (MV/m)	Δ (eV)	Proposed mechanism
423-623	$2\text{-}8 \times 10^{-3}$	$0.43 - 1.55$	Thermally assisted tunneling[9,10]
323-398	≤ 0.4	1.37-1.44	Proton conduction[11]
393-453	0.4-50	1.1	Ionic conduction[12]
273-473	5-45	-	Ionic conduction[13]
323-543	4-40	$0.2 - 0.8$	Schottky injection modified by space charge layer[14]
298-623	$1 - 20$	0.5 eV (298 $\leq$T$\leq$448 K)	Phonon-assisted tunneling
		1.5 eV (448 $\leq$T$\leq$623 K)	theory of Lewis[15]
10-500	0-1.5	0.01-0.85	Phonon assisted hopping[16]
295	>140	-	F-N type tunneling[17]

[9] J. R. Hanscomb and J. H. Calderwood, J. Phys. D: Appl. Phys., 6 (1973) 1093

[10] G. Roberts and J. I. Polanca, Phys. Stat. Solidi, AI (1970) 409

[11] E. Sacher, IEEE Trans. Elect. Insul., **EI-14** (1979) 85-93.

[12] G. Sawa. S. Nakamura, K. Ieda and M. Ieda, Jpn. J. Appl. Phys., **19** (1980) 453.

[13] B. L. Sharma and P. K. C. Pillaai, Polymer, **23** (1982) 17.

[14] G. M. Sessler. B. Hahn, and D. Y. Yoon, J. Phys. D: Appl. Phys. **60** (1986) 318

[15] F. W. Smith. H. J. Neuhaus, S. D. Senturia, Z. Fehn, D. R. Day and T. J. lewis, J. Electron. Mat., **16** (1987) 93-1106

[16] A. A. Alagiri Swamy. K. S. Narayan and G. R. Govinda Raju. To be published

[17] N. R. Tu and K. C. Kao, J. Appl. Phys., **85** (1999) 7267

7.8 CLOSING REMARKS

For a number of years it has been recognized that conduction currents in a polymer are affected by the molecular weight and its distribution, and by density, morphology and crystallinity. In general, electrical conductivity decreases with increasing molecular weight, which in turn is related to a decrease in free volume. Increase in viscosity and intermolecular forces are also associated with the increase in molecular weight and therefore in resistivity. In general conductivity tends to increase with increasing spherulite density and diameter. A regularity of structure and increasing crystallinity will decrease in ionic conductivity (Das Gupta, 1997). Mizutani and Ieda (1984) have proposed a relationship between fluorination and conductivity; increase of fluorine substituent decreases the current. PP, PVF, PE, PVC and PCTFE have no or lower fluorine content and the conduction currents, due to negative carrier, are higher. On the other hand FEP, ETFE, PTFE, PVDF have lower currents, due to positive carriers.

Boggs[61] has provided a collection of conduction formulas which are convenient to apply to experimental results, with the electric field and temperature as parameters. The need for careful and reproducible experimental results cannot be overstressed. Wintle[62] (1994) has drawn attention to the need for caution in interpreting conduction data. Theoretical studies relating the free volume and conductivity help to understand the role of morphology. Further development of models for conduction in cables should be aimed at computation of microscopic parameters in the dielectric so that a comparison with an initial set of values may help to estimate the deterioration of the dielectric (Boggs, 1995).

7.9 REFERENCES

[1] H. L. Saums and W. W. Pendleton, **"Materials for Electrical Insulating and Dielectric Functions"**, Hayden Book Co., 1978.

[2] R. Eisberg and R. Resnick, *"Quantum Physics of atoms, molecules, solids, Nuclei and Particles"*, II edition, John Wiley and Sons, New York, 1985, p. 388.

[3] T. Tanaka and J. H. Calderwood, J. Phys. D., **7** (1974) 5.

[4] S. H. Glarum, J. Phys. Chem. Solids, **24** (1963) 1963.

[5] J. J. Ritzko, in Electronic Properties of Polymers, Chapter 2, Ed: J. Mort and G. Pfister, John Wiley and Sons., New York 1982.

[6] L. A. Dissado and J C. Fothergill, Electrical Degradation and Breakdown in Polymers, Peter Peregrinus Ltd., London, 1992.

[7] J. J. O'Dwyer, *"The Theory of Electrical Conduction and Breakdown in Solid Dielectrics"*, Clarendon Press, Oxford, 1973.

[8] T. Mizutani and Ieda, IEEE Trans. Elec. Insul., **21** (1986) 833.

[9] D. A. Seanor, "Electrical Properties of Polymers", Polymer Science, Vol. 2, Ed: A. D. Jenkins, North Holland Publishing Co., Amsterdam, 1972.

C. C. Ku and R. Liepins, *"Electrical Properties of Polymers"*, Hanser Publishers, Munich, 1987, p. 213.

B. Hilczer and J. Małecki, *Electrets*, Elsevier, Amsterdam (1986).

[10] T. J. Lewis, Proc. Phys. Soc. (London), **B 67** (1956) pp. 187-200.

[11] Y. Miyoshi and K. Chino, Japanese Journal of Applied Physics, **6** (1967) 181.

[12] A. C. Lily, Jr. and J. R. McDowell, J. Appl. Phys., **39** (1968) 141.

[13] J. Frenkel, Phys. Rev., **54** (1938) 647.

[14] D. W. Kim and K. Yoshino, J. Phys. D., Appl. Phys., **33** (2000) 464.

[15] P. Karanja and R. Nath, IEEE Trans. Diel. Elec. Insul., **1** (1994) 213.

[16] R. A. Foss and W. Dannhauser, J. Appl. Poly. Sci., **7** (1963) 1015.

[17] R. Nath, T. Kaura and M. M. Perlman, IEEE Trans. on Elec. Insul., **25** (1990) 419.

[18] M. Ieda, G. Sawa and S. Kato, J. Appl. Phys., **42** (1971) 3733.

[19] R. Nath, T. Kaura and M. M. Perlman, IEEE Trans. Elec. Insul., **25** (1990) 419.

[20] D. K. Das Gupta and M. K. Barbarez, J. Phys. D., Appl. Phys., **6** (1971) 867.

[21] J. Kommandeur and W. G. Schneider, J. Chem. Phys., **28** (1958) 582.

[22] N. F. Mott and R. W. Gurney, "Electronic Processes in Ionic Crystals", Oxford, Clarendon Press, 1940.

[23] A. Rose, Phys. Rev., 97 (1955) 1538.

[24] D. Ma, I. A. Hümmelgen, B. Hu and F. E. Kariasz, J. Phys. D: Appl. Phys., **32** (1999) 2568.

[25] J. Chutia and K. Barua, J. Phys. D: Appl. Phys., **13** (1980) L9.

[26] H. R. Zeller, IEEE Trans. on Elec. Insul., **EI-22** (1987) 215.

[27] Y. Tanaka, N. Ohnuma, K. Katsunami and Y Ohki, IEEE Trans. on Elec. Insul., Vol. 26, pp. 258 - 265, 1991.

[28] M. M. Perlman and A. Kumar, J. Appl. Phys., **72** (1992) 5265.
[29] (a) H. J. Wintle, Engineering Dielectrics, Vol. IIA, Ed. R. Bartnikas and R. M. Eichhorn , ch. 6 (1983) 588.
(b) H. J. Wintle, IEEE Trans. Diel. Elec. Insul., **6** (1999) 1.
[30] M. Ieda, IEEE Trans. Elec. Insul., **21** (1986) 833.
[31] D. K. Das-Gupta, IEEE Trans. Diel. Elec. Insul., **4** (1997) 149.
[32] V. Adamec and J. H. Calderwood, J. Phys.D: Appl. Phys., **14** (1981) 1487.
[33] T. Tanaka, J. Appl. Phys., **44** (1973) 2431.
[34] Y. Tanaka, N. Ohnuma, K. Katsunami and Y Ohki, IEEE Trans. on Elec. Insul., **26** (1991) 258.
[35] A. van Roggen, Phys. Rev. Lett., **9** (1962) 368.
[36] W. G. Lawson, Br. J. Appl. Phys., **16** (1965) 1805.
[37] D. M. Taylor and T. J. Lewis, J. Phys. D: Appl. Phys., **4** (1971) 1346.
[38] D. K. Davies, J. Phys. D: Appl. Phys., **5** (1972) 162.
[39] D. K. Das-Gupta and M. K. Barbarez, J. Phys. D: Appl. Phys., **6** (1973) 867.
[40] T. Tanaka, J. Appl. Phys., **44** (1973) 2430.
[41] T. Tanaka and J. H. Calderwood, J. Phys. D: Appl. Phys., 7 (1974) 1295 .
[42] V. Adamec and J. H. Calderwood, J. Phys. D: Appl. Phys., **14** (1981) 1487.
[43] D. K. Das-Gupta and R. S. Brockley, J. Phys. D: Appl. Phys., **11** (1978) 955-962.
[44] S. Pelissou, H. St. Onge, M. R. Wertheimer, IEEE Trans. on Elec. Insul., **EI-23** (1988), 325.
[45] A. Kumar and M. M. Perlman, J. Appl. Phys., **71** (1992) 735.
[46] S. A. Boggs, IEEE Transactions, **EI-2** (1995) 97.
[47] S. H. Lee, J. Park, C. R. Lee and K. S. Suh, IEEE Trans. Diel. El. Insul., **4** (1997) 425.
[48] A. C. Lily, Jr. and J. R. McDowell, J. Appl. Phys., **39** (1968) 141-168.
[49] W. Vollmann and H. U. Poll, Thin Solid Films, **26** (1975) 201.
[50] G. R. Govinda Raju and M. A. Sussi, CEIDP Annual Report, 1990, pp. 28 – 31.
[51] R. S. Müller, **34** (1963) 2401.
[52] Z. L. Wu and G. R. Govinda Raju, IEEE Trans. Elec. Insul., **2** (1995) 475.
[53] F. W. Smith, H. J. Neuhaus, S. D. Senturia, Z. Fen, D. R. Day and T. J. Lewis, Jour. Electronic Materials, **16** (1987) 93.
[54] Alagiri Swamy, K. S. Narayan and G. R. Govinda Raju: unpublished.
[55] S. Yasufuku, IEEE Electrical Insulation Magazine, **11 (6),** (1995) 27.
[56] G. R. Govinda Raju, IEEE Trans. on Elec. Insul., **27** (1992) 162.
[57] W. O. Baker and W. A. Yager, J. Amer. Chem. Soc., **64** (1942) 2171.
[58] H. J. Wintle, J. Appl. Phys., **63** (1988) 1705.
[59] K. Kaneko, Y. Suzuoki, T. Mizutani and M. Ieda, Jpn. J. Appl. Phys., **29** (1990) 1506.
[60] K. Kaneko, Y. Suzuoki, Y. Yokota, T. Mizutani and M. Ieda, Jpn. J. Appl. Phys., **31** (1992) 3615.
[61] S. A. Boggs, IEEE Trans. Diel. Elec. Insu., **2** (1995) 97.

[62] H. J. Wintle, IEEE Electrical Insulation Magazine, **10, N0. 6,** (1994) 17.

All science is either physics or stamp collecting.
- Ernest Rutherford

8

FUNDAMENTAL ASPECTS OF GASEOUS BREAKDOWN–I

Electrical breakdown in gases has been studied extensively for over one hundred years and the delineation of the various manifestations of discharges has advanced in parallel with a better understanding of the fundamental processes. This vast research area is covered by several excellent books[1, 2]. Fig. 8.1 shows different types of discharge one encounters in practice depending upon the combinations of parameters.

The type of discharge is determined by primary factors such as gas pressure, gas density, electrode shape and distance, polarity of the voltage, type of voltage meaning dc, normal frequency ac, high frequency ac, impulse voltage, etc. Secondary factors are the electrode materials, type and distance to the enclosure, duration of the application of voltage, previous history of the electrodes, etc. Obviously it is not intended to explain all of these phenomena even in a condensed fashion but the fundamental processes that occur are similar, though the intensity of each process and its contribution to the overall discharge process varies over a wide range. In the interest of clarity we limit ourselves to fundamental processes concentrating on the progress that has been achieved during the past twenty five years. However, to provide continuity we recapitulate some fundamental definitions and equations that are relevant to all discharge processes.

8.1 <u>COLLISION PHENOMENA</u>

8.1.1 ELASTIC COLLISION

An electron acquires energy in an electric field and during its acceleration elastic collision occurs between an electron and a molecule. During an elastic collision there is very little exchange of energy with the electron, losing an energy that is proportional to m/M where m and M are masses of the electron and molecule, respectively. The internal

383

energy levels of the molecule do not change during an elastic collision. The main consequence of an elastic collision is that the direction of travel of an electron changes depending upon the angle at which the electron strikes the molecule. A more accurate term for elastic collisions is the **momentum transfer collision** which is an average value that takes into account the angle of approach of the electron.

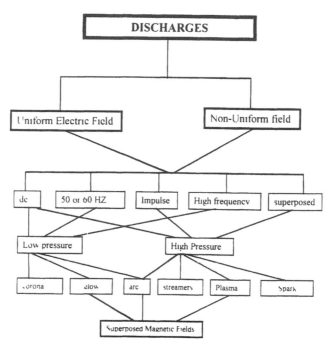

Fig. 8.1 Various manifestations of electrical discharges.

The range of electron energy and the electron density encountered in a wide range of plasmas are shown in fig. 8.2[3]. The parameters chosen to characterize the plasmas are the electron temperature expressed in units of electronvolts and the electron density. One electronvolt (eV) is equal to 11600 degree Kelvin in accordance with $T = (e/k) \varepsilon$ where

ε = energy in eV
e = electronic charge
k = Boltzmann constant

The energy at 300 K is correspondingly 0.026 eV. The electron energy shown in Fig. 8.1 varies over eight decades while the electron density spans a wide range of twenty decades depending upon the type of plasma. Interplanetary space has the lowest number density whereas the fusion plasma has the highest. Ionosphere plasma has the lowest temperature with fraction of an electronvolt energy. Fusion reactors, on the other hand, have an energy ~100,000 eV. Such a wide range of parameters makes the study of plasmas one of the most interesting.

8.1.2 COLLISION CROSS SECTION

The collision between two particles is described by using a fundamental property called the collision cross section, which is defined as the area involved between the colliding particles measured in m^2. Collisions between neutral particles, or between a neutral particle and a charged particle, may be considered to a first approximation as if the particles were hard spheres. The effective target area for such collisions is the collision cross section, Q, which has a value of $10^{-19}\ m^2$ in air for gas molecules. If each particle has a radius a then the collision cross section is given by $4\pi a^2$. Each inelastic process is characterized by a corresponding cross section and the total cross section is the sum of the momentum transfer and inelastic collision cross sections.

8.1.3 PROBABILITY OF COLLISION

The probability of collision is defined as the average number of collisions that an electron makes with a neutral molecule per meter length of drift. The probability is related to the collision cross section according to

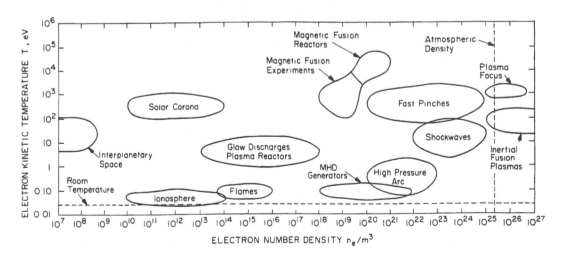

Fig. 8.2 Electron number density and electron energy in various plasmas (Roth, 1995, with permission of Institute of Physics.)

$P = N \ Q_m$ where

Q_m = Momentum transfer cross section (m^2)
P = Number of collisions (m^{-1})
N = Number density of molecules (m^{-3}). The number density is related to gas
pressure at 300 K according to the following relationships.
$N = 3.220 \times 10^{22} \times p$ (Torr) (m^{-3})
$N = 2.415 \times 10^{20} \times p$ (Pa) (m^{-3})
$N = 2.447 \ \times 10^{25} \times p$ (atm) (m^{-3})
1 atmosphere = 760 Torr = 101.33 kPa

The number of collisions, N_c, per second is given by

Number of collisions per second = $N \ Q_m \ v$
v = velocity (m /s)

The probability of collision for electrons in gases is a function of electron energy and is usually plotted against $\sqrt{\varepsilon}$ where ε is the electron energy.

For electrons having energy greater than about 1 eV, P_c increases with ε reaching a maximum in the range of 10-15 eV. For a further increase in energy the probability decreases at various rates depending on the gas (Fig. 8.3). At energies lower than 1 eV the probability of collision is a complicated function due to quantum mechanical effects. In many gases the quantum mechanical wave diffraction of the electron around the atom results in an increased probability with decreasing energy, as found experimentally by Ramsauer during early thirties.

Compilation of cross sections for elastic scattering and momentum transfer for common gases is available in ref.[4]. The cross sections for carbon compounds such as CH_4, CF_4, CF_2Cl_2 etc. which are of interest in plasma processing applications are compiled in ref.[5].

8.1.4 INELASTIC COLLISIONS

Electrons gain energy from the applied electric field and at sufficiently high energy a collision will result in a change of the internal energy level of the molecule. Such a collision is called an inelastic collision. Inelastic collisions with atoms or molecules mainly result in creation of excited species, ionization and metastables (excited molecules with a long life time). Some common inelastic collisions are given in Table 8.1.

8.1.5 MEAN FREE PATH

The **mean free path** of an electron, λ, is the average distance traveled by an electron between collisions with gas molecules. The concept of the mean free path is general and not limited to elastic collisions. We can refer to a mean free path for elastic collisions as distinct from mean free path for ionizing collisions, λ_i. We can also refer to a mean free path for collisions between gas atoms or molecules even in the absence of an electric field. The mean free path for molecules at atmospheric pressure of 101 kPa is approximately 60 nm, a very small distance. As the pressure decreases the mean free path increases and the relationship is

$$\lambda = \frac{1}{P} = \frac{1}{NQ_m} \tag{8.1}$$

Table 8.1
Some commonly observed inelastic collisions

Type	Reaction
Ionization	$e + x \rightarrow x^+ + e + e$
Excitation	$e + x \rightarrow x^* + e$
Dissociative attachment	$e + x\,y \rightarrow x^- + y + e$
Dissociation	$e + x\,y \rightarrow x + y + e$
Recombination	$x^+ + y^- \rightarrow x + y$
Three body recombination	$e + x^+ + y^- \rightarrow x + y$

The distance traveled between collisions is a varying quantity due to the random nature of collisions. The probability of having a free path greater than x is an exponentially decaying function according to $exp(-x/\lambda)$ where λ is the mean free path.

8.1.6 IONIZATION BY COLLISION

When the energy of the electron exceeds the ionization potential of a gas molecule a secondary electron and a positive ion are formed following a collision. The ionization cross section of a electron, Q_i, increases with its energy reaching a peak value at about three times the ionization energy. For a further increase the ionization cross section decreases slowly. Two basic mechanisms for ionization are:

8.1.7 DIRECT IONIZATION

A molecule is directly impacted by an electron of sufficient energy

$$XY + e \rightarrow XY^+ + 2e$$

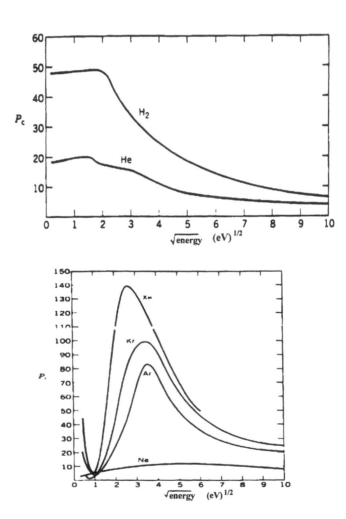

Fig. 8.3 Probability of collision for electrons in gases. Lower figure shows the Ramsauer minima at low energies. The cross section is given by $Q_m = 2.87 \times 10^{-21} P_c \ m^2$.

8.1.8 DISSOCIATIVE IONIZATION

$$XY + e \rightarrow X + Y^+ + 2e$$

X, Y gas atoms or molecules
Y^+ = positive ion
e = electron.

Total ionization cross sections for rare and common molecular gases are available in ref.[6], [7]. Cross sections for dissociative ionization for several molecular gases are reported in ref. [8]. Figure 8.4 shows the ionization cross sections as a function of electron energy for several molecular gases[9].

Photo-ionization occurs when photons of energy greater than the ionization potential of the molecule, ε_i impinge on the molecule. This reaction is represented by:

$$XY + h\nu \rightarrow XY^+ + e; \quad h\nu = \text{photon energy}$$

8.1.9 EXCITATION

The first excitation threshold of an atom is lower than the ionization potential and the excitation cross section is generally higher than the ionization cross section. The excited species returns to its ground state after a short interval, ~10 ns emitting a photon of equivalent energy.

The direct excitation mechanism is

$$X + e \rightarrow X^* \rightarrow e + X + h\nu$$

Where X* denotes an excited atom. Two other types are

$$XY + e \rightarrow XY^* + e$$
$$X^+ + Y \rightarrow X^+ + Y^*$$

where X^+ is a positive ion.

8.1.10 DISSOCIATIVE EXCITATION

The electron impact dissociates the molecule and the excess energy excites one of the atoms

$$XY + e \rightarrow X + Y^* + e$$

8.1.11 PHOTOEXCITATION

$$XY + h\nu \rightarrow XY^*$$

A compilation of excitation cross sections can be found in ref.[10].

8.1.12 ELECTRON ATTACHMENT

Atoms which are electronegative have an affinity for an electron and the process of an electron being captured by such an atom is called attachment. Molecules having an electron attaching atom as a constituent also become electron attaching. Oxygen and halogens are electron attaching elements. Examples of electron attaching molecules are: O_2, CO, CO_2, SF_6 etc. Several processes occur:

A. DIRECT ATTACHMENT

$$XY + e \rightarrow XY^-$$

XY^- = Negative ion

B. DISSOCIATIVE ATTACHMENT

$$XY + e \rightarrow X + Y^-$$

C. THREE BODY ATTACHMENT

Three body attachment occurs in the presence of a molecule that stabilizes and promotes charge transfer

$$XY + Z + e \rightarrow XY^- + Z$$

Total cross sections for negative ion formation in several gases (CO, NO, O_2, CO_2 and SF_6) by electron impact are given in Rapp and Braglia (1963)[11].

8.1.13 ELECTRON DETACHMENT

Electrons may be detached from the negative ions by processes

$$X^- + e \rightarrow X + 2e$$
$$XY^- + e \rightarrow XY + 2e$$
$$XY^- + M \rightarrow XY + M + e$$

The last process is known as three-body attachment and if active, the detachment coefficient is pressure dependent. Electron detachment increases the population of electrons which may further participate in the ionization process.

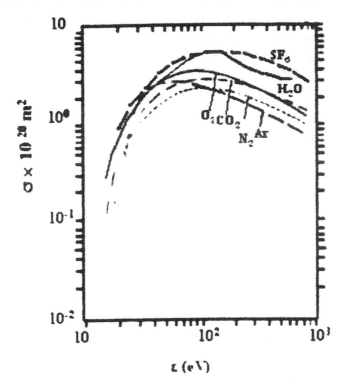

Fig. 8.4 Ionization cross sections for several molecules as a function of electron energy (Pejcev et. al., 1979, with permission of American Chemical Society).

8.1.14 RECOMBINATION

The recombination of positive and negative charges occurs through several different mechanisms. The simplest reaction is

$$A^+ + e \rightarrow A + hv \qquad or \qquad A^+ + e \rightarrow A* + hv$$

These are two body radiative processes and the rate of loss of electrons is given by

$$\frac{\partial n_e}{\partial t} = -\alpha_e n_e^2 \tag{8.2}$$

where n_e is the density of electrons assumed to be equal to the density of positive ions and α_e is the recombination coefficient ($cm^3 s^{-1}$). It has values in the range $1 \times 10^{-14} \leq \alpha_e \leq 1 \times 10^{-7}$ $cm^3 s^{-1}$ depending on the gas, electron density and temperature. Increasing number density and temperature yields higher values of α_e. Several other processes are possible (Meek and Craggs, 1978)

$$(a) A^+ + e + e \rightarrow A* + e \qquad \qquad \textit{Three body recombination}$$
$$(b) A* + e \rightarrow A^+ + e + e \qquad \qquad \textit{impact ionization}$$
$$(c) A* + e \rightarrow A** + e \qquad \qquad \textit{Collisional excitation or de-excitation}$$
$$(d)(AB)^+ + e \rightarrow A* + B* \qquad \qquad \textit{Dissociative recombination}$$
$$(e) A^+ + B^- \rightarrow A + B \; or$$
$$\qquad \qquad A* + B \; or \quad \Big\} \quad \textit{ion-ion recombination}$$
$$\qquad \qquad AB$$

In considering recombination many of these processes have to be taken into account acting in combination and the decrease of electron density due to recombination has led to the term **collisional radiative decay**. At low electron densities, $10^7 \leq n_e \leq 10^{12}$ cm^{-3}, and low electron temperatures ~1 eV the recombination coefficient is relatively constant, independent of n_e. In this region two body recombination processes are dominant. As the temperature of electrons increase the recombination coefficient increases with n_e. This situation arises in considering recombination in hot spark channels, and three body processes become increasingly important.

8.1.15 SECONDARY IONIZATION COEFFICIENT

There are several mechanisms which yield secondary electrons both at the cathode and in the gas. The secondary electrons contribute to the faster growth of current in the discharge gap and the important mechanisms of secondary ionization are classified as below.

A. Impact of Positive Ions on the Cathode

Positive ions which impinge on the cathode liberate secondary electrons provided their energy is equal to or higher than the work function of the cathode. The number of secondary electrons liberated per incident ion is dependent on the surface conditions of the cathode such as oxidation, adsorbed gas layer, etc. The secondary emission is higher for cleaner and non-oxidized surfaces. At low gas pressures the secondary ionization co-efficient is usually of the order of $10^{-2} - 10^{-4}$ per incident electron. In discharges at higher pressures (~ 100 kPa.) the influence of secondary electrons due to positive ions is negligible as the ions do not have sufficient energy.

B. Impact of Metastables on the Cathode

Under certain conditions an excited molecule loses a fraction of its energy by collision and the result is a new excited state from which the excited molecule cannot return to the ground state. The life time of such excited molecules, called metastables, is about 10^{-3} s. Some metastables are destroyed by collision with a gas molecule and some are lost by falling on the anode. However, secondary emission can occur when a metastable strikes the cathode. For a given gas in which the metastables are present (the rare gases and nitrogen are examples) the secondary emission due to positive ions will be more pronounced than that due to metastables. The reason is that, under any particular condition, the positive ions are attracted to the cathode due to Coloumb force where as the metastables, being charge neutral, can only reach the cathode by diffusion. Further, the probability of liberation of a secondary electron due to positive ions is higher than for metastables.

C. Photoelectric Action

Emission of electrons from the cathode due to photoelectric action is an important secondary mechanism and is effective at moderate gas pressures (~10-100 kPa) and gap lengths of a few centimeters. A copious supply of photons of sufficient wave length and a low absorption coefficient in the gas render this secondary process more effective. All of the photons generated will not fall on the cathode since most of them are created near the anode and the number decreases exponentially with gap distance due to photon absorption according to

$$n = n_o \exp(-\mu x) \tag{8.3}$$

where n is the number of photons that survive after a transit distance of x from the avalanche head (anode), n_o, the initial number of number of photons generated and μ the absorption co-efficient. μ is a function of wave length and measurements of absorption co-efficients in gases using monochromatic beams have been published. However these data cannot be used in gas discharge studies because the discharge produces photons having various energies. Govinda Raju et al[12] have measured photon absorption in several gases using a self sustained discharge as source of photons. A corona discharge in a wire-cylinder geometry was also employed[13] and absorption coefficient of 0.4 cm^{-1} at gas number density of ~10^{17} cm^{-3} was measured. This amounts to a decrease of photon intensity of ~33% for a transit of 1 cm and the reduction will be greater at higher gas pressures or longer gap lengths.

The secondary ionization co-efficient observed experimentally is a combination of several of these effects and the relative contributions have been determined for several gases. Generally speaking at low gas number densities (higher E/N) the positive ion action predominates whereas at high gas pressures, ~100 kPa, the photoelectric action predominates. The secondary ionization coefficient also increases with E/N as shown in Fig. 8.5[14].

8.1.16 PHOTO-IONIZATION

Photons having energy equal to or greater than the ionization potential of the gas cause photo-ionization and the process is defined as

$$A + h\nu \rightarrow A^+ + e$$

This process is the converse of radiative electron-ion recombination. Photo-ionization has been invoked in the streamer mechanism (section 8.3) as the principal mechanism in the transition of a primary avalanche into a streamer in gases at atmospheric pressure. The photo-ionization cross sections for a number of gases have been measured by using line spectrum (mono-energy) rather than a continuum. The cross section for this process for several molecules are given in Table 8.2.

While these photo-ionization cross sections are of interest from the point of view atomic and molecular physics, it is not possible to apply them directly to electrical discharges because the discharge produces photons of various energies. The relative intensity of photons generated is also not known. Further, in a strongly photon absorbing gas, the photons generated in the discharge disappear at a very short distance from the detector and are not available at a distance for measurement. Due to these difficulties photon absorption experiments have been carried out at low gas pressures and the results are then applied to high density discharges. Table 8.3 shows the absorption coefficient for photo-ionizing radiations in several gases.

The intensity of photons scattered in a gas is attenuated according to

$$I = I_o \exp(-\mu_p x)$$

where μ_p is the absorption coefficient (m^{-1}) and x is the distance (m). The absorption coefficient is related to the photo-ionization cross section according to

$$\mu_p = NQ_v$$

where N is the gas number density and Q_v the photo-ionization cross section.

8.1.17 ELECTRON SWARM COEFFICIENTS

The cefficients listed in Table 8.4 are usually referred to as swarm parameters in the literature on gas discharges.

Consider a swarm of electrons or ions moving through a reduced electric field E/N and having a drift velocity of W m/s. The mobility is defined as $\mu = W/E$ and in the literature on gas discharges mobility is often referred to standard conditions of pressure and temperature (p = 760 torr and T = 273 K) by expressing

$$\mu_o = W \frac{P_{273}(Pa)}{E} \frac{1}{101.3 \times 10^7} \frac{273}{T(K)} \ \text{m}^2 \text{V}^{-1} \text{s}^{-1} \tag{8.4}$$

where μ_o is called the reduced mobility.

In terms of E/N we have the relationship

$$\mu_o = W \frac{N}{E} \frac{1}{101.3 \times 10^7} \frac{1}{3.54 \times 10^{16}} \ \text{m}^2 \ \text{V}^{-1} \ \text{s}^{-1}$$

(8.5)

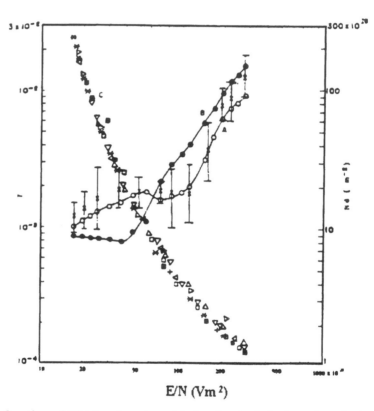

E/N (Vm²)

Fig. 8.5 γ as a function of E/N for air and a gold plated cathode before glow discharge (curve A) and after glow discharge (curve B). The vertical bars indicate the spread of values obtained from log(I, N) plots. The plot of E/N against Nd is included to show the value of N corresponding to each value of E/N (curve c), (Raju and Rao, 1971, with permission of Institute of Physics, UK.)

Experimental and theoretical values for the swarm parameters for a large number of gases are available in the literature. Table 8.5 shows the data for gases which are of practical interest. The swarm coefficients are expressed as a function of the parameter E/N, called **reduced electric field**, which has the dimension of $V \ m^2$. In older literature the reduced electric field is E/p with the dimension of V cm^{-1} Torr^{-1}, the conversion factor being, E/p × 3.1 × 10^{-21} = E/N (V m^2), 1 Townsend (Td) = 1×10^{-21} V m^2.

Table 8.2
Photo-ionization cross sections (Meek & Craggs, 1978).

Molecule	Wave length (nm)	Cross section ($\times 10^{-18}$ cm^2)
	101.9±1 nm	onset
Oxygen	97.9	7.04±1.8
	47.3	23.1 ± 4.1
Nitrogen	78.7	8.8 ± 1.6
	47.3	20.8 ± 3.8
Carbon dioxide	88.3 ± 1.5	onset
	70	30
	98.5 ± 0.5	onset
Water vapor	65	20
	47	12
Hydrogen	80.37	onset
	77.0-66.5	8
Argon	79.0 ± 0.5	onset
	77.0	27

(With permission of John Wiley & Sons)

Table 8.3
Photon absorption coefficients[15]

Gas	N ($\times 10^{16}$ cm^{-3})	μ^{**} (cm^{-1})	Photon efficiency*	source
O_2	32.2-96.6	38	$\sim 10^{-3}$	Cylindrical corona
	3-10	~250		,,
	1-3	~550		,,
	40-220	2.5	$\sim 10^{-5}$	spark
	10-30	38	$\sim 10^{-2}$	,,
N_2	1-5	~750	$\sim 10^{-3}$	Cylindrical corona
Air	10-200	~5	2×10^{-3}	,,
CO_2	1-3	200-800	$\sim 10^{-4}$	,,
CH_4		970		,,
C_2H_5OH		400-700		,,
H_2O		>200		,,

*per ionizing collision, ** Normalized to atmospheric pressure.
(With permission of Springer-Verlag)

Table 8.4
Electron Swarm Parameters

Parameter	Symbol	Units
Drift velocity	W	m s^{-1}
Mean energy	$\bar{\varepsilon}$	eV
Characteristic energy	D/μ	eV
Reduced ionization coefficient	α/N	m^2
Reduced attachment coefficient	η/N	m^2
Reduced detachment coefficient	δ/N	m^2
Recombination coefficient	β	m^3 s^{-1}

Electron swarm parameters are an indispensable set of data for each gas to understand the discharge mechanisms and to develop theoretical approaches. A number of these parameters in analytical form for several gases is provided below.

Table 8.5
Swarm Parameters in Gases

Air[16]

Ionization coefficient

$$\frac{\alpha}{N} \begin{cases} = 2.0 \times 10^{-20} \exp[(-7.248 \times 10^{-19} \times \frac{N}{E})] & m^2; E/N > 1.5 \times 10^{-19} \, Vm^2 \\ = 6.619 \times 10^{-21} \times \exp[(-5.593 \times 10^{-19} \times \frac{N}{E})] & m^2; E/N \leq 1.5 \times 10^{-19} \, Vm^2 \end{cases}$$

Attachment coefficient

$$\frac{\eta_1}{N} = 4.33 \times 10^{-4} \times \frac{E}{N} - 1.0 \times 10^{-23} \, m^2$$

Three body attachment coefficient

$$\frac{\eta_2}{N^2} = 4.778 \times 10^{-69} \times (\frac{E}{N} \times 10^4)^{-1.2749} \, m^5$$

$$\eta = \eta_1 + \eta_2$$

Recombination coefficient

$$\beta = 2.0 \times 10^{-13} \text{ m}^3\text{s}^{-1}$$

A. Oxygen[17]

Electron drift velocity: $W_e = 5.747 \times 10^{16} (E/N)^{0.0604} \text{ ms}^{-1}$

Positive and negative ion mobility: $\mu_i = 5.5 \times 10^{21} (1/N) \text{ V}^{-1}\text{s}^{-1}\text{m}^2$

Recombination coefficient: $\beta = 2.0 \times 10^{-13} \text{ m}^3\text{s}^{-1}$

Electron diffusion coefficient

$$\frac{D}{\mu} = 1.343 \times 10^6 (E/N)^{0.3441} \text{ V; } E/N < 2.0 \times 10^{-21} \text{ Vm}^2$$

$$\frac{D}{\mu} = 1.213 \times 10^{25} (E/N)^{1.2601} \text{ V; } 2 \times 10^{-21} < E/N < 1.23 \times 10^{-20} \text{ Vm}^2$$

$$\frac{D}{\mu} = 1.519 \times 10^9 (E/N)^{0.46113} \text{ V; } E/N > 1.23 \times 10^{20} \text{ Vm}^2$$

Two body attachment coefficient

$$\frac{\eta_1}{N} = 2.0 \times 10^{-20} \text{ m}^2; E/N > 1.835 \times 10^{-20} \text{ Vm}^2$$

$$\frac{\eta_1}{N} = 5.032 \times 10^{-30} (E/N)^{2.655} \text{ m}^2; E/N < 1.835 \times 10^{-20} \text{ Vm}^2$$

Three body attachment rate

$$K_a = \left[\frac{3.0 \times 10^{-42}}{1 + \left(\dfrac{E}{N} \dfrac{1}{4.0 \times 10^{-21}} \right)^{3/2}} \right] m^6 s^{-1}$$

Three body attachment coefficient

$$\frac{\eta_2}{N} = \frac{N K_a}{W_e} \, m^2$$

Total attachment coefficient: $\eta = \eta_1 + \eta_2$

Ionization coefficient

$$\frac{\alpha}{N} = 2.124 \times 10^{-20} \exp(-6.2116 \times 10^{-19} \frac{N}{E}) \, m^2$$

C. Sulphur hexafluoride[18]

Electron drift velocity

$$W_e = 1.027 \times 10^{19} (E/N)^{0.7424} \, ms^{-1}$$

Positive ion mobility (reduced)

$$\mu_{0(+)} = 6.0 \times 10^{-5} m^2 V^{-1} s^{-1}; \, E/N < 1.2 \times 10^{-19} \, Vm^2$$

$$\mu_{0(+)} = 1.216 \times 10^{-5} \ln(E/N) + 5.89 \times 10^{-4} m^2 V^{-1} s^{-1};$$
$$1.2 \times 10^{-19} < E/N < 3.5 \times 10^{-19} \, Vm^2$$

$$\mu_{0(+)} = -1.897 \times 10^{-5} \ln(E/N) - 7.346 \times 10^{-4} m^2 V^{-1} s^{-1}; E/N > 3.35 \times 10^{-19} \, Vm^2$$

Negative ion mobility (reduced)

$$\mu_{0(-)} = 1.69 \times 10^{-32}(E/N)^2 + 5.3 \times 10^{-5}\, \mathrm{m^2 V^{-1} s^{-1}};\ E/N < 5.0 \times 10^{-19}\,\mathrm{Vm^2}$$

The actual mobility is given by

$$\mu = \mu_0 \frac{p_0}{p} \frac{T}{T_0}$$

where $p_0 = 101.325$ kPa and $T_0 = 273.16$ K.

Recombination coefficient

$$\beta = 2.0 \times 10^{-13}(P)^{0.6336}\, \mathrm{m^3 s^{-1}};\quad 1 \le p \le 39 \mathrm{kPa}$$

$$\beta = 2.28 \times 10^{-11}(P)^{-0.659}\, \mathrm{m^3 s^{-1}};\quad 39 \le P \le 270 \mathrm{kPa}$$

$$\beta = 6.867 \times 10^{-10}(P)^{-1.279}\, m^3 s^{-1};\quad 270 \le P \le 2000\ \mathrm{kPa}$$

Negative ion diffusion coefficient

$$\frac{D}{\mu} = \frac{k_T}{39.6}\ \mathrm{V}$$

where k_T, the Townsend's energy factor is given by Table 8.6.

Table 8.6
Mean values of k_T and D/μ for negative ions in SF_6 (Morrow, © 1986, IEEE)

E/N (Td)	30.3	60.7	91.0	121.4	151.7	182	212.3
k_T	1.0	1.2	1.7	2.6	3.5	4.6	5.9
D/μ (V)	0.025	0.030	0.043	0.066	0.086	0.116	0.149

Electron diffusion (longitudinal)

$$\frac{D_L}{\mu} = 8.6488 \times 10^9 (E/N)^{1/2}\ V;\ E/N < 6.5 \times 10^{-19}\ \mathrm{Vm^2}$$

Attachment coefficient: (Fig. 8. 6)[19]

$$\frac{\eta}{N} = 2.0463 \times 10^{-20} - 0.25379(E/N) + 1.4705 \times 10^{18}(E/N)^2 - 3.0078 \times 10^{36}(E/N)^3 \text{ m}^2;$$

$$5.0 \times 10^{-20} < E/N < 2.0 \times 10^{19} \quad \text{Vm}^2$$

$$= 7.0 \times 10^{-21} \exp(-2.25 \times 10^{18} \times E/N) \quad \text{m}^2; \quad E/N > 2.0 \times 10^{-19} \quad \text{Vm}^2$$

Ionization coefficient (Fig. 8.6) (Raju and Liu, 1995)

$$\frac{\alpha}{N} = 3.4473 \times 10^{34}(E/N)^{2.985} \quad \text{m}^2; \quad E/N < 4.6 \times 10^{-19} \quad \text{Vm}^2$$

$$= 11.269(E/N)^{1.159} \quad \text{m}^2; \quad E/N > 4.6 \times 10^{-19} \quad \text{Vm}^2$$

Table 8.7 summarizes the investigations on α and η in SF_6.

8.2 ELECTRON GROWTH IN AN AVALANCHE

Electrons released from the cathode due to an external source of irradiation such as ultraviolet light multiply in the presence of ionization provided the applied electric field is sufficiently high. The multiplication depends on the parameter E/N according to

$$n = n_0 \, e^{\alpha d} \tag{8.6}$$

n_0 = number of initial electrons
α = Townsend's first ionization coefficient (m^{-1})
d = gap length (m)

Equation (8.6) is generally adequate at gas pressures greater than about 15 kPa. At lower gas pressures the electron has to travel a minimum distance before it acquires adequate energy for the onset of the ionization process and equation (8.6) then becomes

$$n = n_0 \exp(d - d_0) \tag{8.7}$$

d_0 = minimum distance for equilibrium (m).

Equations (8.6) and (8.7) apply when the electron multiplication is not large enough to generate secondary electrons. The sources for secondary electrons are photoelectric action and positive ion action at the cathode. The secondary ionization co-efficient is defined as the number of secondary electrons generated per primary electron and usually denoted by the symbol γ. The contribution due to photoelectric action and positive ion impact are denoted by γ_{ph} and γ_+ so that $\gamma = \gamma_{ph} + \gamma_+$. The current growth in the presence of secondary ionization is expressed as

$$i = i_0 \frac{\exp(\alpha d)}{1 - \gamma\left[\exp(\alpha d) - 1\right]} \tag{8.8}$$

The electron growth is also influenced by attachment and detachment processes. The attachment coefficient, η, (m^{-1}), is defined analogous to α as the number of electrons lost due to a primary electron during a drift of unit distance.

The current growth in an electron attaching gas occurs according to

$$i = i_0 \frac{\alpha \exp\{(\alpha - \eta)d\} - \eta}{(\alpha - \eta)} \tag{8.9}$$

The current growth shown in equation (8.9) assumes that the secondary processes are not active. To take into account the secondary electrons released from the cathode the current growth equation should be modified as:

$$i = i_0 \frac{\left(\dfrac{\alpha}{\alpha - \eta} \exp(\alpha - \eta)d - \dfrac{\eta}{\alpha - \eta}\right)}{1 - \left\{\dfrac{\gamma\alpha}{(\alpha - \eta)}\left[\exp(\alpha d) - 1\right]\right\}} \tag{8.10}$$

Electron detachment is also a process that occurs in some electron attaching gases such as oxygen and SF_6. denoting the detachment coefficient as δ an effective attachment co-efficient η^* may be defined as $\eta^* = (\eta - \delta)$ in moderately detaching gases such as O_2. However in the presence of strong detachment the current growth is determined by a more elaborate equation[20] (Dutton, 1978).

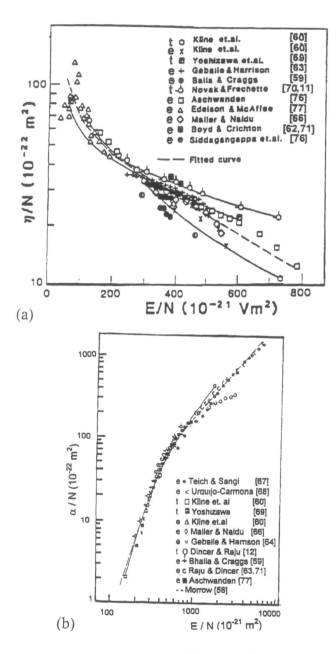

(a)

(b)

Fig. 8.6 Data on (a) the attachment coefficient η/N as a function of E/N, (b) the
ionization coefficient. Fitted curve is the dashed line. Experimental and theoretical
data are indicated by letters 'e' and 't' in the legend. Numbers in the legend are
references in the original paper (Raju and Liu, with permission of IEEE, 1995©)

Table 8.7

Investigations on α AND η in SF_6 (Raju and Liu, 1995) E, Experimental; T, Theory.
(©1995 IEEE)

Authors	E/N (Td)	E/T	Reference
Bhalla and Craggs	$270 \leq E/N \leq 480$	E	Proc. of Phys. Soc., **80** (1962) 151-160
Kline et. al.	$285 \leq E/N \leq 2000$	E/T	J. Appl. Phys., **50** (1979) 6789-6796
Boyd and Crichton	$400 \leq E/N \leq 700$	E	Proc. IEE, **118** (1971) 1872-1877
Geballe and Harrison	$250 \leq E/N \leq 2000$	E	Basic Processes of Gaseous discharges, L. B. Loeb, University of California Press, 1965, 375-476
Maller and Naidu	$350 \leq E/N \leq 4000$	E	IEEE Trans. Plasma Sci., **3** (1975) 205-208; Proc. IEE, **123** (1976) 107-108.
Teich and Sangi	$700 \leq E/N \leq 7000$	E	Proc. Symp. on H. V. Technology, Munich, (1972) 391-395
Urquijo-Carmona	$200 \leq E/N \leq 7000$	E	Ph. D. Thesis, Univ. of Manchester, England, (1980)
Yoshizawa et. al.	$46 \leq E/N \leq 607$	T	J. Phys. D., Appl. Phys., **12** (1979) 1839-1853
Novak and Frechette	$120 \leq E/N \leq 750$	T	J. Appl. Phys., **53** (1982) 8562-8567
Dincer & Govinda Raju	$300 \leq E/N \leq 570$	T	J. Appl. Phys., **54** (1983) 6311-6316
Govinda Raju & Dincer	$300 \leq E/N \leq 1350$	T	J. Appl. Phys., **53** (1982) 8562-8567
Itoh et. al.	$225 \leq E/N \leq 600$	T	J. Phys. D: Appl. Phys., **13** (1979) 1201-1209
Itoh et. al.	$210 \leq E/N \leq 480$	T	J. Phys. D: Appl. Phys., **12** (1979) 2167-2172
Itoh et. al.	$141 \leq E/N \leq 707$	T	J. Phys. D: Appl. Phys., **21** (1988) 922-930
Itoh et. al.	$71 \leq E/N \leq 7656$	T	J. Phys. D: Appl. Phys., **23** (1990) 299-303
Itoh et. al.	$141 \leq E/N \leq 707$	T	J. Phys. D: Appl. Phys., **23** (1990) 415-421
Yousfi et. al.	$100 \leq E/N \leq 960$	T	J. Phys. D: Appl. Phys., **18** (1985) 359-375
Siddagangappa et. al.	$270 \leq E/N \leq 720$	E	J. Phys. D: Appl. Phys., **15** (1983) 763-772
Aschwanden	$150 \leq E/N \leq 918$	E	Fourth Inter. Conf. on Gaseous Electronics, Knoxville, Pergamon Press, (1994) 23-33
Edelson and McAfee	$50 \leq E/N \leq 200$	E	Rev. Sci. Instr. **35** (1964) 187-190
Raju et. al.	$200 \leq E/N \leq 700$	T	J. Appl. Phys., **52** (1981) 3912-3920
Liu & Govinda Raju	$100 \leq E/N \leq 600$	E	IEEE Trans. Plas. Sci., **20** (1992) 515-514

The coefficients α, η, δ and γ are dependent on the applied electric field and the gas number density. It has been found experimentally that the first three coefficients expressed in a normalized way, that is α/N, η/N and δ/N, are dependent on the ratio of E/N only and not on the individual values of E and N. The relationship between α/N and E/N is expressed by the Townsend's semi-empirical relationship

$$\frac{\alpha}{N} = F \exp(-\frac{GN}{E}) = f(E/N) \qquad (8.11)$$

where F and G are characteristics of the gas. It can be shown that the ratio G/F is approximately equal to the ionization potential of the gas. Table 8.8 shows the values of these parameters for several gases.

According to equation (8.6) the current in the gap is a maximum for the maximum value of α at constant d. Differentiating equation (8.11) with respect to N to find the point of maximum α yields

$$\frac{d\alpha}{dN} = \frac{d}{dN}\left[N f(\frac{E}{N})\right] = 0 \tag{8.12}$$

Differentiating the middle term yields

$$f(\frac{E}{N}) - \frac{E}{N} f'(\frac{E}{N}) = \frac{\alpha}{N} - \frac{E}{N} f'(\frac{E}{N}) = 0 \tag{8.13}$$

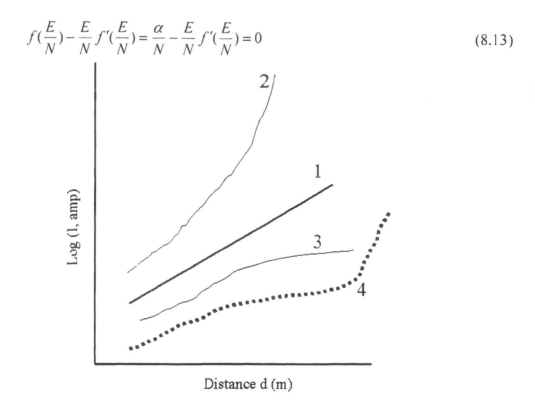

Fig. 8.7 Schematic current growth curves in a gap at various E/N. Curves 1 and 2 are observed in a non-attaching gas. Curves 3 & 4 are observed in an electron attaching gas.

This equation is rearranged to yield the condition for the maximum value of α as a function of N,

$$\frac{\alpha / N}{E / N} = f'\left(\frac{E}{N}\right) = \tan \theta \tag{8.14}$$

Equation (8.14) predicts that the current will be a maximum at the point of intersection of the tangent from the origin to the α/N versus E/N curve (Fig. 8.8). This point is called the **Stoletow point**. At this point the value of E/N is equal to G in equation (8.11). Table 8.9 lists the values of $(E/N)_{Sto}$ and $(\alpha/N)_{Sto}$ at the Stoletow point for several gases in addition to the G/F and the ionization potential. The co-efficient η, defined as the ionization per volt (α/E), is sometimes found in the literature in place of α. The expression for maximum α may be obtained as follows. Differentiation of equation (8.11) yields

$$\frac{d\alpha}{dN} = F\left[1 - \frac{NG}{E}\right]\exp\left(-\frac{GN}{E}\right) = 0 \tag{8.15}$$

Table 8.8
Gas constants in expression (8.11)

Gas	$F \times 10^{22}$ (m^2)	$G \times 10^{22}$ (V m^2)
Argon	372.7	6211
Air	378.9	11335
CO_2	621.1	14472
H_2	329.2	10869
HCl	776.4	11801
He	56.5	1553
Hg	621.1	11491
H_2O	400.6	8975
Kr	450.3	6832
N_2	329.2	10621
Ne	124.2	3105
Xe	689.4	9627

Equating the term in square brackets to zero yields the condition for N at which the current is a maximum as

$$N_{max} = \frac{E}{G} \tag{8.16}$$

The Stoletow point corresponds to the minimum of the Paschen breakdown curve for gases, discussed in the next section. The ratio G/F is known as the effective ionization potential and in every case it is greater than the actual ionization potential due to the fact that the latter is obtained from beam experiments in which an electron beam of the specific energy passes through the gas. The physical significance of the Stoletow point is that at this value of E/N the energy for generating an ion-pair is a minimum resulting in a minimum sparking potential of the gas.

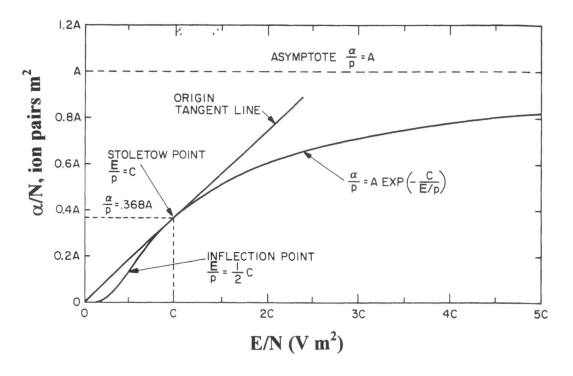

Fig. 8.8 Townsend's first ionization co-efficients plotted against the reduced electric field, (8.11) with major features of the curve indicated (Roth, 1995, with permission of Institute of Physics, UK.)

8.3 CRITERIA FOR BREAKDOWN

Electron multiplication at low gas pressures under sufficiently high electric fields results in a regeneration of secondary electrons, each primary electron resulting in a secondary electron. This condition is known as electrical breakdown of the gas and the discharge is said to be self sustaining. The current due to the electrons will be maintained even though the external agency, which provides the initiating electrons, is switched off.

Mathematically the condition is expressed by equating to zero the denominator of the growth equations (8.8) to (8.10).

A. Non-attaching gases

$$\{\gamma(e^{\alpha d} - 1) - 1\} = 0 \tag{8.17}$$

Due to the large exponential term $e^{\alpha d}$ the condition $e^{\alpha d} >> 1$ holds true and equation (8.17) simplifies to

$$\gamma e^{\alpha d} = 1 \tag{8.18}$$

This equation is known as the Townsend's criterion for breakdown.

B. Attaching gases

The condition for breakdown in electron-attaching gases is given as

$$1 - \left\{\frac{\gamma \alpha}{(\alpha - \eta)}[\exp(\alpha d) - 1]\right\} = 0 \tag{8.19}$$

Taking the natural logarithm of equation (8.19) we obtain

$$\ln(1 + \frac{1}{\gamma}) = \alpha d \tag{8.20}$$

These breakdown conditions lead to Paschen's law.

8.4 PASCHEN'S LAW

Paschen discovered experimentally that the sparking potential of a uniform field gap is dependent on the product Nd rather than, on individual values of N and d. According to equation (8.11) α/N is a function of E/N which may be rewritten as $\alpha/N = \phi(V_s/Nd_s)$

where the suffix s denotes quantities when sparking occurs. If γ is also a function of E/N then $\gamma = F(V_s/Nd_s)$. Substituting these values in equation (8.18) we may write

$$FNd_s \left[\exp\left(\frac{-GNd_s}{V_s} \right) \right] = \ln(1 + \frac{1}{\gamma}) \tag{8.21}$$

This expression leads to

$$V_s = \frac{GNd_s}{\ln\left[\dfrac{FNd_s}{1 + 1/\gamma} \right]} \tag{8.22}$$

which may be rewritten as

Table 8.9

Stoletow Constants for gases (Roth, 1995)

Gas	$(E/N)_{Sto} = G$ $(V\ m^2)$	$(\alpha/N)_{Sto}$ $\times 10^{22}\ m^2$	$\eta = E/\alpha$ V	G/F	ε_i (eV)
Argon	6211	137.6	45	16.7	15.7
Air	11335	139.4	81	29.9	-
CO$_2$	14472	228	63	23.3	13.7
H$_2$	10869	121	90	33.0	15.4
HCl	11801	286	41	15.2	-
He	1553	21	74	27.5	24.5
Hg	11491	228	50	18.5	10.4
H$_2$O	8975	147	61	22.4	12.6
Kr	6832	165	41	15.2	14
N$_2$	10621	121	88	32.2	15.5
Ne	3105	46	68	25.0	21.5
Xe	9627	254	38	14.0	12.1

(with permission from Institute of Physics, Bristol)

$$V_s = f(Nd_s) \tag{8.23}$$

The nature of the function f must be determined experimentally for each gas.

Each gas has a minimum sparking voltage, called the Paschen minimum, below which breakdown cannot occur. To obtain an expression for the minimum voltage we use equation (8.22). By differentiating this expression with respect to the product Nd and equating the derivative to zero the value of Nd_{min} can be shown to be

$$Nd_{min} = \frac{e}{F} \ln(1 + \frac{1}{\gamma}) = \frac{2.718}{F} \ln(1 + \frac{1}{\gamma}) \tag{8.24}$$

where the parameter F is already listed in Table 8.8 for several gases. A dimensionless reduced Nd may be defined according to

$$x \equiv (Nd)_{red} = \frac{Nd}{(Nd)_{min}} \tag{8.25}$$

A dimensionless reduced sparking voltage may also be defined according to

$$y \equiv \frac{V_s}{V_{s.min}} \tag{8.26}$$

Substituting equations (8.25) and (8.26) in (8.22) yields

$$y \equiv \frac{V_s}{V_{s.min}} = \frac{x}{1 + \ln x} \tag{8.27}$$

A plot of y versus x gives a universal Paschen curve and it has the following features. At low values of $x \ll 1$ the curve rises asymptotically at the left reaching a value of infinity at $x = -1$. The co-ordinates of the lowest point on the curve are of course (1,1). On the right hand side the curve rises somewhat linearly on the right due to the logarithmic term in the denominator.

The breakdown voltage on the left of the minimum increases because the mean free path for ionizing collisions increases and the number of ionizing collisions is correspondingly reduced. The number of ionizing collisions is also small because of the reduced number of neutral gas molecules for ionization. On the other hand, at higher gas pressures, on the right hand side of the minimum, the number of neutral molecules is so great that frequent collisions occur. Such increased frequency of collisions does not allow the electron energy to build up to the ionization limit, thus requiring higher voltage for breakdown. Theoretically, breakdown is impossible at $\ln x = -1$ in equation (8.27), but this is not true in practice because other mechanisms such as field emission come into play at such low values of *Nd*.

8.5 BREAKDOWN TIME LAGS

Electrical breakdown of a gap does not occur instantly although the applied electrical field has the critical magnitude to satisfy the criterion, equation (8.20). The time interval from the instant of application of voltage to complete breakdown is called the time lag and arises out of two reasons:

1. The time required for one or more initial electrons to appear in a favourable position in the gap to lead to the necessary avalanche. This is called the statistical time lag.
2. The time required for these electrons to build up the primary avalanche and succeeding generations that leads to a current rise at breakdown. This is called the formative time lag.

8.5.1 THE STATISTICAL TIME LAG

The statistical time lag depends upon the nature and degree of irradiation, cathode material and surface conditions such as the presence of oxidized layer. If β is the rate at which electrons are produced in the gap by external irradiation, P_1 the probability of an electron appearing in a region of the gap where it can lead to a breakdown and P_2 the probability that it will actually lead to a spark, the average statistical time lag is given by

$$\tau_s = \frac{1}{\beta P_1 P_2} \tag{8.28}$$

If the gap has not broken down for a time interval t the probability that it will breakdown in the next interval of time dt is $\beta P_1 P_2\, dt$.

Let us consider n_o number of voltage applications out of which n_t is the number of breakdowns with a time lag greater than t. In the next interval dt the number of breakdowns will be

$$dn = -\beta P_1 P_2\, n\, dt \tag{8.29}$$

The total number of breakdowns is obtained by integrating this expression as

$$\int_{n_0}^{n_t} \frac{dn}{n} = -\int_0^t \beta P_1 P_2 \, dt \tag{8.30}$$

Assuming that P_1 and P_2 are independent of t we get

$$\log\left(\frac{n_t}{n_0}\right) = -\beta P_1 P_2 t \tag{8.31}$$

which yields

$$n_t = n_0 \exp\left(-\frac{t}{\tau_s}\right) \tag{8.32}$$

τ_s is called the mean statistical time lag and equation (8.32) is known as the Laue equation. A plot of $\ln(n_t/n_0)$ yields a straight line having a slope of $-1/\tau_s$ as shown in Fig. 8.9. The time lags decrease with an increasing over voltage and an increasing level of irradiation. The time lag corresponding to $n_t/n_0 = 1$ gives the formative lag for the particular conditions.

8.5.2 FORMATIVE TIME LAGS IN UNIFORM FIELDS

The formative time lag is determined mainly by the fundamental processes and depends upon the gas. Other parameters that influence the formative times are the electrode geometry and the magnitude of the over-voltage. Measured formative time lags can theoretically be interpreted in terms of the avalanche build up and this requires calculation of the electron multiplication as a function of time and space. Let us consider a uniform field gap at moderate pressures of 10-100 kPa. The primary and secondary ionization mechanisms are considered to be the growth components. The continuity equations for the electrons and positive ions may be expressed as

$$\frac{\partial n_e(x,t)}{\partial t} = W_e \, \alpha n_e(x,t) - \frac{W_e \, \partial n_e(x,t)}{\partial x} \tag{8.33}$$

$$\frac{\partial n_p(x,t)}{\partial t} = W_p \alpha n_p(x,t) + \frac{W_p \partial n_p(x,t)}{\partial x} \tag{8.34}$$

Here n_e and n_p are the number of electrons and positive ions, respectively; α the Townsend's first ionization co-efficient; W_e and W_p the drift velocity of electrons and positive ions, respectively, x the spatial variable and t the temporal variable. The left hand terms in equations (8.33) and (8.34) denote the temporal rate at which the electrons and the positive ions increase. The first term on the right hand side is the increase due to ionization by collision and the second term is due to the electrons (positive ions) leaving (arriving) the element of distance dx. Note the negative sign for the electrons leaving element dx.

The analytical solution of these two equations is obtained by Davidson[21] by applying the initial and boundary conditions:

i. $n_e(x,t) = 0$ $t \leq 0$
ii. $n_p(x,t) = 0$ $t \leq 0$
iii. $n_p(d,t) = 0$ $t > 0$

iv. $n_e(0,t) = n_0 + \gamma\, n_p(0,t) + \delta \int_0^d n_e(x,t)e^{-\mu x}$ (8.35)

where n_0 is the number of initial electrons, γ the secondary ionization co-efficient due to positive ion action at the cathode, and δ the photo-electric efficiency (number of secondary electrons generated per primary electron). Initial conditions (i) and (ii) apply when there are no charge carriers in the gap before the application of the voltage, boundary condition (iii) applies because at the anode there ($x = d$) are no positive ions. Boundary condition (iv) shows that both secondary mechanisms (positive ions and photons) are taken into account.

The number of electrons at (x,t) is given by

$$n_e(x,t) = n_e\left(0, t - \frac{x}{W_e}\right) e^{\alpha x}$$ (8.36)

Application of Laplace transforms to equations (8.33) and (8.34) gives

$$\frac{s}{W_e} n_e(x) = \alpha n_e(x) - \frac{\partial n_e(x)}{\partial x}$$ (8.37)

$$\frac{s}{W_p} n_p(x) = \alpha n_e(x) + \frac{\partial n_p(x)}{\partial x}$$ (8.38)

Applying the transformed boundary conditions yields the solution of these equations,

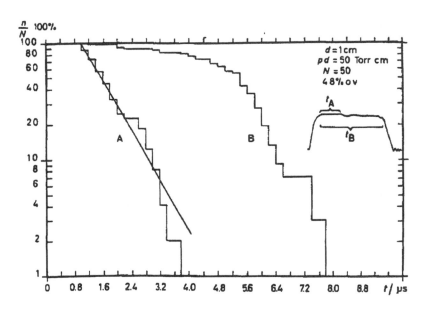

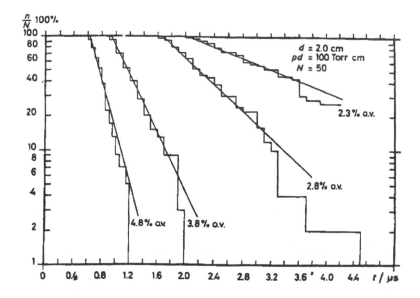

Fig. 8.9 Linear and non-linear time lag distributions in SF_6 (Meek and Craggs, 1978, with permission of John Wiley and Sons, New York).

$$n_e(x) = Ae^{\psi x} \tag{8.39}$$

$$n_p(x) = \frac{A\alpha}{\phi} e^{\psi x} + Be^{sx/W_p} \tag{8.40}$$

where $\phi = \alpha - s/W_e$, $\psi = \alpha - \mu - (s/W_p)$ and A and B are constants given by

$$A = \frac{n_o}{s\left[1 - (\alpha\gamma/\phi)(e^{\phi d} - 1) + (\delta/\psi)(e^{\psi d} - 1)\right]} = \frac{n_o}{s\,f(s)} \tag{8.41}$$

and

$$B = \frac{\alpha A e^{\phi d}}{\phi} \tag{8.42}$$

where n_o is the initial number of electrons released from the cathode. Substituting equations (8.41) and (8.42) in (8.39) and (8.40) we get the solution for the number of electrons and positive ions as

$$n_e(x) = n_o \frac{e^{\psi x}}{s\,f(s)} \tag{8.43}$$

$$n_p(x) = \frac{\alpha\, n_o e^{sx/W_p}}{s\,f(s)} \left\{ \frac{e^{\phi d} - e^{\phi x}}{\phi} \right\} \tag{8.44}$$

The solution is obtained by finding the inverse Laplace transforms of equations (8.43) and (8.44). The details of computation are explained in the following section.

8.5.3 Formative Time Lags in Cylindrical Geometry

We consider a cylindrical geometry[22] with the inner and outer electrodes having a radii of R_1 and R_2, respectively because this geometry is more relevant to practical applications. Let r be the spatial variable, and as in the previous case, we consider a source of external irradiation which liberates uniformly a constant initial current of i_o

from the cathode and at t = 0 a voltage is applied to the gap. The current increases temporally and spatially according to

$$\frac{1}{W_e}\frac{\partial i_-(r,t)}{\partial t} = \alpha(r)i_-(r,t) - \frac{\partial i_-(r,t)}{\partial r} \tag{8.45}$$

$$\frac{1}{W_p}\frac{\partial i_+(r,t)}{\partial t} = \alpha(r)i_-(r,t) + \frac{\partial i_+(r,t)}{\partial r} \tag{8.46}$$

where i_- and i_+ are currents due to electrons and positive ions respectively. They replace n_e and n_p respectively. The initial and boundary conditions are:

    i. $i_-(r,t) = 0, \quad t \le 0$
    ii. $i_+(r,t) = 0, \quad t \le 0$
    iii. $i_+(R_2,t) = 0$

$$\text{iv.} \quad i_-(R_1,t) = i_0 + \gamma_+ i_+(R_2,t) + \gamma_{ph}\int_{R_2}^{R_1}\alpha(r)i_-(r,t)dr \tag{8.47}$$

where γ_+ and γ_{ph} are the relative contributions of positive ions and photons, respectively, to the total secondary ionization co-efficient, γ_T. Hence

$$\gamma_T = \gamma_+ + \gamma_{ph} \tag{8.48}$$

The Townsend's first ionization co-efficient is a function of E/N according to equation (8.11) and, since the field varies radially according to

$$E(r) = \frac{V}{r\ln(R_2/R_1)} \tag{8.49}$$

the radial variation of α is taken into account according to

$$\alpha(r) = FN\exp\left[-\frac{GN}{E(r)}\right] \tag{8.50}$$

Substituting equation (8.49) in (8.50) we get

$$\alpha(r) = FN\exp(-cr) \tag{8.51}$$

where

$$c = \frac{GN}{V} \ln(R_2 / R_1)$$

(8.52)

If we wish to use equation (8.47) as a boundary condition the solutions of equations (8.45) and (8.46) become extremely complicated and therefore we make an approximation that at voltages below breakdown and at the threshold voltage at least 90% of the positive ions are generated near the anode. This yields the equation

$$i_{-(R_1 .t)} = i_0 + i_{+(R_2 .t - t_i)} A$$

(8.53)

in which t_i is the ion transit time and A is called the regenerative factor, defined by

$$A = \gamma_T \left\{ \left[\exp \int_{R_1}^{R_2} \alpha(r) dr \right] - 1 \right\}$$

(8.54)

The ion transit time is given by

$$t_i = \left(\frac{R_2 - R_1}{W_p} \right)$$

(8.55)

For $t \gg t_i$ we can make the approximation

$$i_{+(R_2 t - t_i)} = i_{+(R_2 .t)} - t_i \frac{d\left[i_{+(R_2 .t)} \right]}{dt}$$

(8.56)

Substituting equation (8.56) in (8.53) and omitting the suffixes for the sake of simplicity we obtain the differential equation

$$i = i_0 + A\left[i - t_i\, i' \right]$$

(8.57)

the solution of which is given by

$$i = \frac{i_o}{A-1} \left\{ \exp\left[\left(\frac{A-1}{A} \right) \frac{t}{t_i} \right] - 1 \right\} \tag{8.58}$$

If we use equation (8.58) as the boundary condition in place of (8.47) we can obtain the analytical solutions of (8.45) and (8.46).

Taking Laplace transforms the currents are expressed as

$$I_{-(r.s)} = K_1 e^{\frac{-sr}{W_e} - \frac{FN}{c} \exp - cr} \tag{8.59}$$

$$I_{+(r.s)} = K_3 e^{\frac{sr}{W_p}} + \frac{FNK_2 e^{\frac{sR_1}{W_e}} e^{-r(c+\frac{s}{W_e})W_{eff}}}{At_1 s \left[s + \frac{(1-A)}{At_i} \right] \left[s + cW_{eff} \right]} \tag{8.60}$$

in which K_1 and K_2 are constants to be determined from the transformed boundary conditions, and

$$K_2 = i_0 e^{\frac{FN}{c} [\exp(-cR_1) - (\exp - cR_2)]} \tag{8.61}$$

In equations (8.60) and (8.61) we have defined the following quantities

$$\frac{1}{W_{eff}} = \frac{1}{W_e} + \frac{1}{W_p} \tag{8.62}$$

$$\frac{1}{R} = \frac{1}{R_1} + \frac{1}{R_2} \tag{8.63}$$

Hence

$$K_1 = \frac{i_0 e^{\frac{sR_1}{W_e} + \frac{FN}{c} \exp(-cR_1)}}{s \left[1 + At_1 s - A \right]} \tag{8.64}$$

$$K_3 = -\frac{FNK_2 W_{eff} e^{\frac{sR_1}{W_e}} e^{-(c+\frac{s}{W_e})(R_2-R_1)-\frac{s}{W_p}(R_2-R_1)}}{At_i s \left[s + \frac{(1-A)}{At_i} \right] \left[s + c W_{eff} \right]}$$

(8.65)

Substituting equations (8.61), (8.64) and (8.65) in equations (8.59) and (8.60) and taking inverse Laplace transforms yields

$$i_{-(r,t)} = \frac{K_2}{(A-1)} \left\{ \exp\left[\left(\frac{A-1}{At_i} \right) \left(t + \frac{R_1}{W_e} - \frac{r}{W_e} \right) \right] - 1 \right\} ; t \ge \left(\frac{r}{W_e} - \frac{R_1}{W_e} \right)$$

$$= 0 : t \le \left(\frac{r}{W_e} - \frac{R_1}{W_e} \right)$$

(8.66)

$$i_{+(r,t)} = FNK_2 \exp(-cR_2) \left\{ \frac{1}{c(A-1)} + \frac{At_i W_{eff}}{(A-1)(1-A-At_i c W_{eff})} e^{\left(\frac{A-1}{At_i} \right) \left\{ t - \left[\frac{(R_2-R_1)}{W_e} + \frac{R_2}{W_t} - \frac{r}{W_t} \right] \right\}} \right.$$

$$\left. - \frac{1}{c\left(A-1+cW_{eff} At_i \right)} e^{-cW_{eff} \left[t - \frac{(R_2-R_1)}{W_e} + \frac{(R_2-r)}{W_t} \right]} \right\}$$

$$- FNK_2 \exp(-cr) \left\{ \frac{1}{c(1-A)} + \frac{A W_{eff} t_i}{\left(At_i c W_{eff} + A-1 \right)(A-1)} e^{\left(\frac{A-1}{At_i} \right) \left[t - \left(\frac{r}{W_e} - \frac{R_1}{W_e} \right) \right]} \right.$$

$$\left. - \frac{1}{c\left(1-A-Ac W_{eff} t_i \right)} \exp\left[-cW_{eff} \left(t - \frac{r}{W_e} + \frac{R_1}{W_e} \right) \right] \right\}$$

(8.67)

If we substitute

$$i_{+(r,t)} = P_1 + P_2$$

then

$$
\left.
\begin{aligned}
P_1 &= 0; t < \frac{(R_2 - R_1)}{W_e} + \frac{R_2}{W_i} - \frac{r}{W_i} \\
P_2 &= 0; t < \frac{r}{W_e} - \frac{R_1}{W_e}
\end{aligned}
\right\}
\tag{8.68}
$$

Equation (8.67) gives the positive ion current at (r,t) subject to the condition imposed by equation (8.68). The total current at (r,t) is given by

$$
i_{(r,t)} = i_{-(r,t)} + i_{+(r,t)}
\tag{8.69}
$$

The current from the cathode at time t is given by

$$
i_{-(R_1,t)} = i_0 + \gamma_+ i_{+(R_1,t)} + \gamma_{ph} i_{-(R_2,t)}
\tag{8.70}
$$

This is the true value of i_0 which should be used in the calculation of current multiplication which is given by

$$
M = \frac{i_{-(R_2,t)}}{i_0} = \frac{i_{-(R_1,t)}}{i_0(A-1)} \left\{ \exp\left[\frac{(A-1)\left(t + \frac{R_1}{W_e} - \frac{r}{W_e}\right)}{A t_{eff}} \right] - 1 \right\}
\tag{8.71}
$$

in which the effective transit time is defined according to Heylen's[23] derivation

$$
\frac{1}{t_{eff}} = \frac{\gamma_+}{\gamma_T} \frac{1}{t_i} + \frac{\gamma_{ph}}{\gamma_T} \frac{1}{t_e}
\tag{8.72}
$$

The formative time lag is given by that value of t in equation (8.71) for which a specified multiplication of current occurs for breakdown to become inevitable.

To demonstrate the applicability of the theory the following experimental conditions are employed: $R_1 = 0.7$ cm, $R_2 = 3.1$ cm, $N = 3.54 \times 10^{21}$ m^{-3} and 1.75×10^{21} m^{-3}. Very good agreement between measured and calculated[24] time lags are obtained as shown in Fig. 8.10.

8.6 THE STREAMER MECHANISM

The theory of electrical breakdown based on the regenerative mechanism of secondary electrons at the cathode, mainly due to the photoelectric action at the cathode, is applicable to low gas pressures. Experimental evidence suggests that at high pressures (~ 1 atm. and above) streamers develop in the gap due to photo-ionization in the gas and breakdown occurs at time lags in the range of 1 ns-10μs, depending upon the over-voltage, gas pressure, gap length, etc.

The origin of the streamer inception is the number of electrons at the head of the advancing primary avalanche. When the number reaches a value in the range of $10^7 - 10^8$ the electric field due to the space charge is sufficiently strong to allow secondary electrons due to photo-ionization to develop into the avalanche head. The streamer theory explains many details observed in non-uniform field breakdown in long gaps and electrical corona. The dependence of the corona inception voltage and breakdown voltage on the polarity of the applied voltage may also be explained using the streamer criterion. The streamer mechanism was first proposed by Meek (Meek and Craggs (1978) and independently by Raether[25] and many details have been clarified in the literature since then.

The streamer mechanism of electrical breakdown is based on the following considerations. The primary avalanche is initiated by a single electron and as the avalanche moves close to the anode there is a distribution of positive ions which is most intense at the head of the avalanche. The positive ions produce a field distortion in both the radial and axial direction. The axial component will add to the applied field, whereas the radial component will allow the electrons generated due to photo-ionization to grow into it. The radial component of the field should be of the same order of magnitude as the applied field for the successful inception of a streamer.

Fig. 8.11 shows the development of a secondary avalanche into the head of the advancing streamer. The secondary electrons generated due to photo-ionization in the gas should be within the active region which is defined as the region that allows for electron multiplication. In electron attaching gases the boundary of the active region is determined by the criterion $\alpha = \eta$. In air at atmospheric pressure this criterion is satisfied at 26 kV/cm and in SF_6 at 91 kV/cm. In low electric fields the streamer may still advance because of its own field provided the number of ions is approximately 10^8 and the avalanche head has a radius of 30 μm[26]. The field in the active region may be calculated as

$$E = E_a + \frac{N_s e}{4\pi\varepsilon_0 (z_0 + R_s)^2} \tag{8.73}$$

where E_a is the applied field, N_s the number of charges in the avalanche head of radius R_s and Z_0 the starting point of the secondary avalanche along the axis (Fig. 8.11).

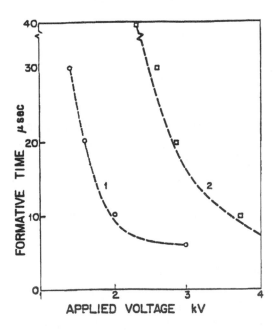

Fig. 8.10 Formative time lags in nitrogen at $N = 3.54 \times 10^{21}$ m^{-3} (curve 1) and $N = 1.57 \times 10^{21}$ m^{-3} (curve 2). Broken lines: theory, symbols: experimental. (Govinda Raju, 1983, with permission of American Institute of Physics).

Substituting typical values for SF$_6$ 91 kV/cm, $N_s = 10^8$, $z_0 = 80$ μm, $R_s = 30$ μm the field in the active region is E = 75 kV/cm which is of the same order of magnitude as the applied voltage.

The secondary avalanche starts to grow in the active region as long as the energy gained by the drifting electrons from the applied field is balanced by the energy lost in collisions with gas molecules (ionization, attachment, momentum transfer etc.) and the energy stored as potential energy in the streamer tip. The energy balance equation is

$$\varepsilon_e = \varepsilon_a + \Delta\,\varepsilon_{pot} \tag{8.74}$$

where ε_e is the energy lost in collisions (momentum transfer, electronic excitation, dissociation, attachment, vibrational excitation and rotation, ionization), and ε_{pot} the potential energy because of the concentration of the charges and their position in the applied field. Figure 8.12 shows the agreement between theoretical values based on the model and experimental results (Gallimberti, 1981).

The criterion for the initiation of the streamers from a single electron avalanche in high pressure electron attaching gases (p > 1 bar) has been expressed as $\alpha = \eta$. This condition was initially derived by Harrison and Geballe (Loeb, 1965) and used by many authors to calculate or predict breakdown voltages. The physical significance of this criterion is that the electron avalanche ceases to grow at this limiting condition, often expressed as $(E/N)_{lim}$ and therefore, strictly speaking, it does not lead to breakdown. The condition implies that if the voltage exceeds that given by $(E/N)_{lim}$ by a small fraction (~0.1%)the electron avalanche multiplies so rapidly that breakdown ensues. The limiting value is 86.4 V/m Pa for SF_6 , 89.5 V/m Pa for CCl_2F_2 and 23.6 V/m Pa for air.

8.6.1 Leader Mechanism

The mechanisms of breakdown in long gaps, as in spark-over in a transmission line, discharges along bushings and in lightning are different from the streamer mechanism in uniform electric fields. Because of the very high voltages applied and the non-uniformity of the electric field, the field near the anode is very high. Intense ionization occurs close to the electrode and the electrons collide with neutral molecules transferring energy.

The local temperature may rise to as high as 10^4 K. Thermal ionization occurs and the area surrounding the electrode becomes a hot plasma channel, called the leader channel. The thermal ionization renders the channel highly conductive and the axial field is quite small. However the radial variation of the field is sharp. The head of the channel is approximately at the same potential as the electrode. The electric field ahead of the channel is intensified and several streamers are launched. The electrons from the new streamers flow back into the head of the channel and lengthen the leader until it extends to the cathode. The velocity of propagation of a leader can reach values as high as 2×10^6 m/s.

Leader initiated breakdown may actually begin as corona at the high voltage electrode. Corona formation results in a pre-ionized region making possible the transition to a leader or facilitating the propagation of the leader. Figure (8.13) shows a beautiful picture of corona formation in the gap while the leader has bridged the gap partially. The corona region adjacent to the leader is very similar to a glow discharge in characteristics. The region is almost charge neutral, with a low concentration of electrons and a high concentration of equal positive and negative ions. The resulting conductivity is quite low. Several models have been proposed for the transition of corona into a leader, one of them being the thermal ionization briefly referred to above.

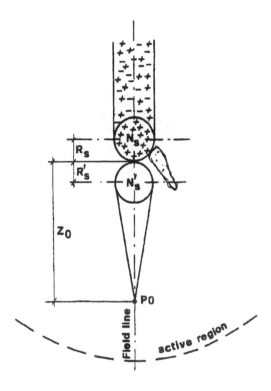

Fig. 8.11   Development of secondary avalanche into advancing streamer, adopted from (Gallimberti, 1981, with permission from Plenum Press, New York).

8.7 FIELD DISTORTION DUE TO SPACE CHARGE

When the electron multiplication is large the difference in mobility of electrons and positive ions introduces a space charge which distorts the applied field. To calculate the inhomogeneity that arises due to space charge we consider a parallel plane gap with an applied voltage V, and the electric field is uniform initially across the gap length d. Multiplication of electrons occurs and the electric field at dx at a distance of x from the cathode is obtained by applying Poisson's equation,

$$\frac{dE}{dx} = -\frac{\rho}{\varepsilon_0} \qquad (8.75)$$

where ρ is the volume charge density. As a simplification we assume that the charge density is proportional to $e^{\alpha x}$ where x is the distance over which the growth of avalanche has occurred.

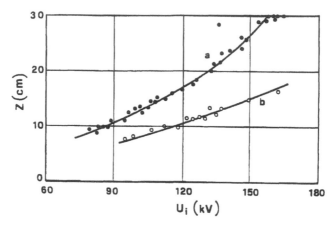

Fig. 8.12 The streamer length as a function of the inception voltage for gap lengths of 30 cm (curve a) and 150 cm (curve b); the solid lines represent the computed values; The points are experimental results (square cut rod of 1 cm radius). (Gallimberti, 1978, with permission of Plenum Press, New York.)

Substituting $\rho = Ke^{\alpha x}$ where K is a constant having the dimension of C m^{-3} and integrating equation (8.75) we get the electric field at x as

$$E_x = -\frac{K}{\varepsilon_0}\frac{e^{\alpha x}}{\alpha} + C \qquad (8.76)$$

where C is the integration constant. At the cathode $x = 0$ and $E_x = E_c$. Substituting these values in equation (8.76) we get

$$C = E_c + \frac{K}{\varepsilon_0 \alpha} \qquad (8.77)$$

Substituting equation (8.77) in (8.76) we get

$$E_x = \frac{K\rho}{\varepsilon_0 \, \alpha} - \frac{K\rho}{\varepsilon_0} \frac{e^{\alpha x}}{\alpha} + E_c \tag{8.78}$$

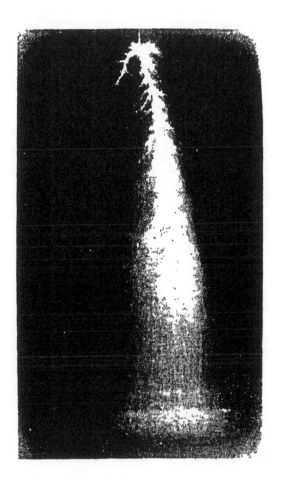

Fig. 8.13 Partially developed leader in 1.5 m gap. The leader ahead of the streamer can be clearly seen (Gallimberti, 1981, with permission of Plenum Press, New York).

The multiplication of electrons required to distort the applied electric field should be large, i.e., $e^{\alpha x} >> 1$ and we can make the approximation

$$E_x = E_c - \frac{K\rho}{\varepsilon_0 \, \alpha} e^{\alpha x} \tag{8.79}$$

Since the applied voltage is constant we may express it as

$$\int_0^d E_x dx = V \tag{8.80}$$

Integration yields

$$-\frac{K\rho}{\varepsilon_0 x^2}[e^{\alpha x}]_0^d + E_c d = V \tag{8.81}$$

Substituting the limits and approximating $(e^{\alpha d} - 1) \approx e^{\alpha d}$ we get

$$\frac{K\rho}{\varepsilon_0 x^2}[e^{\alpha d}] + E_c d = V \tag{8.82}$$

combining equations (8.82) and (8.79) we get

$$E_x = \frac{V}{d} + \frac{K\rho}{\varepsilon_0 d^2}\frac{e^{\alpha d}}{d} - \frac{K}{\varepsilon_0 \alpha}e^{\alpha x} \tag{8.83}$$

This equation may be rewritten as

$$E_x = E_0 + \frac{K\rho}{\varepsilon_0}\left(\frac{e^{\alpha d}}{\alpha d} - e^{\alpha x}\right) \tag{8.84}$$

where E_0 is the applied field. The electric field at the cathode, i. e., at $x = 0$ is

$$E_c = E_0 - \frac{K}{\varepsilon_0 \alpha}\frac{e^{\alpha d}}{\alpha d} \tag{8.85}$$

The field at the anode, i. e., at $x = d$ is

$$E_a = E_0 - \frac{K}{\varepsilon_0 \alpha}e^{\alpha d} \qquad \text{because } \alpha d \gg 1$$

We can calculate the current which increases the cathode field significantly, say by 10%. Let us assume that the applied voltage is 30 kV across 1 cm gap length in air at atmospheric pressure. The multiplication appropriate to these conditions is $\alpha d \sim 18$. Using equation (8.85) the constant K is evaluated as 13.5×10^{12} C/m^3. The average diameter of the electron avalanche is approximately 0.1 mm and an approximate volume of 5×10^{-12} m^3. Hence $K = 1.6 \times 10^{-19} /(5 \times 10^{-12}) = 3 \times 10^{-6}$ C/m^3 and $\rho = K e^{\alpha x} = 3 \times 10^{-6} \times e^{18 \times 0.01} = 3 \times 10^{-6}$ C/m^3. The current density corresponding to 10% field distortion at the cathode is given by $j = \rho \mu E_c$ where μ is the mobility of positive ions (10^{-3} m^2 $V^{-1}S^{-1}$) and $E_c = 1.1 \times 3 \times 10^6$ V/m. Substituting these values we obtain a current density of 10 mA/m^2. A current of this magnitude is observed experimentally at higher pressures emphasizing the need for taking space charge distortion into account.

8.8 SPARKOVER CHARACTERISTICS OF UNIFORM FIELD GAPS IN SF$_6$

A large volume of experimental data is available in the literature on the sparkover characteristics of uniform field gaps[27] in various gases and we will not attempt to present these data. However, data in SF$_6$ are plotted in figure (8.14) and compared with those in nitrogen. SF$_6$ has a higher breakdown voltage when compared with nitrogen and at atmospheric pressure the ratio of breakdown voltages is approximately 2.95. The higher dielectric strength is attributed to the electron attachment in SF$_6$.

The ratio of V_s/Nd gives the value of E_s/N at which breakdown occurs and this quantity is also plotted in figure 8.9 for both gases. For a non-attaching gas such as N$_2$ the ratio E_s/N decreases at first rapidly but at larger values of Nd the decrease is rather more gradual. However for an electron attaching gas there is a limiting value of E_s/N below which breakdown does not occur. At this limiting value the condition $\alpha/N = \eta/N$ holds and for SF$_6$ the limiting E/N = 361 (Td). The proof for the existence of a limiting value of E/N for electron attaching gases may be obtained as follows:

For electron-attaching gases the breakdown criterion is given by equation (8.19). Rearranging terms we get

$$\gamma = \frac{(\alpha - \eta)}{\alpha} \frac{1}{\exp[(\alpha - \eta)d] - 1} \tag{8.86}$$

Since α/N and η/N are both functions of E/N the breakdown voltage is dependent on α as well as $(\alpha-\eta)$. As the product Nd increases the reduced field E_s/N decreases as demonstrated in figure 8.14, thereby decreasing α/N and increasing η/N.

As $\alpha \to \eta$ at constant N equation (8.86) becomes

$$\gamma = \underset{(\alpha-\eta) \to 0}{Lt} \frac{(\alpha-\eta)}{\alpha[\exp(\alpha-\eta)d-1)]} \tag{8.87}$$

$$= \frac{1}{\alpha d} \tag{8.88}$$

When $\alpha = \eta$ breakdown occurs for a value of γ given by equation (8.88).

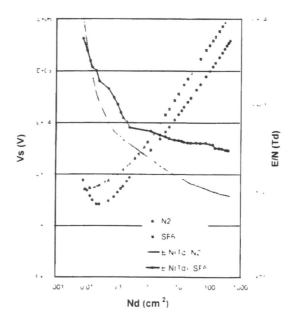

Fig. 8.14 Breakdown voltages in N_2 and SF_6 [data from Meek and Craggs, 1978]. (With permission of John Wiley and Sons, New York.)

If $\alpha \ll \eta$ equation (8.86) becomes

$$\gamma = \frac{\eta}{\alpha} \frac{1}{(1-\exp-\eta d)} \tag{8.89}$$

Both the terms on the right side are greater than unity and therefore breakdown occurs only if $\gamma > 1$. In electron attaching gases the secondary ionization co-efficient is usually quite small, for example in SF_6 it is of the order of 10^{-5} and in oxygen $\sim 10^{-4}$. Therefore breakdown does not occur for $\alpha < \eta$.

8.9 SPARKOVER CHARACTERISTICS OF LONG GAPS

Due to its engineering importance corona inception and breakdown of long gaps ≥ 1m have been studied extensively in various electrode configurations, under d.c. and impulse voltages of various wave shapes. There is generally good understanding of the various features of discharges observed and a general picture may be drawn to characterize the influence of the gap length and the polarity on the corona inception (V_c) and breakdown voltages (V_s). Fig. (8.15) represents schematically the general characteristics in a point plane gap and sphere-plane gaps which present non-uniform electric fields of varying degree, depending upon the diameter of spheres[28].

At small gap lengths on the order of a cm or less, it is difficult to detect corona before the sparkover takes place, and the curve for corona inception and sparkover coincide irrespective of the polarity of the point electrode. As the gap length is increased the corona inception voltage may be distinguished from the sparkover voltage and the difference between the two voltages increases as the gap length increases, making it easier to observe the polarity dependence of V_c. The gap length at which the onset of corona is observed is called the critical gap length.

The corona inception voltage for the negative polarity voltage is lower than that due to the positive polarity for all electrode geometries, possibly due to the fact that the electric field at the tip of the point is several times the average field in the gap. Initiating electrons are therefore more successfully launched into the gap as the avalanche grows from the high field electrode. Availability of a fortuitous electron to launch an avalanche, due to dissociation of a negative ions close to the negative point electrode, is also a contributing factor. The corona inception voltage for both positive and negative point electrode remains essentially the same as the gap length is increased. This is due to the fact that the gap length has relatively minor influence in lowering the electric field at the tip of the point electrode.

The sparkover voltage for positive polarity (curve 1, 3, 5 & 7 in Fig. 8.15) increases with the gap length and the difference between V_c and V_s increases as well. The initial phase of discharge in this region is an intense corona discharge from the point which develops into a leader. The positive leader is associated with a more diffuse discharge (Fig. 8.13) at its tip and is termed "leader-corona".

After the passage of the leader tip the luminosity falls, until complete bridging of the gap causes the narrow channel to be reilluminated by the main discharge current (Meek & Craggs, 1978). The return stroke which is sustained by the main current does not illuminate the branches. If the earthed electrode is a point or a rod then a negative leader propagates towards the positive leader. The junction of the two leaders occurs with a small region surrounding the junction point being brighter than the rest of the gap. The junction of the two leaders launches the return stroke. The negative leader velocity is about 30% of the positive leader.

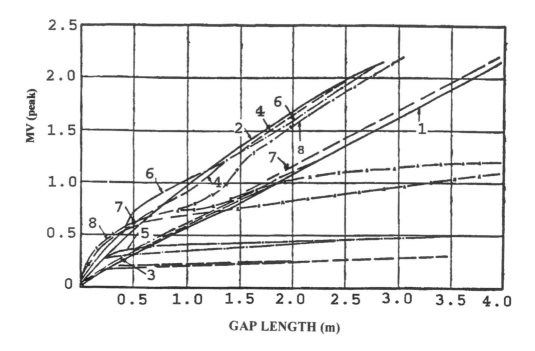

Fig. 8.15 Impulse breakdown voltage characteristics of large air gaps [Author's measurements]. The electrode geometry is: Breakdown voltage:1&2, point-plane gap, positive and negative polarity, 3&4, 6.25 cm diameter sphere-plane gap, positive and negative polarity, 5&6, 12.5 cm diameter sphere-plane gap, positive and negative polarity, 7&8. 50 cm diameter sphere plane gap, positive and negative polarity. Corona inception voltage: 9&10, point plane gap, positive and negative polarity, 11&12, 6.25 cm diameter sphere-plane gap, positive and negative polarity, 13 &14, 12.5 cm diameter sphere-plane gap, positive and negative polarity.

The sparkover voltage for a negative point electrode is higher than that for positive polarity. The difference in the sparking voltage increases with gap length as shown in Curve *d* of Fig. 8.15. In the case of negative polarity the avalanche growth is quenched at the edge of the active region and thereafter the electrons drift in a low electric field towards the earthed electrode. The electrons get attached to the neutral molecules forming a negative ion barrier that drifts very slowly (~ms) due to the low ionic mobility.

This barrier reduces the electric field in the region between the cathode and the negative ion space charge, suppressing the growth of the leader channel. Replacing the earthed electrode by a point or rod permits the propagation of vigorous positive leader channels from a point of contact of negative corona with the anode. The hatched region in Fig. 8.9 shows that a reduction of sparkover voltage with increasing gap length may be observed under certain combination of electrode geometry and atmospheric conditions such as pressure and humidity. The role of ion space charge is quite complex in these situations and a detailed analysis will not be attempted here.

8.10 BREAKDOWN VOLTAGES IN AIR WITH ALTERNATING VOLTAGES

Fig. 8.16 shows the RMS value of the corona inception and breakdown voltage for various gaps in the range of gap length, $11\mu m \leq d \leq 0.1\ m$ covering a voltage range of $0.5 \leq V \leq 90\ kV$[29]. The gaps were irradiated with ultraviolet radiation to reduce scatter in both V_c and V_s. The scatter in observed voltages was much greater in the region where corona inception was not distinguishable from breakdown voltage. However in the region where corona inception preceded the breakdown by a significant voltage difference, the effect of irradiation was reduced, possibly due to the fact that corona provides considerable number of initiatory electrons.

The breakdown voltage of non-uniform field gaps under 50 Hz voltages is given by the empirical formulae[30]:

$$V_s = 25 + 4.55\ d \qquad \text{for rod-plane gaps}$$
$$V_s = 10 + 5.25\ d \qquad \text{for rod-rod gaps}$$

where V_s is expressed in kilovolts and the gap length d in centimetres. For much larger gap lengths, in the range of $1 \leq d \leq 9$ meters, the breakdown voltage follows the equation (Meek and Craggs, 1978)

$$V_s = 0.0798 + 0.4779d - 0.0334d^2 + 0.0007d^3 \qquad \text{rod-plane gaps}$$
$$V_s = -0.990 + 0.6794d - 0.405d^2 \qquad \text{rod-rod gaps}$$

8.11 CONCLUDING REMARKS

In this chapter we have presented the essentials of gas discharge phenomena covering a wide range of experimental conditions. The details of the mechanisms operative in gas discharges depend upon so many parameters that an exhaustive discussion is not possible in a limited space. In addition to the dc and alternating voltages, equipments in the electrical power sector encounter transient voltages. The transient voltages may have a rise time that varies from a few (~10-50) nanosecond to as long as 250 μs. Lightning impulses have a rise time of the order of 1.5 μs and the switching impulses have a rise time upto 300 μs. A significant of failures of gas insulated equipments have failed in practice due to fast rising transients (≤ 10 ns)[31]. A notable contribution has been made in predicting the breakdown in SF_6 under impulse conditions by Xu et. al.[32] and the developed model yields results that are in agreement with the measured breakdown probabilities (Fig. 8.17).

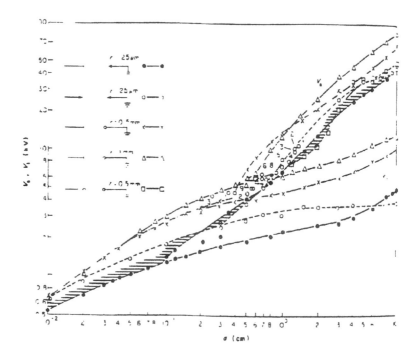

Fig. 8.16 RMS value of the 50 Hz corona inception voltage V_c and breakdown voltage V_s for different electrode radii, r, in atmospheric air. The points 1-13 refer to gaps that occur in printed circuits (Meek and Craggs, 1978, with permission of John Wiley & Sons, New York)

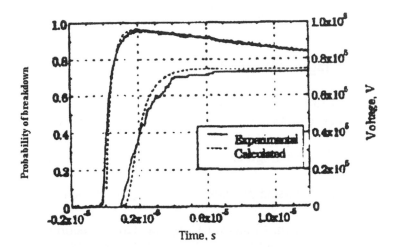

Fig. 8.17 Measured and calculated breakdown probabilities for the lightning impulse at 100 kV crest voltage along with the test voltage wave form (upper trace), (Xu et. al, 1996, ©IEEE)

The more fundamental aspects of determining the nature of corona and swarm parameters from the electron energy distribution functions will be dealt with in the next chapter.

8.12 REFERENCES

[1] J. M. Meek and J. D. Craggs, "Electrical breakdown in Gases", John Wiley & Sons, 1978.

[2] L. B. Loeb, "Electrical Coronas", University of California Press, 1965, R. S. Sigmond and M. Goldman, "Electrical Breakdown and Discharges in Gases, Part B., Edited by Kunhardt and L. H. Leussen pp. 1-64, Plenum Press, New York, 1983.

[3] J. R. Roth, "Industrial Plasma Engineering, Vol. 1, Principles", Institute of Physics Publishing, Bristol, 1995.

[4] S. C. Brown, Basic Data of Plasma physics, Technology Press and Wiley, New York, 1959.

[5] W. M. Huo and Yong-Ki Kim, "Electron Collision Cross-Section Data for Plasma Modeling", IEEE Transactions on Plasma Science, **27** (1999) 125.

[6] D. Rapp and P. Golden-Englander, "Total Cross Sections for Ionization and Attachment in Gases by Electron Impact. I Positive Ionization", J. Chem. Phys., Vol. **43** (1965) 1464.

[7] L. J. Kieffer, G. H. Dunn, "Electron Impact Ionization Cross-section Data for Atoms, Atomic Ions and Diatomic Molecules: I. Experimental Data", Rev. Mod. Phy., **38** (1966) 1.

[8] D. Rapp, P. Golden-Englander and D. D. Braglia, "Cross sections for Dissociative Ionization of Molecules by Electron Impact", J. Chem. Phys., **42** (1965) 4081.

[9] V. M. Pejcev, M. V. Kurepa and I. M. Cadez, Chem. Phys. Letters, **63** (1979) 301.

[10] B. L. Moseiwitch and S. J. Smith, "Electron impact excitation of atoms", Rev. Mod. Phys., **40** (1968) 238.

[11] D. Rapp and D. D. Braglia, "Total Cross Sections for Ionization and Attachment in Gases by Electron Impact. II Negative Ion Formation", J. Chem. Phys., **43** (1963) 1480.

[12] G. R. Govinda Raju, J. A. Harrison and J. M. Meek, Brit. J. Appl. Phys., **16** (1965) 933.

[13] G. R. Govinda Raju, J. A. Harrison and J. M. Meek, Proc. IEE., **111** (1964) 2097.

[14] G. R. Govinda Raju and C. Raja Rao, J. Phys. D., Appl. Phys., **4** (1971) 494.

[15] A. Przybylyski, Zeits. fur Physi., **168**, 1962.

[16] D. K. Gupta, S. Mahajan and P. I. John, J. Phys.D., Appl. Phys., **33** (2000) 681.

[17] R. Morrow, Physical Review A, **32** (1985) 1799.

[18] R. Morrow, IEEE Trans. on Plasma Science, **PS-14** (1986) 234.

[19] G. R. Govinda Raju and J. Liu, IEEE Trans. on DI & EI, **2** (1995) 1004.

[20] J. Dutton, p. 219, Ch. 3 in ref. 1.

[21] P. M. Davidson, Brit. J. Appl. Phys., **4** (1953) 170.

[22] G. R. Govinda Raju, J. Appl. Phys., **54** (1983) 6745.

[23] A. E. D. Heylen, Proc. IEE, **120** (1972) 1565.

[24] A. D. Mokashi and G. R. Govinda Raju, IEEE Trans. on Electrical Insulation, **18** (1983) 436.

[25] H. Raether, "Electron Avalanches and Breakdown in Gases", Butterworth, London, 1964.

[26] I. Gallimberti, "Long Air Gap Breakdown Processes", in "Electrical Breakdown and Discharges in Gases", Ed: E. E. Kunhardt and L. H. Luessen, Plenum Press, New York, 1981.

[27] D. T. A. Blair, "Breakdown Voltage Characteristics", in , "Electrical breakdown in Gases", Ed; J. M. Meek and J. D. Craggs, John Wiley & sons, 1978.

[28] Present author's measurements (1958).

[29] W. Hermstein, Elektrotech. Z. Ausg., **A90** (1969) 251.

[30] K. Feser, Energie und Technik, **22** (1970) 319.

[31] S. A. Boggs, F. Y. Chu, N. Fujimoto, A. Krenicky, A. Plessel and D. Schlicht, IEEE Transactions on Power Apparatus and Systems, **101** (1982) 3593.

[32] X. Xu, S. Jayaram and S. A. Boggs, IEEE Transactions on Dielectrics and Electrical Insulation, **3** (1996) 836.

Common sense is nothing more than a deposit of prejudices laid down in the mind before you reach eighteen.
                                        - Albert Einstein

9

FUNDAMENTAL ASPECTS OF GASEOUS BREAKDOWN–II

We continue the discussion of gaseous breakdown shifting our emphasis to the study of phenomena in both uniform and non-uniform electrical fields. We begin with the electron energy distribution function (EEDF) which is one of the most fundamental aspects of electron motion in gases. Recent advances in calculation of the EEDF have been presented, with details about Boltzmann equation and Monte Carlo methods. The formation of streamers in the uniform field gap with a moderate over-voltage has been described. Descriptions of Electrical coronas follow in a logical manner. The earlier work on corona discharges has been summarized in several books[1,2] and we shall limit our presentation to the more recent literature on the subject. However a brief introduction will be provided to maintain continuity.

9.1 ELECTRON ENERGY DISTRIBUTION FUNCTIONS (EEDF)

One of the most fundamental aspects of gas discharge phenomena is the determination of the electron energy distribution (EEDF) that in turn determines the swarm parameters that we have discussed briefly in section (8.1.17). It is useful to recall the integrals that relate the collision cross sections and the energy distribution function to the swarm parameters. The ionization coefficient is defined as:

$$\frac{\alpha}{N} = \frac{1}{W}\left(\frac{2e}{m}\right) E \int_{\varepsilon_i}^{\infty} \varepsilon^{1/2} Q_i(\varepsilon) F(\varepsilon) d\varepsilon \tag{9.1}$$

in which e/m is the charge to mass ratio of electron, $F(\varepsilon)$ is the electron energy distribution function, ε the electron energy, ε_i the ionization potential and $Q_i(\varepsilon)$ the ionization cross section which is a function of electron energy. Other swarm parameters are similarly defined. It is relevant to point out that the definition of (9.1) is quite general and does not specify any particular distribution. In several gases $Q_i(\varepsilon)$ is generally a function of ε according to (Fig. 8.4),

$$Q_i = \frac{Q_i^0 (\varepsilon - \varepsilon_i)^n}{\varepsilon} \tag{9.2}$$

Substitution of Maxwellian distribution function for $F(\varepsilon)$, equation (1.92) and equation (9.2) in eq. (9.1) yields an expression similar to (8.11) thereby providing a theoretical basis[3] for the calculation of the swarm parameters.

9.1.1 EEDF: THE BOLTZMANN EQUATION

The EEDF is not Maxwellian in rare gases and large number of molecular gases. The electrons gain energy from the electric field and lose energy through collisions. In the steady state the net gain of energy is zero and the Boltzmann equation is universally adopted to determine EEDF. The Boltzmann equation is given by[4]:

$$\frac{\partial}{\partial t} F(r,v,t) + a \cdot \nabla_v F(r,v,t) + v \cdot \nabla_r F(r,v,t) = J\left[F(r,v,t)\right] \tag{9.3}$$

where F is the EEDF and J is called the collision integral that accounts for the collisions that occur. The solution of the Boltzmann equation gives both spatial and temporal variation of the EEDF. Much of the earlier work either used approximations that rendered closed form solutions or neglected the time variation treating the equation as integro-differential. With the advent of fast computers these are of only historical importance now and much of the progress that has been achieved in determining EEDF is due to numerical methods.

The solution of the Boltzmann equation gives the electron energy distribution (EEDF) from which swarm parameters are obtained by appropriate integration. To find the solution the Boltzmann equation may be expanded using spherical harmonics or the Fourier expansion. If we adopt the spherical harmonic expansion then the axial symmetry of the discharge reduces it to Legendre expansion and in the first approximation only the first two terms may be considered. The criterion for the validity of the two term expansion is that the inelastic collision cross sections must be small with respect to the elastic collision cross sections or that the energy loss during elastic collisions should be small. These assumptions may not be strictly valid in molecular gases where inelastic collisions occur with large cross sections at low energies due to vibration and rotation. The two-term solution method is easy to implement and several good computer codes are available[5].

The Boltzmann equation used by Tagashira et. al.[6] has the form

$$\frac{\partial}{\partial t} \int_0^\infty n(\varepsilon', z, t) d\varepsilon' = N_c + N_E + N_z \tag{9.4}$$

where n (ε',z,t) is the electron number density with (ε', z, t) as the energy, space and time variables, respectively, N_c, N_E and N_z are the change rate of electron number density due to collision, applied electric field and gradient, respectively. Equation (9.4) has a simple physical meaning: the electron number density is conserved. The solution of equation (9.4) may be written in the form of a Fourier expansion[7]:

$$n_s(\varepsilon, z, t) = e^{isz} e^{-w(s)t} H_o(\varepsilon, s) \tag{9.5}$$

where s is the parameter representing the Fourier component and

$$w(s) = -w_o + w_1(is) - w_2(is)^2 + w_3(is)^3 ... \tag{9.6}$$

$$H_o(\varepsilon, s) = f_o(\varepsilon) + f_1(\varepsilon)(is) + f_2(\varepsilon)(is)^2 + ... \tag{9.7}$$

where w_n (n = 0, 1, 2, ...) are constants. The method of obtaining the solution is described by Liu [7]. The method has been applied to obtain the swarm parameters in mercury vapor and very good agreement with the Boltzmann method is obtained. The literature on Application of Boltzmann equation to determine EEDF is vast and, as an example, Table 9.1 lists some recent investigations in oxygen[8].

9.1.2 EEDF: THE MONTE CARLO METHOD

The Monte Carlo method provides an alternative method to the Boltzmann equation method for finding EEDF (Fig. 9.1). and this method has been explored in considerable detail by several groups of researchers, led by, particularly, Tagashira, Lucas and Govinda Raju. The Monte Carlo method does not assume steady state conditions and is therefore responsive to the local deviations from the energy gained by the field. Different methods are available.

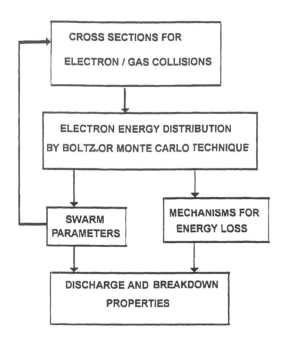

Fig. 9.1 Methods for determining EEDF and swarm parameters

A. MEAN FREE PATH APPROACH

In a uniform electric field an electron moves in a parabolic orbit until it collides with a gas molecule. The mean free path λ (m) is

$$\lambda = \frac{1}{N Q_t(\varepsilon)} \qquad (9.8)$$

where Q_t is the total cross section in m^2 and ε the electron energy in eV. Since Q_t is a function of electron energy, λ is dependent on position and energy of the electron. The mean free path is divided into small fractions, $ds = \lambda / a$, where a is generally chosen to be between 10 and 100 and the probability that an electron collides with gas molecules in this step distance is calculated as $P_1 = ds/\lambda$. The smaller the ds is chosen, the longer the calculation time becomes although we get a better approximation to simulation. The collision event is decided by a number of random numbers, each representing a particular type.

B. MEAN FLIGHT TIME APPROACH

The mean flight time of an electron moving with a velocity $W(\varepsilon)$ is

$$T_m = \frac{1}{NQ_T(\varepsilon)W(\varepsilon)} \tag{9.9}$$

where $W(\varepsilon)$ is the drift velocity of electrons. The time of flight is divided into a number of smaller elements according to

$$dt = \frac{T_m}{K} \tag{9.10}$$

where K is a sufficiently large integer.

The collision frequency may be considered to remain constant in the small interval dt and the probability of collision in time dt is

$$P = 1 - \exp\left[-\frac{dt}{T_m}\right] \tag{9.11}$$

For each time step the procedure is repeated till a predetermined termination time is reached. Fig. (9.2) shows the distribution of electrons and energy obtained from a simulation in mercury vapour.

C. NULL COLLISION TECHNIQUE

Both the mean free path and mean collision time approach have the disadvantage that the CPU time required to calculate the motion of electrons is excessively large. This problem is simplified by using a technique known as the null collision technique. If we can find an upper bound of collision frequency v_{max} such that

$$v_{max} = \max[NQ_T(\varepsilon)W(\varepsilon)] \tag{9.12}$$

and the constant mean flight time is $1/v_{max}$ the actual flight time is

$$dt = -\frac{\ln R}{V_{max}} \qquad (9.13)$$

where R is a random number between 0 and 1.

Table 9.1

Boltzmann Distribution studies in oxygen. The parameters calculated are indicated. W = Drift velocity of electrons, ε_m = mean energy, ε_k = characteristic energy, η = attachment co-efficient, α = ionization co-efficient, $f(\varepsilon)$ = electron energy distribution, × denotes the quantities calculated [Liu and Raju, 1995].

Author	E/N (Td)	W	ε_m	ε_k	η	α	$f(\varepsilon)$
Hake et. al.[9]	0.01-150	×	×	×	×	×	×
Myers[10]	10^{-3}-200		×	×			
Wagner[11]	90-150				×	×	
Lucas et. al.[12]	15-152	×	×		×	×	×
Masek[13]	1-140	×	×	×		×	×
Masek et. al[14]	1-200	×	×	×			×
Masek et. al.[15]	10-200				×		×
Taniguchi et. al.[16]	1-30				×		
Gousset et. al.[17]	0.1-130	×	×	×		×	×
Taniguchi et. al.[18]	0.1-20	×			×		×
Liu and Raju (1993)	20-5000	×	×	×	×	×	×

The assumed total collision cross section Q_t is

$$Q_T = Q_T + Q_{null} \qquad (9.14)$$

where Q_{null} is called the null collision cross section.

We can determine whether the collision is null or real after having determined that a collision takes place after a certain interval dt. If the collision is null we proceed to the next collision without any change in electron energy and direction. In the mean free path and mean flight time approaches, the motion of electrons is followed in a time scale of T_m / k while in the null collision technique it is on the T_m scale. The null collision technique is computationally more efficient but it has the disadvantage that it cannot be used in situations where the electric field changes rapidly.

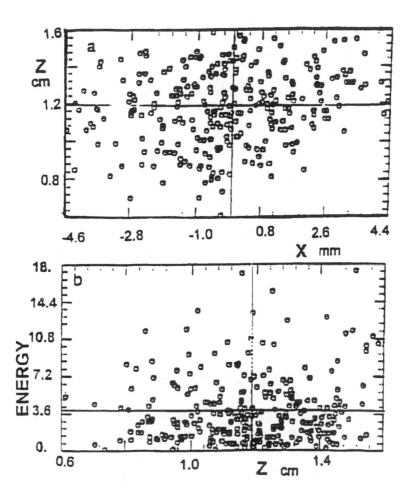

Fig. 9.2 Distribution of electrons and energy in mercury vapour as determined in Monte-Carlo simulation, E/N = 420 Td. T = 40 ns [Raju and Liu, 1995, with permission of IEEE ©.)

D. MONTE CARLO FLUX METHOD

In the techniques described above, the electron trajectories are calculated and collisions of electrons with molecules are simulated. The swarm parameters are obtained after following one or a few electrons for a predetermined period of distance or time. A large number of electrons are required to be studied to obtain stable values of the coefficients, demanding high resolution and small CPU time, which are mutually contradictory. The problem is particularly serious at low and high electron energies at which the distribution function tends to have small values. To overcome these difficulties Schaffer and Hui[19] have adopted a method known as the Monte Carlo flux method which is based on the concept that the distribution function is renormalized by using weight factors which have changing values during the simulation. The low energy and high energy part of the distribution are also redetermined in a separate calculation.

The major difference between the Monte Carlo flux method and the conventional technique is that, in the former approach, the electrons are not followed over a long period of time in calculating the transition probabilities, but only over a sampling time t_s. One important feature of the flux method is that the number of electrons introduced into any state can be chosen independent of the final value of the distribution function. In other words, we can introduce as many electrons into any phase cell in the extremities of the distribution as in other parts of the distribution.

The CPU time for both computations is claimed to be the same as long as the number of collisions are kept constant. The conventional method has good resolution in the ranges of energy where the distribution function is large, but poorer resolution at the extremities. The flux method has approximately the same resolution over the full range of phase space investigated. Table 9.2 summarizes some recent applications of the Monte Carlo method to uniform electric fields.

9.2   STREAMER FORMATION IN UNIFORM FIELDS

We now consider the development of streamers in a uniform field in SF_6 at small overvoltages ~ 1-10%. In this study 1000 initial electrons are released from the cathode with 0.1 eV energy[20]. During the first 400 time steps the space charge field is neglected. If the total number of electrons exceeds 10^4, a scaling subroutine chooses 10^4 electrons out of the total population. In view of the low initial energy of the electron, attachment is large during the first several steps and the population of electrons increases slowly. At electron density of 2×10^{16} m^{-3} space charge distortion begins to appear. The electric field behind and ahead of the avalanche is enhanced, while in the bulk of the avalanche the field is reduced.

In view of the large attachment the number of electrons is less than that of positive ions, and the field behind the avalanche is enhanced. On the other hand, the maximum field enhancement in a non-attaching gas occurs at the leading edge of the avalanche. The development of streamers is shown in Fig. 9.3. As the first avalanche moves toward the anode, its size grows. The leading edge of the streamer propagates at a speed of 6.5×10^5 ms^{-1}. The trailing edge has a lower velocity ~ 2.9×10^5 ms^{-1}. At t = 1.4 ns, the primary streamer slows down (at A) by shielding itself from the applied field.

Table 9.2
Monte Carlo Studies in Uniform Electric Fields (Liu and Raju, (©1995, IEEE)

GAS	AUTHORS	RANGE (Td)	REFERENCE
N_2	Kucukarpaci and Lucas	$14 \leq E/N \leq 3000$	J. Phys. D.: Appl. Phys. **12** (1979) 2123-2138
	Schaffer and Hui	$50 \leq E/N \leq 300$	J. Comp. Phy. **89** (1990) 1-30
	Liu and Raju	$20 \leq E/N \leq 2000$	J. Frank. Inst. **329** (181-194) 1992; IEEE Trans. Elec. Insul. **28** (1993) 154-156.
	Lucas & Saelee	$14 \leq E/N \leq 3000$	J. Phys. D.: Appl. Phys. **8** (1975) 640-650.
	Mcintosh	$E/N = 3$	Austr. J. Phy. **27** (1974) 59-71.
	Raju and Dincer	$240 \leq E/N \leq 600$	IEEE Trans. on Plas. Sci., **17** (1990) 819-825
O_2	Liu and Raju	$20 \leq E/N \leq 2000$	IEEE Trans. Elec. Insul. **28** (1993) 154-156.
	Al Amin et. al	$25.4 \leq E/N \leq 848$	J. Phys. D.: Appl. Phys. **18** (1985) 1781-1794
air	Liu and Raju	$20 \leq E/N \leq 2000$	IEEE Trans. Elec. Insul. 28 (1993) 154-156.
CH_4	Al Amin et. al	$25.4 \leq E/N \leq 848$	J. Phys. D.: Appl. Phys. **18** (1985) 1781-1794
Ar	Kucukarpaci and Lucas	$141 \leq E/N \leq 566$	J. Phys. D.: Appl. Phys. **14** (1981) 2001-2014.
	Sakai et. al.	$E/N = 141. 283. 566$	J. Phys. D.: Appl. Phys. **10** (1995) 1035-1049.
Kr	Kucukarpaci and Lucas	$141 \leq E/N \leq 566$	J. Phys. D.: Appl. Phys. **18** (1985) 1781-1794
CO_2	Kucukarpaci and Lucas	$14 \leq E/N \leq 3000$	J. Phys. D.: Appl. Phys. **12** (1979) 2123-2138
H_2	Hunter	$1.4 \leq E/N \leq 170$	Austr. J. Phys., **30** (1977) 83-104
	Read & Hunter	$0.5 \leq E/N \leq 200$	Austr. J. Phys., **32** (1979) 255-259
	Blevin et. al	$40 \leq E/N \leq 200$	J. Phys. D.: Appl. Phys. **11** (1978) 2295-2303.
	Hayashi	$3 \leq E/N \leq 3000$	J. de Physique. **C740** (1979) 45-46
Hg	Liu & Raju	$10 \leq E/N \leq 2000$	J. Phys. D.: Appl. Phys. **25** (1992) 167-172
	Nakamura and Lucas	$0.7 \leq E/N \leq 50$	J. Phys. D.: Appl. Phys. **11** (1978) 337-345.
SF_6	Dincer and Raju	$300 \leq E/N \leq 540$	J. Appl. Phys., 54 (1983) 6311-6316
He	Braglia and Lowke	$E/N = 1$	J. de Physique **C740** (1979) 17-18
	Liu and Raju	$200 \leq E/N \leq 700$	IEEE Trans. on Plas. Sci., **20** (1992) 515-524
	Lucas	$30 \leq E/N \leq 150$	Int. J. Electronics, 32 (1972) 393-410
Na	Lucas	$0.7 \leq E/N \leq 50$	J. Phys. D..Appl. Phy. **11** (1978) 337-345

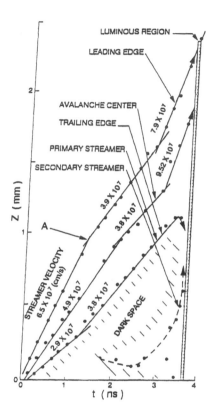

Fig. 9.3 Streamer development and calculated luminosity vs position and time in a uniform electric field at 7% over voltage (Liu and Raju, © 1993, IEEE)

The velocity of the leading edge decreases to 3.9×10^5 ms^{-1}; however the trailing edge propagates faster than before, at 3.8×10^5 ms^{-1}. The enhanced field between the cathode and the primary streamer is responsible for this increase in velocity. The secondary streamer, caused by photo-ionization, occurs at t = 2 ns and propagates very fast in the maximum enhanced field between the two streamers. The secondary streamer moves very fast and connects with the primary streamer within ~ 0.2 ns. The observed dark space exists for ~ 2 ns. These results explain the experimentally observed dark space by Chalmers et. al.[21] in the centre of the gap at 4% over-voltage. Between the primary and secondary streamer there is a dark space, shown hatched in Fig. 9.3.

The theoretical simulation of discharges that had been carried out till 1985 are summarized by Davies[22]. The two dimensional continuity equation for electron, positive ion and excited molecules in He and H$_2$ have been considered by Novak and Bartnikas[23, 24, 25,26]. Photoionization in the gap was not considered, but photon flux, ion flux and metastable flux to cathode as cathode emission were included. The continuity equations were solved by finite element method. Because of the steep, shock-like density gradients the solution by ordinary finite difference method is difficult and is limited to the early stages of streamer formation. Dhali and Pal[27] in SF_6 and Dhali and Williams[28] in N_2 handle the steep density gradients by using flux-corrected transport techniques which improved the numerical method for the two dimensional continuity equations.

9.3 THE CORONA DISCHARGE

Due to the technological importance of corona in electrophotography, partial discharges in cables, applications in the treatment of gaseous pollutants, pulsed corona for removing volatile impurities from drinking water etc. (Jayaram et. al., 1996), studies on corona discharge continue to draw interest. Corona is a self sustained electrical discharge in a gas where the Laplacian electric field confines the primary ionization process to regions close to high field electrodes. When the electric field is non-uniform, as exists in an asymmetrical electrode geometry (Fig. 9.4), the collision processes near the smaller electrode will be more intense than in other regions of the gap.

The non-uniformity of the electric field results in a partial breakdown of the gap. This phenomenon is called the electrical corona. The inter electrode gap may be divided into several regions[29], viz., (1) Glow region very close to the active (high voltage) electrode, (2) The drift region where ionization does not occur because of the low electric field and charge carriers drift in the field (3) charge free region which is separated by the active region by the Laplacian boundary.

The domain of the individual regions varies depending on the configuration of the electrode geometry, the characteristics of the insulating gas and the magnitude of the applied voltage which may or may not be time dependent. The existence of a charge-free region is not certain in all electrode configurations; for example in a concentric cylinder geometry we have only the glow and drift region. Depending upon the polarity of the smaller electrode the discharge that occurs in different manifestations though the Laplacian electric field is independent of the polarity.

In the active region, ionization by collision occurs and a self maintained discharge exists at sufficiently high voltage when according to the Townsend's criterion, $\gamma \exp(\int_0^{d_0} \alpha dx) = 1$ where d_0 is the edge of the glow region. The integral is used because the electric field is spatially varying and therefore the Townsend's first ionization coefficient is not constant.

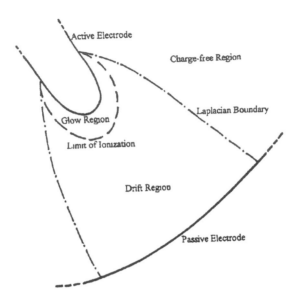

Fig. 9.4 Schematic description of regions in a corona discharge. The boundaries of the three regions vary depending upon the electrode dimension and shape of the electrode [Jones, 2000] (With permission of the Institute of Physics, England).

9.4 BASIC MECHANISMS : NEGATIVE CORONA

Negative coronas in gases have been studied quite extensively and there is general agreement on the broad characteristics of corona discharge in common gases like air, oxygen, etc. and rare gases such as argon and helium (Loeb, 1965). In electronegative gases the negative corona is in the form of regular pulses, called Trichel pulses. The pulses have a very fast rise time (~ ns) and short duration with a long period of relatively low current between the pulses. In oxygen and air the pulses are extremely regular, increasing in frequency with the corona discharge current. Sharper points have a higher frequency for the same corona current.

In SF_6 the initial corona current in a negative point plane gap flows in the form of intermittent pulses[30]. The frequency of the pulses depends on the magnitude of the corona current and not on the gap length. Further, for the same corona current the frequency is higher for a sharper point. The frequency increases approximately linearly with the average corona current and at very high currents the pulses occur so rapidly that they merge into each other forming a glow; the current now becomes continuous.

An early explanation for the high frequency pulses in a negative point-plane was given by Loeb (1965) in terms of the space charge, and it is still valid in its broad features. Near the point electrode the electron avalanches produce a positive ion space charge that increases exponentially with distance from the point electrode. At some finite distance from the tip of the point electrode the electric field, which decreases with increasing distance, becomes low enough to make attachment a dominant process. The resulting negative ion space charge thus formed chokes off the current during the time necessary for most of the ions to be swept away to the positive electrode.

After the negative ions are cleared a new pulse is initiated at the negative electrode with the process repeating itself. Figure 9.5 shows the space charge and potential near the tip of the electrode at the instant the corona pulse is extinguished, (a), and the instant at which the negative space charge is nearly cleared, (b). The top Figure shows the presence of positive ions closer to the electrode and the negative ions farther away. The distortion of the Laplacian electric field due to the space charges are also shown. Close to the point there is intensification of the field due to the positive ions and a corresponding reduction in the field in the region of negative ion space charge. At (b), conditions just before clearing of the space charge and reinstatement of the Laplacian field are shown.

In view of its technological importance, the corona in SF_6 has attracted considerable interest for understanding the mechanisms, and a typical experimental set up used by Van Brunt and Leap (1981) is shown in Figure 9.6. A significant observation is that the negative corona critically depends on the point electrode surface which is not surprising because the initiatory electrons originate from the electrode surface.

Fig. (9.7) shows the measured corona pulse repetition rates and pulse-height distributions at the indicated voltages for electrodes conditioned by prior discharges. The pulses appear intermittently with a low frequency of 100 Hz carrying a charge < 10 pC. Increasing the voltage increases the frequency to above 100 kHz as shown at 15.15 kV. The three distinct peaks evident at lower voltages (9.7 and 10.5 kV) probably correspond to discharges from the different spots or regions of the electrode. In general the negative corona was observed to be less reproducible than the positive corona.

The condition to be satisfied for corona inception is given by

$$\int_0^{d_0} (\alpha - \eta)dx = k \qquad (9.15)$$

where k is a constant (~10) and d_0 is the distance at which the ionization coefficient equals the attachment coefficient, $\alpha = \eta$. Once corona is initiated it could be further enhanced by secondary electron emission from the cathode or photon emission in the gas. Thus, even at voltages much higher than that required to satisfy equation (9.15) the corona consists of predominantly small pulses of magnitude ~ 1 pC.

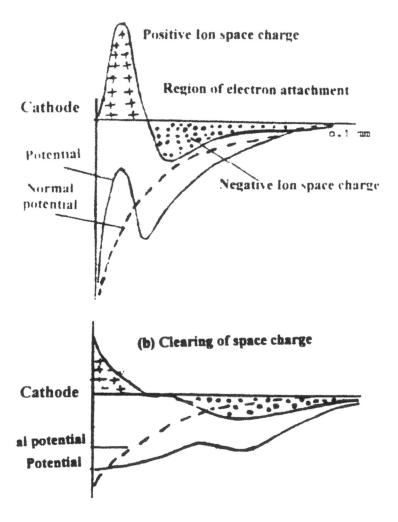

Fig. 9.5 Mechanism for the formation of the Trichel pulses from a negative point. (a) The top Figure shows the presence of positive ions closer to the electrode and the negative ions farther away. (b) Just before clearing of the space charge and restoration of the Laplacian field.

As in other gases, electron-attaching or not, the negative corona inception voltage is smaller than the inception voltage for positive corona. The reason for this phenomenon is partially due to the fact that the initiatory electrons for negative corona are found in a well defined high field region on the surface of the electrode. In contrast the initiatory electrons for positive corona originate in the volume, this volume being very small at the onset voltage. As the voltage is increased the volume increases with an increase in the detachment coefficient contributing to greater number of initiatory electrons.

9.5 BASIC MECHANISMS : POSITIVE CORONA

The corona characteristics are extremely polarity dependent and we have already explained that the positive corona inception voltage is higher than the negative inception voltage. The difference between the inception voltages increases with increasing divergence of the electric field. The corona from a positive point is predominantly in the form of pulses or pulse bursts corresponding to electron avalanches or streamers. This appears to be true in SF_6 with gas pressures above 20 kPa and from onset to breakdown voltages [van Brunt, 1981].

At low pressures < 50 kPa the intermittent nature of corona does not permit a definite frequency to be assigned and only an average corona current can be measured in the range of 0.1 nA-1 µA. As the pressure is increased predominantly burst pulses are observed with a repetition rate of 0.1-10 kHz. The charge in an individual pulse is higher than that in a negative corona pulse. At higher pressures the time interval between pulse bursts becomes less than 2 µs.

Near inception positive corona appears in the form of infrequent electron avalanches of low charge, (q < 1 pC). The initiatory electrons are probably due to collisional detachment of negative ions, though field detachment has also been proposed. At higher voltages and lower gas pressures the bursts occur rapidly forming a train of corona pulses. Fig. 9.7 compares the influence of polarity and ultraviolet radiation on the pulse discharge repetition rate for both polarities at 400 kPa.

Van Brunt and Leep (1981) draw the following conclusions for positive corona in SF_6.

(1) As voltage is increased, positive corona appears as low-level electron avalanches of low repetition rate (f < 1Hz) and then develops into avalanches or relatively large (10-100pC) streamer pulses that act as precursors to burst pulses.

(2) At corona currents above about 0.1nA corona pulses and/or pulse bursts have a repetition rate and mean amplitude that increases with increasing voltage (Fig. 8.20).

(3) The average duration of positive corona pulses tends to increase with decreasing gas pressure and increasing applied voltage.

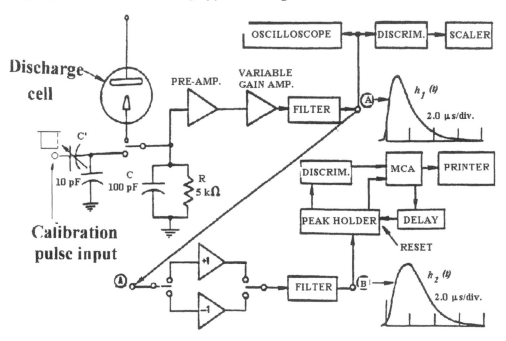

Fig. 9-6. System for measuring electrical characteristics of corona pulses. Shown also are the measured impulse responses $h_1(t)$ and $h_2(t)$ at points A and B where the pulse repetition rates and pulse height distributions are measured (Van Brunt and Leap,1981; with permission of the American Physical Society.)

Under appropriate conditions of parameters like gas pressure, gap length, electrode dimensions etc. the burst pulses form into a glow region usually called Hermstein glow (Loeb, 1965). Though Hermstein thought that the positive glow occurs only in electron attaching gases, recent investigations have disclosed that non-attaching gases such as Ar, He and N_2 also exhibit the same phenomenon[31].

The current due to the positive glow increases with the applied voltage (Fig. 9.8) and consists of nonlinear oscillations of high frequency ($10^5 - 10^6$ Hz). The positive glow can be sustained only in the presence of a fast replacement of electrons leaving the ionization region, and the mechanism of this fast regeneration is photoelectric action at the cathode and photo-ionization. Photoelectric action is the preferred mechanism in non-attaching gases, whereas photo-ionization in the gaseous medium plays a dominant role in electronegative gases such as air. Soft x-rays have also been detected in N_2 (Yu et. al., 1999) contributing to the ionization of the gas.

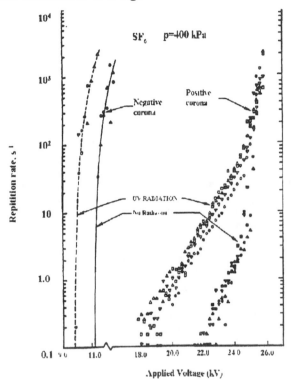

Fig. 9.7 Partial discharge repetition rate vs applied voltage for positive and negative point-plane dc corona in SF_6 at an absolute pressure of 400 kPa. Open symbols correspond to data obtained for a gap irradiated with UV radiation. Included are all pulses with a charge in excess of 0.05 pC (Van Brunt and Leep, 1981; with permission of American Institute of Physics).

9.6 MODELING OF CORONA DISCHARGE: CONTINUITY EQUATIONS

General comments on the methods available for modeling a discharge will be considered in section 9.7. Focusing our attention to the literature published since about 1980, the contributions of Morrow and colleagues[32] and Govinda Raju and colleagues will be considered, because of the different approaches adopted for the theoretical study. The use of continuity equations provides the starting point for the method of Morrow; the temporal and spatial growth of charge carriers, namely electrons, positive ions, negative ions and excited species of metastables coupled with Poisson's equation are solved. Ionization, attachment, recombination, and electron diffusion are included for the growth

size, either one-dimensional form or radial coordinates, both for spherical and cylindrical coordinates are solved. It is assumed that there is no variation in the other coordinate directions.

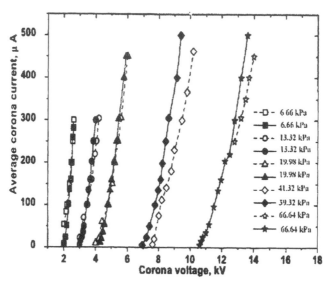

Fig. 9.8 Measured corona-voltage characteristics for positive corona in nitrogen at various gas pressures. Open and full symbols correspond to meshy and solid cathodes. Lines are guide to the eye. (Akishev et. al., 1999; with permission of Institute of Physics.)

The one-dimensional continuity equations for corona in oxygen are Kunhardt and Leussen, 1981):

$$
\left.
\begin{aligned}
\frac{\partial N_e}{\partial t} &= N_e \alpha W_e - N_e \eta W_e - N_e N_p \beta - \frac{\partial (N_e W_e)}{\partial x} + \frac{\partial^2 (D N_e)}{dx} \\
\frac{\partial N_p}{\partial t} &= N_e \alpha W_e - N_e N_p \beta - N_n N_p \beta - \frac{\partial (N_p W_p)}{\partial x} \\
\frac{\partial N_n}{\partial t} &= N_e \eta W_e - N_n N_p \beta - \frac{\partial (N_n W_n)}{\partial x}
\end{aligned}
\right\} \qquad (9.16)
$$

Here t is time, x the distance from the cathode, N_e, N_p, and N_n are the electron, positive ion and negative ion densities, respectively, and W_e, W_p, and W_n are the drift velocities of these particles, respectively. The swarm parameters of the gas are, ionization (α), attachment (η), recombination (β), and diffusion (D) coefficients. The mobilities are considered to be positive in sign for all particles. The continuity equations are coupled to the Poisson's equation via the charge density. It is important to solve the Poisson's equations in three dimensions to allow for the radial extent of the charge distribution.

The last two equations of the set (9.16) are both first order and require only one boundary condition each. These are: (i) $N_p = 0$ at $x = d$, (ii) $N_n = 0$ at $x = 0$. The first equation of the set (9.16) is for electrons and it is of the second order requiring two boundary conditions. These are: (i) At the anode, i.e. at $x = d$, $N_e = 0$ and (ii) at the cathode, i.e. at $x = 0$, $N_e = N_e^P + N_e^i$ where N_e^P is the number of secondary electrons released due to the photoelectric action at the cathode and N_e^i is the number of secondary electrons released due to positive ion bombardment.

The current I in the external circuit due to the motion of electrons and ions between the electrodes is calculated according to

$$I = \frac{\pi r^2 e}{V_A} \int_0^d \left(N_p W_p - N_n W_n - N_e W_e + \frac{\partial (D N_e)}{\partial x} \right) E_L \, dx \tag{9.17}$$

The results of computations are discussed below.

The negative corona current at inception occurs in the form of a pulse having a rise time of about 10 ns (Fig. 9.9 a). The current in the pulse then decreases over a much longer period of about 1300 ns. Initially the electron multiplication in the Laplacian field is the dominant process and the influence of space-charge or the negative ions is negligible. At the end of this period the influence of space charge in distorting the electric field begins to be felt. Near the cathode the field is slightly enhanced due to the presence of positive ions, and further out it is somewhat depressed due to the accumulation of negative ion space charge. The electrons leaving the ionization zone are fast replenished ensuring the growth of the discharge. The light output increases with a weak diffuse form.

A prominent cathode fall region is formed immediately adjacent to the cathode, and almost zero field is formed within the plasma. The decrease of the field nearer the anode is attributed to space charge, which intensifies as the current in the pulse reaches a maximum. The current does not rise indefinitely because the feed back mechanism of the secondary electrons due to photoelectrons declines after reaching a maximum. The trailing edge of the corona pulse is due to the progressive decay of the discharge which allows the build up of the electric field in the low field region (Fig. 9.9 b). The current pulse is quenched because the low electric field in the plasma reduces the energy of the electrons to such an extent that three body attachment becomes dominant. Furthermore, the cathode fall region becomes reduced to such a short distance that insignificant current the cathode fall region becomes reduced to such a short distance that insignificant current is generated from this region. Because of the low mobility of the negative ions, the current remains low and the structure of the space-charge fields changes only slowly with time between pulses.

The development and decay of a discharge may be summed up in the following steps:

1. $t \leq 26$ ns: Current multiplication in the Laplacian field initiates the negative corona pulse (A to B in Fig. 9.9).
2. $26 \leq t \leq 39$ ns: The feed back of secondary electrons, predominantly due to photoelectric action reaches a maximum and then decreases. The current pulse has a shape that is related to this mechanism (B to C). Photons and positive ions are generated in the same ionization region but the ions arrive at the cathode at a much later time. The secondary emission mechanism due to ions comes into play only during the decay phase of the pulse, at times of the order of $\sim$ 100ns.

The initial rise of the corona pulse has been experimentally observed to have a step[33] and the step is attributed to the longer time required for ions to drift to the cathode. In a recent theoretical investigation, however, Gupta, et al[34] have proposed that two mechanisms, namely photoelectric action and ion bombardment are not required to explain the observed step. A negative corona current pulse with the step on the leading edge is observed in the presence of ion-impact electron emission feed back source only. The step is explained in terms of the plasma formation process and enhancement of the feed back source. Ionization wave-like movement is observed after the step.

3. $39 \leq t \leq 70$ ns: The cathode sheath forms during the decay period of the current pulse (C to D). Due to a decrease in the replenishment of electrons falling below the critical value the discharge progressively decays. In addition to the decrease of replenishment of electrons, a number of other mechanisms are possible for the decay of current. The primary electrons move away from the negative point to the low field region where they form negative ions, thus reducing the current. Positive ions are absorbed by the cathode and reduce the space charge near the cathode and the current.
4. $70 \leq t \leq 131$ ns: The electron removal phase due to the formation of negative ions (D to E). At the end of this period the densities of electrons and negative ions are comparable.
5. $t > 131$ ns: Electrons are lost due to three body attachment, and the discharge current continues to decay for longer periods.

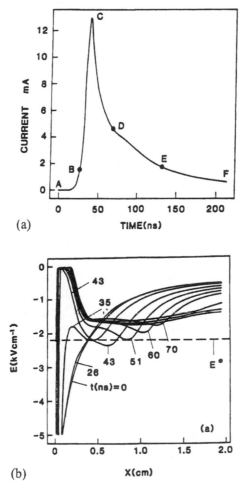

(a)

(b)

Fig. 9.9 (a) Computed current in the external circuit vs time in a negative point-plane gap in oxygen. (b) Electric field vs. position at the times shown (ns) after the release of primary electrons. x = 0 is the cathode. E* is the field at which the ionization coefficient is equal to the attachment coefficient (Morrow, 1985; with permission of American Physical Society.)

Morrow has added significantly for our present knowledge of corona phenomena. Use of continuity equations has an implied equilibrium assumption, i.e., the electron, ion transport coefficients and rate coefficients are only a function of the local reduced electric field E/N. Sigmond and Goldman (1978) have pointed out that this assumption becomes questionable in very high or very inhomogeneous fields. To investigate these effects we have to adopt a different simulation technique, which is possible by the use of the Monte Carlo method.

9.7 NON-EQUILIBRIUM CONSIDERATIONS

Generally, in a uniform electric field remote from boundaries, the macroscopic parameters, such as the drift velocities and ionization and attachment coefficients, are dependent on the ratio E/N because the energy gain from the field is balanced by the energy lost in collisions. However when the field changes rapidly with position or time, this energy balance is disturbed and the transport parameters and coefficients may differ from the predictions made on the basis of the local field. Because of the complexity of the non-equilibrium behaviour the swarm parameters have been analyzed in non-uniform fields in several gases[35] both by the Monte Carlo method and the diffusion flux equations. Table 9.1 summarizes some recent investigations.

Table 9.3
Monte Carlo studies in non-uniform fields [35]

Author	Gas	Field Configuration	Field Slope
Boeuf & Marode	He	Decreasing	Two
Sato & Tagashira	N_2	Decreasing	one
Moratz et. al.	N_2	Decreasing & increasing	Four
Liu & Govinda Raju	SF_6	Decreasing & increasing	one

(Liu and Raju, 1992; with permission of IEEE ©.)

In considering the effects of non-uniformity on the swarm parameters, a distinction has to be made on decreasing electric field or increasing electric field along the direction of electron drift. Liu and Govinda Raju [1992] have found that α/N is lower than the equilibrium value for increasing fields and is higher in decreasing fields (Fig. 9.7), with the relaxation rate depending upon the field slope, β. The relaxation rate has been found to depend upon not only the field slope but also on the number gas density. The effects of changes in field slope and pressure, acting alone or together are discussed below.

1) *Effect of Field slope (β^2)* : To ensure that the electrons are in equilibrium before entering the non-uniform region, the simulation includes a region of uniform electric field at both the cathode and the anode. The initial electrons released from the cathode are required to travel a minimum distance before attaining equilibrium as they enter the non-uniform region. The ionization coefficients are higher than the uniform field values for decreasing field slope and vice versa for increasing fields. The deviations from the equilibrium values are also higher, particularly in the mid gap region for higher values of

β^2. Both increasing fields and decreasing fields influence the swarm parameters in a consistent way, though they are different from the uniform field values.

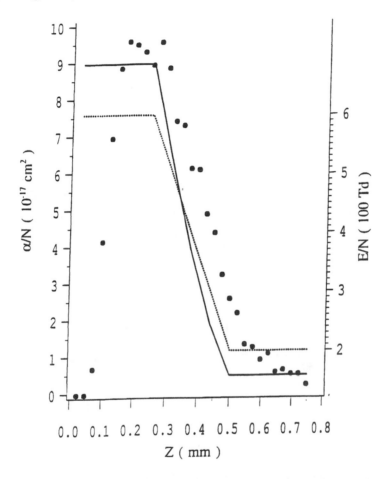

Fig. 9.10 Ionization coefficients in SF_6 in non-uniform field gap in a decreasing field slope of $\beta^2=16$ kTd/cm at N= 2.83 × 10^{23} m^{-3} Symbols are computed values. Closed line for equilibrium conditions. The reduced electric field is shown by broken lines (Liu and Raju, 1997; IEEE ©.)

In non-uniform fields, although the energy gain from the field changes instantly with the changing field, the energy loss governed by collisions is a slower, non local process. Also the collisional energy loss is dependent on the electron energy and different parts of the electron energy distribution will readjust to changing field with different rates. Fig. (9.11 a) shows the energy distribution function , $f(\varepsilon)$, at 400 Td in uniform, decreasing and increasing fields at the midpoint of the gap [(Liu and Raju, 1997). It is clear that relatively more electrons are distributed in the high energy range for the decreasing field, and fewer electrons for the increasing field, compared with the uniform situation. The lower portion of Fig. (9.11) shows the energy distribution function along the non-uniform gap for a decreasing field at $N = 2.83 \times 10^{23}$ m^{-3}, $\beta^2 = 8$ kTd/cm.

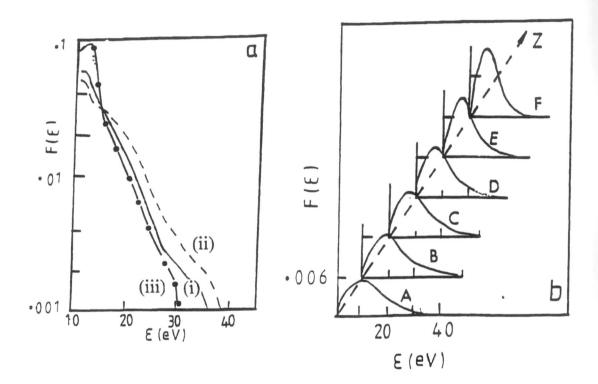

Fig. 9.11. (a) Energy distribution function in uniform and non-uniform fields at E/N = 400 Td corresponding to the mid point of the gap, N= 2.83 × 10^{23} m^{-3}, (i) uniform field (ii) decreasing field (iii) increasing field. (b) Energy distribution function along the non-uniform gap in a decreasing field. A: z = 0.5 mm, 600 Td; B: z = 0.6 mm, 520 Td; C: z = 0.7 mm, 440 Td; D: z = 0.8 mm, 360 Td; E: z = 0.9 mm, 280 Td; F: z = 1mm, 200 Td [Liu and Raju, 1992]. (With permission of IEEE).

As z increases from 0.5 mm to 1 mm, E/N decreases from 600 Td to 200 Td, and the peak of $f(\varepsilon)$ moves toward low energy range. From such analyses of the distribution function at various reduced electric fields, E/N, we can explain the fact reduced ionization coefficients, α_d/N, in decreasing fields are higher than the equilibrium values, i. e., $\alpha_d/N > \alpha/N > \alpha_i/N$. where α_i is the ionization coefficient in increasing fields.

We have already referred to the polarity dependence of the corona inception voltage, V_c, (Fig. 8.15) in non-uniform fields. The corona inception voltage for negative polarity is lower than that for positive polarity and the results previously discussed support the following mechanism for corona. With negative polarity, the electrons experience decreasing fields as they move toward the anode and the actual ionization coefficients, α_d, are higher than the equilibrium values. Furthermore, the magnitude of the difference between the actual and equilibrium values, $(\alpha_d - \alpha)$, is higher for the negative polarity. The corona inception voltage which is dependent on the exponential of α_d is therefore higher for the positive polarity voltage than that for the negative polarity.

The attachment coefficients are also affected by the changing field when compared with equilibrium values. With changing field, the electrons cannot catch up with the faster decreasing field and have energies greater than those dictated by the local field. This means that relatively more electrons are distributed in the high energy tail, causing more ionizing collisions (Fig. 9.11). These ionization collisions generate new electrons with low energy and these electrons attach to SF_6 molecules and increase the attachment coefficient. The observed sharp decrease of the attachment coefficients near the anode is probably caused by the gradual disappearance of the backward directed electrons arising from electron absorption by the anode. With regard to the other swarm parameters the mean energy and drift velocity rise near the anode and the attachment coefficients fluctuate, particularly with higher β^2, possibly due to the small number of attachment coefficients.

2)   *The effect of gas number density (N)*: The gas number density is expected to have an effect on the non-equilibrium behavior due to the fact that the relaxation rate is dependent on the collision rate. At higher gas densities more collisions occur within the sampling distance and the electron swarm will readjust itself faster as appropriate to the applied field. This results in a coefficient that deviates from the equilibrium value by a smaller amount. The mean energy at the first part of the non-uniform gap is lower than the equilibrium value because the larger number of ionization collisions results in loss of more energy. The anode effect is observed by a sharp decrease of η/N and a sharp rise in the mean energy.

3) *The effect of simultaneous change in β^2 and N*: Low N and high field slopes enhance the non-equilibrium behaviour of the electron swarm. An important consideration is whether the rate of change of the coefficients remains the same if β^2/N remains the same. By considering the three sets of individual values of β^2 and N such that their ratio remains the same, it has been found that individual changes in each parameter do not affect α/N and η/N as long as the ratio is not affected. The same observations are also

qualitatively true for the mean energy and drift velocity though the change in them is not as significant as the change in α/N and η/N. This is reasonable because the change in the energy distribution affects the latter two parameters more than the former.

4) Electron and ion distribution: Space charge plays an important role in determining the electric field in the gap. as the previous discussions have clearly demonstrated. Fig. 9.9 shows the distribution of electrons and ions in the gap, both in a uniform field and a non-uniform field. The essential features of the differences in the distribution of charges may be summarized as follows:

a. In a uniform field gap (Fig. 9.12a) the electron density increases exponentially uneventfully. However in decreasing fields (Fig. 9.9b) the electron density reaches a peak at approximately the mid gap region. This is due to the combination of two opposing factors: (1) The electric field, and therefore the ionization coefficient, decreases. (2) The number of electrons increases exponentially. In increasing field, both these factors act cumulatively and N_e increases initially slowly but very rapidly as the electrons approach the anode. It is easy to recognize that the two situations of non-uniform field correspond to negative and positive coronas.

b. The total number of negative ions is denoted by N^- and the individual contribution of each species are identified in Fig. (9.12a). In both uniform and non-uniform electric fields the ion density oscillates and is possibly due to relaxation process of the electron energy distribution as suggested by Itoh, et. al.[36].

c. The negative ion density in the non-uniform field shows a peak, first increasing till the end of the non-uniform region, then decreasing in the low field region because of the decrease in the total number of electrons in the low field region. There is a shift between the peak of the electron density and the negative ion density. Though the field has attained the critical value corresponding to $\alpha=\eta$ multiplication continues to occur because of the non-uniform equilibrium effects, thereby increasing N^-. Beyond the peak the reduced number of electrons yield reduced negative ions.

9.8 MONTE CARLO SIMULATION: NEGATIVE CORONA IN SF$_6$

The Monte Carlo technique has been extended to the study of negative corona and development of streamers[37]. A flow diagram used for the purpose is shown in Fig. 9.13. The initial electrons are emitted from the cathode according to a cosine distribution because the emission occurs on only one side of the electrode. The Laplacian field in the gap is

$$E_L(z) = \frac{2Vd}{Ln\left[\dfrac{4d}{r_c}\right]\left[d(2z + r_c) - z^2\right]}$$ (9.18)

where z is the space coordinate , V the applied voltage and r_c the radius of the cathode tip. The energy gain of the electrons in a small time interval Δt is governed by the equations of motion.

Whether a collision between an electron and a gas molecule occurs or not is decided by generating a random number R uniformly distributed between 0 and 1. A collision is deemed to occur at the end of a time step if P > R where P is the probability of collision. In the event of an elastic collision the fractional loss of energy is 2m/M' where m and M' are masses of electron and SF_6 molecule. The direction of electron motion after a collision is determined according to equations of motion. To reduce the CPU time of the simulation of the motion of a large group of electrons during ~50ns, Liu and Govinda Raju (1994) assume that the electrons move in one dimension in space with a velocity which has three components (V_x, V_y, V_z). In the simulation of a corona discharge time step and the cell size should be paid great attention since the electric field is steep close to the smaller electrode. Also the accumulation of the space charge causes the field near the cathode to change abruptly; therefore smaller cell size and time steps are required.

The length of field intensive region varies with various voltages and gap separations and it is varied for each situation.

$$T_m = \frac{1}{NQ_T W_e}$$ (9.19)

where Q_T is the total collision cross section and W_e is the electron drift velocity. The time step Δt used to calculate the motion of electrons and space charge field is chosen to be very small ($\Delta t \ll T_m$) and Δt may be different for different applied voltages. In negative coronas Δt is in the range of 0.5 to 1 ps.

At each time step, the new position and energy is calculated according to the equation of motion. New electrons, positive and negative ions may be produced by ionization, photo-ionization and attachment collisions. At the end of each time step, the space charge field is calculated from the Poisson's equation as a function of charge distribution and is stored for use in the next time step.

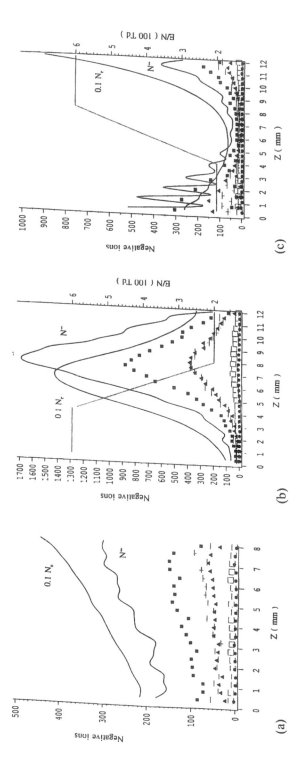

Fig. 9.12 Electron and ion distribution in the gap in SF$_6$. (a) Uniform electric field (400Td). Ion species are: closed square- SF$_6^-$, closed triangle-SF$_5^-$, Plus-SF$_4^-$, closed circle-F$_2^-$; Arbitrary units. The electron density is multiplied by 0.1. (b) decreasing electric field, symbols same as (a). (c) increasing electric field, symbols same as (a). (Liu and Raju. 1994; IEEE ©.)

The space charge field in k^{th} cell is given by

$$E_k = E_{(k-1)} + \frac{1}{2}[\Delta E_{(k-1)} + \Delta E_k); \qquad 1 \le k \le M \tag{9.20}$$

where M is the total number of cells in the low field region. The integration of the space charge field over all cells should be zero (space charge produces no external voltage on the gap). The final field is the sum of the applied Laplacian field and the space charge field.

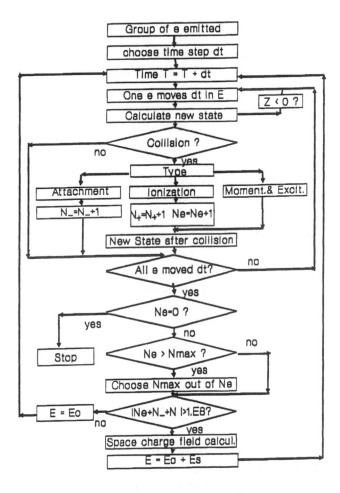

Fig. 9.13 Flow diagram of Monte Carlo simulation of corona discharge . e=electron, N_e = number of electrons, N_+ = number of positive ions, N_- = number of negative ions (Liu and Raju, 1994, IEEE ©.)

The current density J in the external circuit due to the motion of electrons, and ions between the electrodes is calculated using Sato's equation

$$J = \frac{e}{V} \int [N_p W_p - N_n W_n - N_e W_e] E_L \, dz \tag{9.21}$$

where the subscripts e, n and p stand for electron, negative ion and positive ion, respectively. Fig. (9.14a) shows a single current pulse at the onset of corona and the current is of avalanche type with a single pulse.

With increasing voltage the current density increases and the current now becomes a streamer type with more than one pulse contributing to the current. The time at which the peak occurs becomes shorter due to the faster development of the electron avalanche in a higher field. The pulse shape obtained (Fig. 9.14a) is similar to those observed by Van Brunt and Leap (1981) and Morrow[38]. The larger pulses are usually followed by a long tail, or a burst of lower level pulses which is characteristic of positive corona.

The total field distribution in the gap is shown in Fig. (9.14b). The space charge distortion begins at 2.8 ns and increases up to 5.95 ns. It then remains unchanged for ~5 ns (not shown) because no more ions are produced after this time. The smallest electric field is only 0.4 V/cm. A comparison with Fig. (9.14b) shows that the space charge distortion occurs in SF_6 much earlier than in oxygen because the attachment in the former gas is much more intensive.

Fig. (9.14c) shows the development of subsequent avalanches. The second avalanche gives rise to the third avalanche due to photo-ionization at 1.6 ns and in the time interval 1.6-2.3 ns the third avalanche grows much faster because of intense field near the cathode. After about 4.5 ns all the electrons are in the low field region and the avalanches die out soon thereafter.

9.9 MONTE CARLO SIMULATION: POSITIVE CORONA IN SF$_6$

As a preliminary to the discussion we again refer to the experimental measurements of Van Brunt and Leap (1981) who have contributed significantly to the advancement of understanding positive corona. The studies were carried out at relatively high pressure of 50 ~ 500 kPa in SF_6. Positive corona pulses appear at onset as low-level electron avalanches that, contain relatively low charge (q< 1 pC). The corona inception current is less than 0.1 nA. The frequency of the pulses vary from about 0.3-5 kHZ depending upon the voltage and corona current. At higher voltages, the pulses develop into large streamers usually followed by a burst of many small pulses. The burst characteristics of positive corona show a definite dependence on pressure and voltage which is evident in the pulse height distribution data. Comparing with negative corona studies, positive corona develops much more slowly and has a higher inception voltage as expected (Fig. 9.7).

The simulation technique shows that the initial current pulse is very fast with a half maximum amplitude of ~0.3 ns. The electrons are absorbed by the anode very fast; during collision photo-ionization occurs and the secondary electrons are produced leading to further avalanches. The positive ions are left behind leading to a fast build up of space charge[39]. The space charge field depresses the applied field at the anode and a 'spike' is produced in the electric field as fast as ~0.5 ns. The spike is attributed to the high density of positive charges in the streamer head which propagates toward the cathode. The net positive charge behind the streamer front extends to the anode.

The streamer front moves very rapidly with a velocity of $\sim 8 \times 10^6$ m/s and as it progresses into the gap the electric field remains at the critical value, E*. There is less voltage remaining at the streamer head and eventually there is insufficient voltage at the streamer tip to sustain its progress. The electron density is constant through the length of the streamer but farther away from the tip it declines rapidly. As the streamer progresses into the gap the region of constant electron density also extends into the gap. Outside this region, which is towards the cathode, ionization slows down and the recombination causes the positive ion, negative ion and electron densities to fall rapidly from their previously high values.

A further elucidation of the positive corona mechanism is obtained by the Monte-Carlo simulation[40]. The corona inception occurs in the form of pulses. The shape of the current pulse obtained is similar to the experimental observations of Van Brunt and Leap (1981) with multiple peaks except that their duration of pulse is much longer: ~30 ns compared to the present duration of ~ 5ns.

The development of the secondary avalanches is due to photo-ionization and occurs on the cathode side of the electron avalanche. Here the field is intensified whereas the field between the avalanche head and the point electrode is decreased. This is clearly shown in Fig. (9.14c) where avalanche #2, the parent avalanche, is smaller than the progeny avalanche, #3. The influence of the polarity in secondary avalanche development may be discerned from comparing with Fig. (9.15); the field intensification regions are different.

The differences between positive and negative coronas are shown in Table 9. 4.

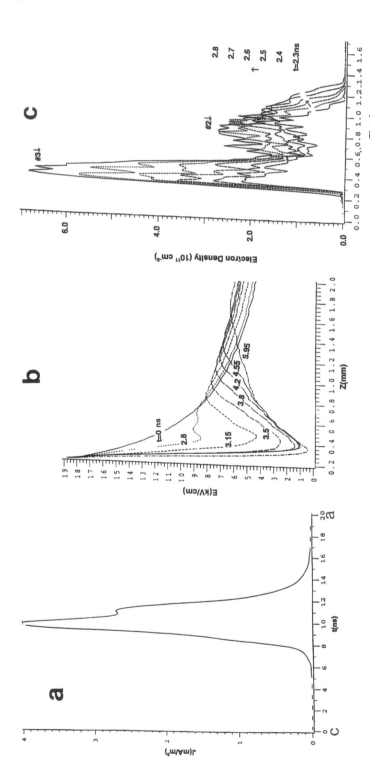

Fig. 9.14 Development of negative corona pulses in SF$_6$ at N = 2.12 × 10^{24} m^{-3} (a) current due to a single pulse, V = 2.4 kV (b) electric field in the gap at various times, V = 3.0 kV (c) electron density distribution at 2.3 ≤ t ≥ 2.8 ns, V = 3.5 kV. Liu and Raju. 1994; with permission of IEEE ©.)

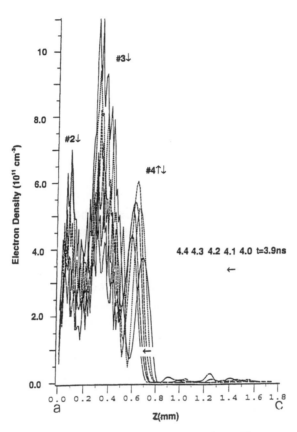

Fig. 9.15 Electron density distribution in positive corona pulses. The sequence of development of secondary pulses development should be compared with figure 14(c) for negative polarity. Applied voltage is 4.0 kV and $3.9 \leq t \leq 4.4$ ns. (Liu and Raju, 1994; with permission of IEEE.)

Table 9.4

A Comparison of Positive and Negative Coronas in SF_6

	POSITIVE CORONA	NEGATIVE CORONA
Effect of space charge on total field	(a) Depressed near anode (b) Enhanced in the rest of region	(a) Enhanced near the cathode (b) Enhanced near the anode (c) Depressed in the midgap region
Second avalanche	(a) Anode side of the primary avalanche	(a) Cathode side of the primary avalanche
Subsequent avalanches	Cathode side of the previous avalanche	Cathode side of the previous avalanche
Concentration of negative ions in the gap	Negligible	Appreciable
Concentration of positive ions in the gap	Appreciable	Appreciable

9.10 CONCLUDING REMARKS

Comparing the continuity equations method with the Monte Carlo method, the former is concise and consumes less CPU time. However more assumptions are implicit and it is not as straight forward as the Monte Carlo method. Further, the continuity equation method has the disadvantage that all the swarm parameters should be known prior to the analysis. Usually these parameters are assumed to be in equilibrium with the field. Since the streamer mechanism is the result of space charge distortion with the total electric field may be as large as 5 times the average field. The field slope, which has been shown to be significant, may be quite large. These aspects need to be viewed against the longer CPU time required by the Monte Carlo method.

The range of experimental conditions and electrode geometries that are encountered in practice or under laboratory experimental conditions are so large that it is not possible to take into account all the parameters that determine the details of the discharge development. Further, the modelling process requires the initial and boundary conditions to be set up which may not be exact and require inordinate long computation time. However the advances in modern computing techniques and faster processing of various steps have led to success in simulating the discharge process, which go hand in hand with the experimental results.

9.11 REFERENCES

[1] R. S. Sigmond, "Corona Discharges", Chapter 4 in J. M. Meek and J. D. Craggs, "Electrical Breakdown of Gases", John Wiley & Sons, New York, 1978.

[2] L. B. Loeb, "Electrical coronas", University of California Press, 1965.

[3] G. R. Govinda Raju and R. Hackam, J. Appl. Phys., **52** (1981) 3912.

[4] J. Liu, Ph. D. Thesis, University of Windsor, 1993.

[5] E. E. Kunhardt and L. H. Luessen, (Ed), article by L. C. Pitchford, "Electrical Breakdown and Discharges in Gases", Plenum Press, New York, 1981, pp. 313.

[6] H. Tagashira, Y. Sakai and S. Sakamoto, J. Phys. D; Appl. Phys. **10** (1977) 1051.

[7] J. Liu and G. R. Govinda Raju, Can. J. Phys., **70** (1993) 216.

[8] J. Liu and G. R. Govinda Raju, IEEE Trans. on DIE & EI, **2** (1995) 1004.

[9] R. D. Hake Jr., A. V. Phelps, Phys. Rev., **158** (1962) 70. Only valid for low E/N.

[10] H. Myers, J. Phys.B., At. Mol. Phys., **2** (1969) 393-401. Attachment cross section is derived.

[11] H. K. Wagner, Z. Physik., **241** (1971) 258. Curve fitting to experimental data.

[12] J. Lucas, D. A. Price and J. Moruzzi, J. Phys. D.: Appl. Phys., **6** (1973) 1503. Valid for low E/N.

[13] K. Masek, Czech. J. Phys., 25 (1975) 686. Valid for low E/N.

[14] K. Masek, T. Ruzicka and L. Laska, Czech. J. Phys., B., **27** (1977) 888. Neglects attachment and dissociation.

[15] K. Masek, L. Laska and T. Ruzicka, J. Phys. D.,: Appl. Phys., **10** (1977) L25.

[16] T. Taniguchi, H. Tagashira and Y. Sakai, J. Phys. D.: Appl. Phys., **11** (1978) 1757. Only three body attachment calculated.

[17] G. Gousset, C. M. Ferrir, M. Pinheiro, P. A. Sa, M. Tanzeau, M. Vialle and J. Loureiro, J. Phys. D.: Appl. Phys., **24** (1980) 290. No attachment co-efficient, emphasis on heavy particle.

[18] T. Taniguchi, H. Tagashira, I. Okada and Y. Sakai, J. Phys.D.: 11 (1978) 2281.

[19] G. Schaffer and P. Hui, J. Comp. Phys., **89** (1990) 1.

[20] J. Liu and G. R. Govinda Raju, IEEE Trans. on EI, **28** (1993) 261.

[21] I. D. Chalmers, H. D. Duffy and D. J. Tedford, Proc. Roy. Soc., London, A, **329** (1972) 171.

[22] A. J. Davies, Proc. IEE, Pt. A., **133** (1986) 217.

[23] J. P. Novak and R. Bartnikas, J. Appl. Phys., **62** (1987) 3605.

[24] J. P. Novak and R. Bartnikas, J. Phys. D.: Appl. Phys. **21** (1988) 896.

[25] J. P. Novak and R. Bartnikas, J. Appl. Phys., **64** (1988) 1767.

[26] J. P. Novak and R. Bartnikas, IEEE Trans. on Plasma Sci., **18** (1990) 775.

[27] S. K. Dhali and A. K. Pal, J. Appl. Phys., **63** (1988) 1355.

[28] S. K. Dhali and P. F. Williaams, J. Appl. Phys., **62** (1987) 4696.

[29] (a) J. E. Jones, J. Phys. D., Appl. Phys., **33** (2000) 389.

(b) A. A. Al-Arainy, S. Jayaram, J. D. Cross, 12th International Conference on Conduction and Breakdown in Liquids, Italy, July 15-19, 1996.

[30] R. J. Van Brunt and D. Leap, J. Appl. Phys. **52** (1981) 6588.

[31] Yu. S. Akishev, M. E. Grushin, A. A. Deryugin, A. P. Napartovich, M. V. Pan'kin and N. I. Trushkin, J. Phys. D., Appl. Phys., **32** (1999) 2399.

[32] R. Morrow, J. Phys. D., Appl. Phys., **30** (1997), 3099.

[33] M. Cernak and T. Hosokawa, Japan. J. Appl. Phys., **27** (1988) 1005.

[34] D. K. Gupta, S. Mahajan and P. I. John, J. Phys. D: Applied Phys. **33** (2000) 681.

[35] For a list of references see J. Liu and G. R. Govinda Raju, IEEE Trans. on Plas. Sci., **20** (1992) 515.

[36] H. Itoh, M. Kawaguchi, K. Satoh, Y. Miura, Y. Nakao and H. Tagashira, J. Phys. D; Appl. Phys. **23** (1990) 299.

[37] J. Liu and G. R. Govinda Raju, IEEE Trans. on DIE & EI, **1** (1994) 520-529.

[38] R. Morrow, Phys. Rev., A, **32** (1985) 1799.

[39] R. Morrow, IEEE Trans. on DIE. & EI, **26** (1991) 398.

[40] J. Liu and G. R. Govinda Raju, IEEE Trans. on DIE & EI, **1** (1994) 530.

If we do not find anything pleasant at least we shall find something new.
-Voltaire

10

THERMALLY STIMULATED PROCESSES

Thermally stimulated processes (TSP) are comprised of catalyzing the processes of charge generation and its storage in the condensed phase at a relatively higher temperature and freezing the created charges, mainly in the bulk of the dielectric material, at a lower temperature. The agency for creation of charges may be derived by using a number of different techniques; Luminescence, x-rays, high electric fields corona discharge, etc. The external agency is removed after the charges are frozen in and the material is heated in a controlled manner during which drift and redistribution of charges occur within the volume. During heating one or more of the parameters are measured to understand the processes of charge generation. The measured parameter, in most cases the current, is a function of time or temperature and the resulting curve is variously called as the glow curve, thermogram or the heating curve. In the study of thermoluminescence the charge carriers are generated in the insulator or semiconductor at room temperature using the photoelectric effect.

The experimental aspects of TSP are relatively simple though the number of parameters available for controlling is quite large. The temperature at which the generation processes are catalyzed, usually called the poling temperature, the poling field, the time duration of poling, the freezing temperature (also called the annealing temperature), and the rate of heating are examples of variables that can be controlled. Failure to take into account the influences of these parameters in the measured thermograms has led to conflicting interpretations and in extreme cases, even the validity of the concept of TSP itself has been questioned.

In this chapter we provide an introduction to the techniques that have been adopted in obtaining the thermograms and the methods applied for their analysis. Results obtained in specific materials have been used to exemplify the approaches adopted and indicate the limitations of the TSP techniques[1, 2, 3]. To limit the scope of the chapter we limit ourselves to the presentation of the Thermally Stimulated Depolarization (TSD) Current.

In what follows we adopt the following terminology:

The electric field, which is applied to the material at the higher temperature, is called the poling field. The temperature at which the generation of charges is accelerated is called the poling temperature and, in polymers, mostly the approximate glass transition temperature is chosen as the poling temperature.

The temperature at which the electric field is removed after poling is complete is called the initial temperature because heating is initiated at this temperature.

The temperature at which the material is kept short circuited to remove stray charges, after attaining the initial temperature and the poling field is removed, is called the annealing temperature. The annealing temperature may or may not be the initial temperature.

The current released during heating is a function of the number of traps (n_t). If the current is linearly dependent on n_t then first order kinetics is said to apply. If the current is dependent on n_t (Van Turnhout, 1975) then second order kinetics is said to apply.

10.1 TRAPS IN INSULATORS

The concept of traps has already been introduced in chapter 7 in connection with the discussion of conduction currents. To facilitate understanding we shall begin with the description of thermoluminiscence (Chen and Kirsch, 1981). Let us consider a material in which the electrons are at ground state G and some of them acquire energy, for which we need not elaborate the reasons, and occupy the level E (Fig. 10.1). The electrons in the excited state, after recombinations, emit photons within a short time interval of 10^{-8} s and return to the ground state. This phenomenon is known as floroscence and emission of light ceases after the exciting radiation has been switched off.

The electrons may also lose some energy and fall to an energy level M where recombination does not occur and the life of this excited state is longer. The energy level corresponding to M may be due to metastables or traps. Energy equivalent to ε needs to be imparted to shift the electrons from M to E, following which the electrons undergo recombination. In luminescence this phenomenon is recognized as delayed emission of light after the exciting radiation has been turned off. In the study of TSDC and TSP the energy level corresponding to M is, in a rather unsophisticated sense, equivalent to traps of single energy. The electrons stay in the trap for a considerable time which results in delayed response.

The probability of acquiring energy, ε_a, thermally to jump from a trap may be expressed as

$$\frac{n}{n_o} = s \exp\left(-\frac{\varepsilon_a}{kT}\right) \tag{10.1}$$

where n is the number of electrons released at temperature T, n_o the initial number of traps from which electrons are released ($n/n_o \leq 1$), s is a constant, k the Boltzmann constant and T the absolute temperature. Eq. (10.1) shows that the probability increases with increasing temperature. The constant s is a function of frequency of attempt to escape from the trap, having the dimension of s^{-1}. A trap is visualized as a potential well from which the electron attempts to escape. It acquires energy thermally and collides with the walls of the potential well. s is therefore a product of the number of attempts multiplied by the reflection coefficient. In crystals it is about an order of magnitude less than the vibrational frequency of the atoms, $\sim 10^{12}\ s^{-1}$.

The so called first order kinetics is based on the simplistic assumption that the rate of release of electrons from the traps is proportional to the number of trapped electrons. This results in the equation

$$\frac{dn}{dt} = -\alpha\, n(t) \tag{10.2}$$

where the constant, α represents the decrease in the number of trapped electrons and has the dimension of s^{-1}. The solution of eq. (10.2) is

$$n(t) = n_0 \exp(-\alpha t) \tag{10.3}$$

where n_o is the number of electrons at $t = 0$.

In terms of current, which is the quantity usually measured, equation (10.3) may be rewritten as

$$I = -C\left(\frac{dn}{dt}\right) = C\frac{n}{\tau}\exp\left(-\frac{\varepsilon_a}{kT}\right) \tag{10.4}$$

where τ is called the relaxation time, which is the reciprocal of the jump frequency and C a proportionality constant. Let us assume a constant heating rate β. We then have

$$T = T_o + \beta t \tag{10.5}$$

where T_0 and t are the initial temperature and time respectively.

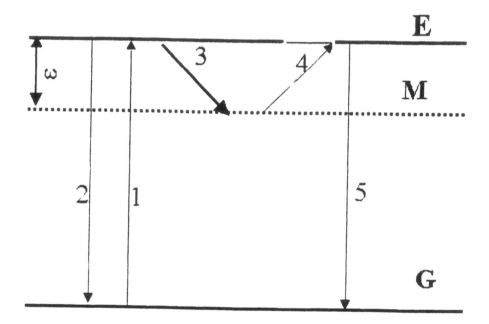

Fig. 10.1 Energy states of electrons in a solid. G is the ground state, E the excited level and M the metastable level. Excitation shifts the electrons to E via process 1. Instantaneous return to ground state-process 2 results in fluorescence. Partial loss of energy transfers the electron to M via process 3. Acquiring energy ε the electron reverts to level E (process 4). Recombination with a hole results in the emission of a photon (Process 5) and phosphorescence. Adopted from Chen and Kirsch (1981), (with permission of Pergamon Press)

The solution of equation (10.4) is given as

$$I(T) = \frac{n_o}{\tau} \exp\left(-\frac{\varepsilon_a}{kT}\right) \exp\left[\left(-\frac{1}{\beta\tau}\right)\int_{T_o}^{T}\exp\left(-\frac{\varepsilon_a}{kT'}\right)dT'\right] \tag{10.6}$$

This equation is known as the first order kinetics. The second order kinetics is based on the concept that the rate of decay of the trapped electrons is dependent on the population of the excited electrons and vacant impurity levels or positive holes in a filled band. This leads to the equation

$$I = -\frac{dn}{dt} = -\frac{n^2}{\tau'}\exp\left(-\frac{\varepsilon_a}{kT}\right)$$

(10.7)

where τ' is a constant having the dimension of $m^3\ s^{-1}$. The solution of this equation is

$$I(T) = \frac{n_o^2}{\tau'}\exp\left(-\frac{\varepsilon_a}{kT}\right)\left[1 + \left(\frac{n_o}{\tau'\beta}\right)\int\exp\left(-\frac{\varepsilon_a}{kT'}\right)dT'\right]^{-2}$$

(10.8)

without losing generality, equations (10.6) and (10.8) are expressed as

$$I(t) = -\frac{dn}{dt} = \left(\frac{n}{n_o}\right)^b\frac{n_o}{\tau}\exp\left(-\frac{\varepsilon_a}{kT}\right)$$

(10.9)

where the exponent b is the order of kinetics.

Returning to Fig. (10.1) the metastable level M may be equated to traps in which the electrons stay a long time relative to that at E. The trapping level, having a single energy level ε, and a single retrapping center (Fig. 10.1) shows a single peak in the measured current as a function of the temperature. The trap energy level is determined, according to equation (10.1) for n_T, by the slope of the plot of ln (I) versus l/T.

A polymer having a single trap level and recombination center is a simplified picture, used to render the mathematical analysis easier. In reality, the situation that one obtains is shown in Fig. 10.2 where n_c denotes the number of electrons in the conduction band. The trap levels, T_1, T_2, ... are situated closer to the conduction band and the holes, H_1, H_2, ... are retrapping centers. Introduction of even this moderate level of sophistication requires that the following situations should be considered (Chen and Kirsch, 1981).

1. The trap levels have discreet energy differences in which case each level could be identified with a distinct peak in the thermogram. On the other hand the trap levels may form a local continuum in which case the current at any temperature is a contribution of a number of trap levels. The peak in the thermogram is likely to be broad.
2. The traps are relatively closer to the conduction band so that thermally activated electron transfer can occur. The holes are situated not quite so close to the valence

band so that the holes do not contribute to the current in the range of temperatures used in the experiments. The reverse situation, though not so common, may obtain; the holes are closer to the valence band and traps are deep. The electron traps now become recombination centers completing the "mirror image".

3. The essential feature of a thermogram is the current peak, which is identified with the phenomenon of trapping and subsequent release due to thermal activation. The sign of the carrier, whether it is a hole or electron, is relatively of minor significance. In this context the trapping levels may be thermally active in certain ranges of temperature while, in other ranges, they may be recombination centers.

4. An electron which is liberated from a trap may drift under the field before being trapped in another center that has the same energy level. The energy level of the new trap may be shallower, that is closer to the conduction band. This mode of drift has led to the term "hopping".

The development of adequate theories to account for these complicated situations is, by no means, straight forward. However, certain basic concepts are common and they may be summarized as below:

1) The intensity of current is a function of the number of traps according to equation 10.4. The implied condition that the number density of traps and holes is equal is not necessarily true. To render the approach general let us denote the number of traps by n_t and the number of holes by n_h. The number of holes will be less if a free electron recombines with a hole. Equation (10.7) now becomes

$$I = -\frac{dn_h}{dt} = f_{rc} n_h n_c \tag{10.10}$$

where n_c is the number density of free carriers in the conduction band and f_{rc} is the recombination probability with the dimension of $m^3\ s^{-1}$. The recombination probability is the product of the thermal velocity of free electrons in the conduction band, v, and the recombination cross section of the hole, σ_{rc}.

2) The electrons from the traps move to the conduction band due to thermal activation. The change in the density of trapped carriers, n_t, is dependent on the number density of trapped charges and the Boltzmann factor. Retrapping also reduces the number that moves into the conduction level. The retrapping probability is dependent on the number density of unoccupied traps. Unoccupied trap density is given by $(N-n_t)$ where N is the concentration of traps under consideration. The rate of decrease of electrons from the traps is given by

$$-\frac{dn_t}{dt} = sn_t \exp\left(-\frac{\varepsilon}{kT}\right) - n_c (N - n_t) f_n \qquad (10.11)$$

where f_n is the retrapping probability ($m^3\ s^{-1}$). Similar to the recombination probability, we can express f_n as the product of the retrapping cross section, σ_n, and the thermal velocity of the electrons in the conduction band.

3) The net charge in the medium is zero. Accordingly

$$n_c + n_t - n_h = 0 \qquad (10.12)$$

This equation transforms into

$$\frac{dn_h}{dt} = \frac{dn_t}{dt} + \frac{dn_c}{dt} \qquad (10.13)$$

Substituting equation (10.11) in (10.13) gives

$$\frac{dn_c}{dt} = sn \exp\left(-\frac{\varepsilon}{kT}\right) - n_c \left(n_h f_n + (N - n_t) f_n\right) \qquad (10.14)$$

Equations (10.10), (10.13) and (10.14) are considered to be generally applicable to thermally stimulated processes, with modifications introduced to take into account the specific conditions. For example the charge neutrality condition in a solid with number of trap levels, T_1, T_2, ... etc., and a number of hole levels H_1, H_2 ... etc., (Fig. 10.2) is given by

$$\sum n_{ti} + n_c + \sum n_{hj} = 0 \qquad (10.15)$$

Numerical solutions for the kinetic equations governing thermally stimulated are given by Kelly, et al. (1972) and Haridoss (1978)[4, 5]. Invariably some approximations need to be made to find the solutions and Kelly, et al. (1971) have determined the conditions under which the approximations are valid. The model employed by them is shown in Fig. 10.3. Let N number of traps be situated at depth E below the conduction band, which has a density of states, N_c. On thermal stimulation the electrons are released from the traps to the conduction band with a probability lying between zero and one, according to the Boltzmann factor $e^{-E/kT}$.

The electrons move in the conduction band under the influence of an electric field, during which event they can either drop into recombination centers with a capture coefficient γ or be retrapped with a coefficient of β. The relative magnitude of the two coefficients depend on the nature of the material; retrapping is dominant in dielectrics whereas recombination with light output dominates in thermoluminiscent materials.

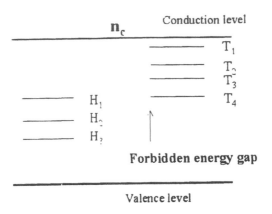

Fig. 10.2 Electron trap levels (T) and hole levels (H) in the forbidden gap of an insulator. N_c is the number density of electrons in the conduction band. The number density of electrons in T_1 is denoted by the symbol n_{t1} and hole density in H_1 is n_{h1}. Charge conservation is given by equation (10.15). Adopted from (Chen and Kirsch, 1981, with permission of Pergamon Press, Oxford).

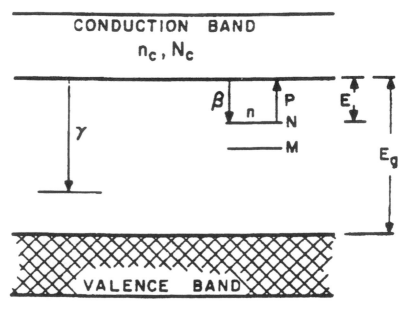

Fig. 10.3 Energy level diagram for the numerical analysis of Kelly et. al. (1972), (with permission of Am. Inst. of Phy.)

In the absence of deep traps, the occupation numbers in the traps and the conduction band are n and n_c respectively. Note that the energy diagram shown in fig. 10.3 is relatively more detailed than those shown in Figs. 10.1 and 10.2.

10.2 CURRENT DUE TO THERMALLY STIMULATED DEPOLARIZATION (TSDC)

We shall focus our attention on the current released to the external circuit during thermally stimulated depolarization processes. To provide continuity we briefly summarize the polarization mechanisms that are likely to occur in solids:

1) Electronic polarization in the time range of $10^{-15} \le t \le 10^{-17}$ s
2) Atomic polarization, $10^{-12} \le t \le 10^{-14}$ s
3) Orientational polarization, $10^{-3} \le t \le 10^{-12}$ s
4) Interfacial polarization, $t > 0.1$ s
5) Drift of electrons or holes in the inter-electrode region and their trapping
6) Injections of charges into the solid by the electrodes and their trapping in the vicinity of the electrodes. This mechanism is referred to as electrode polarization.

Considering the orientational polarization first, generally two experimental techniques are employed, namely, single temperature poling (Fig. 10.4) and windowing[6] (Fig. 10.5). The windowing technique, also called fractional polarization, is meant to improve the method of separating the polarizations that occur in a narrow window of temperature. The width of the window chosen is usually 10°C. Even in the absence of windowing technique, several techniques have been adopted to separate the peaks as we shall discuss later on. Typical TSD currents obtained with global thermal poling and window poling are shown in figs. 10-6[7] and 10-7[8], respectively.

Bucci et. al.[9] derived the equation for current due to orientational depolarization by assuming that the polar solid has a single relaxation time (one type of dipole). Mutual Interaction between dipoles is neglected and the solid is considered to be perfect with no other type of polarization contributing to the current. It is recalled that the orientational polarization is given by

$$P_\mu = \frac{1}{3}\frac{N\mu^2 E_p}{kT_p} \tag{2.51}$$

where E_p is the applied electric field during poling and T_p the poling temperature.

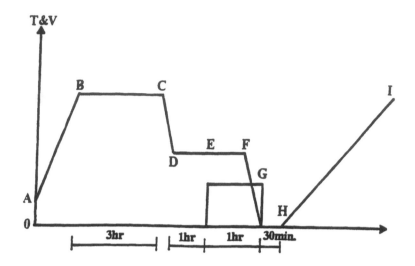

Fig. 10.4 Thermal protocol for TSD current measurement. AB-initial heating to remove moisture and other absorbed molecules, BC-holding step, time duration and temperature for AB-BC depends on the material, electrodes etc. CD-cooling to poling temperature, usually near the glass transition temperature, DE-stabilizing period before poling, electrodes short circuited during AE, EF-poling, FG-cooling to annealing temperature, GH-annealing period with electrodes short circuited, HI-TSD measurements with heating rate of β.

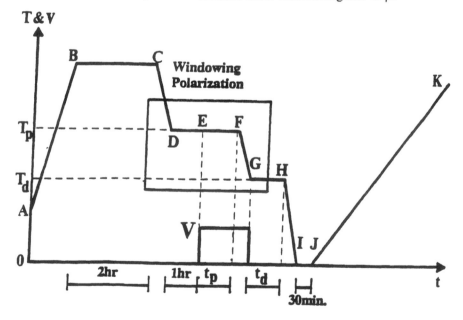

Fig. 10.5 Protocol for windowing TSD. Note the additional detail in the region EFGH. T_p and T_d are poling and window temperatures respectively. t_p and t_d are the corresponding times, normally $t_p = t_d$.

The relaxation time which is characteristic of the frequency of jumps of the dipole is related to the temperature according to

$$\tau = \tau_o \exp\frac{\varepsilon}{kT} \tag{10.16}$$

where $1/\tau$ (s^{-1}) is the frequency of a single jump and τ_o is independent of the temperature. Increasing the temperature decreases the relaxation time according to equation (2.51), as discussed in chapters 3 and 5.

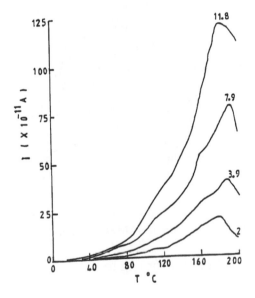

Fig. 10.6 TSD currents in 127 μm thick paper with β = 2K/min. The poling temperature is 200°C and the poling field is as shown on each curve, in MV/m.

The decay of polarization is assumed proportional to the polarization, yielding the first order differential equation

$$-\frac{dP(t)}{dt} = \frac{P(t)}{\tau(T)} = \frac{P(t)}{\tau_o}\exp\left(-\frac{\varepsilon}{kT}\right) \tag{3.10}$$

The solution of equation (3.10) may be written in the form[10]

$$P(T) = P_o \exp\left(-\frac{1}{\beta}\int_{T_o}^{T}\frac{dT'}{\tau(T')}\right) \tag{10.17}$$

The current density is given by

$$J = -\frac{dP}{dt}$$

(10.18)

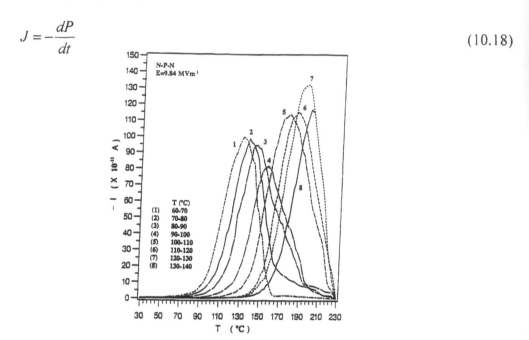

Fig. 10.7 TSD currents in a composite dielectric Nomex-Polyester-Nomex with window poling technique (Sussi and Raju. With permission of SAMPE Journal)

Assuming that the heating rate is linear according to

$$T = T_o + \beta t$$

(10.19)

Using equation (10.16) this gives the expression for the current

$$J = \frac{P_\infty}{\tau} \exp\left(-\frac{1}{\beta\tau} \int_{T_o}^{t} \exp\left(-\frac{\varepsilon}{kT} \right) dT' \right)$$

(10.20)

where T_o is the initial temperature (K), β the rate of heating (Ks⁻¹), t the time (s). Substituting equation (2.51) in (10.20) Bucci et. al (1966) derive the expression for the current density as:

$$J = \frac{N\mu^2 E_p}{3kT\tau_0} \exp\left(-\frac{\varepsilon}{kT}\right) \exp\left[-\left(\frac{1}{\beta\tau_0}\right) \int_{T_0}^{T} \exp\left(-\frac{\varepsilon}{kT'}\right) dT'\right] \tag{10.21}$$

Equation (10.21) is first order kinetics and has been employed extensively for the analysis of thermograms of solids. By differentiation the temperature T_m at which the current peak occurs is derived as

$$T_m = \left[\varepsilon \ \tau_0 \ \exp\left(\frac{\varepsilon}{kT_m}\right)\right]^{1/2} \tag{10.22}$$

T_m is independent of the poling parameters E_p and T_p but dependent on β. The number density of the dipoles is obtained by the relation

$$N = \frac{3kT_p}{\mu^2 E_p} \int_{T_0}^{\infty} J(T') dT' \tag{10.23}$$

where the integral is the area under the J-T curve.

According to equation (10.21) the current density is proportional to the poling field at the same temperature and by measuring the current at various poling fields dipole orientation may be distinguished from other mechanisms.

The concentration of charge carriers n_t for the case of mono-molecular recombination (τ is constant) and weak retrapping is given by[11]

$$n_t = \frac{2.7 J_m kT_m^2}{eL\beta\varepsilon_a} \tag{10.24}$$

where J_m is the maximum current density.

10.3 TSD CURRENTS FOR DISTRIBUTION OF ACTIVATION ENERGY

Bucci's equation (10.21) for TSD currents assumes that the polar materials possess a single relaxation time or a single activation energy. As already explained, very few materials satisfy this condition. Before considering the distributed activation energies, it is simpler to consider the analysis of TSD current spectra using the theory developed by Frohlich[12, 13] .

$$J(T) = \frac{P_o}{\tau_o} \exp\left[-\frac{\varepsilon_a}{kT} - \frac{1}{\beta\tau_o} \int_{T_o}^{T} \exp(-\frac{\varepsilon_a}{kT'})dT' \right]$$ (10.25)

in which P_o is a constant related to the initial polarization. Using a single value of ε_a in equation (10.25) generates a spectrum which is asymmetric as a function of the temperature while the experimental data are symmetric (Sauer and Avakian, 1992). This is remedied by assigning a slight breadth to the distribution of energies. An alternative is to express the TSD current in the form

$$J(T) = \sum_{i=1}^{n} a_i J(T, \varepsilon_{a_i})$$ (10.26)

The mean value of ε_{ai} is found to give a good fit to the data as that obtained by Bucci method.

Fig. 10.8 shows the TSD current in poly(ethyl methacrylate) (PEMA) in the vicinity of T_g using a poling temperature of 30°C (Sauer and Avakian, 1992). Application of the Bucci equation gives an activation energy of 1.4 eV. Application of equation (10.25) with $\varepsilon_a = 1.2$ eV gives a poor fit. Application of equation (10.26) with n = 3 and $\varepsilon_{a1} =$ 1.35 eV, $\varepsilon_{a2} = 1.37$ eV, $\varepsilon = 1.41$ eV, $a_1 = 47\%$, $a_2 = 41\%$, $a_3 = 12\%$ gives a fit which is comparable to a Bucci fit. A gaussian distribution shows considerable deviation at low temperatures indicating a slower relaxing entity. Non-symmetrical dipoles and dipoles of different kinds (bonds) as in polymers are reasons for a solid to have a distribution of relaxation times. Interacting dipoles result in a distribution of activation energies. In both cases of distribution of activation energies and distribution of relaxation times, the thermogram is much broader than that observed for single relaxation time and the TSD current is

$$J = \frac{N\mu^2 E_p}{3kT\tau_o} \int_{0}^{\infty} F(\varepsilon) \quad \exp\left[-\frac{\varepsilon}{kT} - \frac{1}{\beta\tau_o} \int_{T_o}^{T} \exp\left(-\frac{\varepsilon}{kT'}\right)dT' \right] d\varepsilon$$ (10.27)

where $F(\varepsilon)$ is the distribution function of activation energies. A similar equation for continuous distribution of pre-exponentials may be derived from equation (10.21).

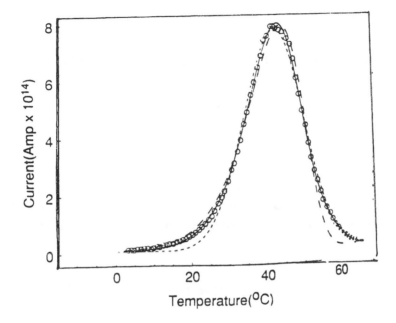

Fig. 10.8 TSD current in poly(ethyl methacrylate) poled near T_g. Fit to Frohlich's equation (10.25) is shown. Open circles: Experimental data; broken line: Frohlich's equation, single energy; full line: Frohlich's equation, distributed energy, G-Gaussion distribution.

10.4 TSD CURRENTS FOR UNIVERSAL RELAXATION MECHANISM

The dipolar orientation of permanent dipoles according to Debye process results in a TSD current according to equation (10.21). However we have seen in chapters 3 and 5 that the assumption of non-interacting dipoles has been questioned by Jonscher[14] who has suggested a universal relaxation phenomenon according to fractional power laws:

(a) In the frequency range $\omega > \omega_p$:

$$\chi''(\omega) \, \alpha \, \omega^m$$

(b) In the frequency range $\omega < \omega_p$:

$$\chi' = \chi_s - \tan(m\pi/2)\chi''(\omega)$$

$$\chi''(\omega) = \cot(\frac{n\pi}{2})\chi'(\omega) \, \alpha \, \omega^{n-1}$$

where the exponents m and n fall in the range (0,1). The TSD current in materials behaving in accordance with these relaxations is given by:

$$J(t) = aP_\mu \omega_p \left[\int_0^t \omega_p dt \right]^{-n} \quad \text{where } t < (1/\omega_p) \qquad (10.28)$$

In the case of distribution of relaxation times and single activation energy the initial slopes are equal whereas a distribution of activation energies results in different slopes. Peak cleaning technique is employed to separate the various activation energies as will be demonstrated below, while discussing experimental results.

10.5 TSD CURRENTS WITH IONIC SPACE CHARGE

The origin of TSD currents is not exclusively dipoles because the accumulated ionic space charge during poling is also released. The decay of space charge is generally more complex than the disorientation of the dipoles, and Bucci, et al. bring out the following differences between TSD current characteristics due to dipoles and release of ionic space charge;

1. In the case of ionic space charge the temperature of the maximum current is not well defined. As T_p is increased T_m increases.
2. The area of the peak is not proportional to the electric field as in the case of dipolar relaxation, particularly at low electric fields.
3. The shape of the peak does not allow the determination of activation energy (Chen and Kirsch, 1981).

The derivation of the TSD current due to ionic space charge has been given by Kunze and Müller[15]. Let us suppose that the dark conductivity varies with temperature according to

$$\sigma = \sigma_0 \exp\left(-\frac{\varepsilon}{kT}\right) \qquad (10.29)$$

and the heating rate is reciprocal according to

$$\frac{1}{T} = \frac{1}{T_o} - \beta t \tag{10.30}$$

$$-\frac{d(T^{-1})}{dt} = \frac{1}{T^2}\frac{dT}{dt} = a = \text{constant} \tag{10.31}$$

where T_o and β are the initial temperature and heating rate ($K^{-1}s^{-1}$) respectively. The TSD current density is given by

$$J(t) = \frac{\sigma_o}{\varepsilon\varepsilon_o} Q_o \exp\left(-\frac{\varepsilon}{kT}\right)\exp\left[-\frac{\sigma_o}{\varepsilon\varepsilon_o\beta}\int_{T_o}^{T}\exp\left(-\frac{\varepsilon}{kT'}\right)dT'\right] \tag{10.32}$$

where Q_o is the charge density on the electrodes at temperature T_o.

10.6 TSD CURRENTS WITH ELECTRONIC CONDUCTION

Materials which possess conductivity due to electrons or holes present additional difficulties in analyzing the TSD currents. The theory has been worked out by Müller[16] who considered a dielectric (ε_1) under investigation sandwiched between insulating foils of dielectric constant ε_2, and thickness d_2. This arrangement prevents the superposition of current due to electron injection from the TSD current. The theory is relevant to these experimental conditions.

$$J(T) = \frac{f\sigma_o P_o}{\varepsilon_o\varepsilon_2}\exp\left[-\frac{\varepsilon}{kT} - \frac{f\sigma_o k}{a\varepsilon\varepsilon_o\varepsilon_2}\exp\left(-\frac{\varepsilon}{kT}\right)\right] \tag{10.33}$$

The current peak is observed at

$$\frac{f\sigma_o k}{a\varepsilon\varepsilon_o\varepsilon_2}\exp\left(-\frac{\varepsilon}{kT_m}\right) = 1 \tag{10.34}$$

where T_m is the temperature at the current maximum. The current peak is asymmetrical. The activation energy is determined by the equation

$$\ln \varepsilon = \ln\left(\frac{J_m}{aP_\mu}\right) + 1 \qquad (10.35)$$

where J_m is the maximum current density and P_μ is determined from the area of the J-t curve.

10.7 TSD CURRENTS WITH CORONA CHARGING

The mechanism of charge storage in a dielectric may be studied by the measurements of TSD currents, and the theory for the current has been given by Creswell and Perlman[17]. Sussi and Raju[18] have applied the theory to corona charged aramid paper. Let us assume a uniform charge density of free and trapped charge carriers, the charge in an element of thickness at a depth x and unit area is

$$dQ = \rho dx \qquad (10.36)$$

in which ρ is the surface charge density. The contribution to the current in the external circuit due to release of this element of charge is [Creswell, 1970]

$$dJ = \frac{WdQ}{s} = \frac{J(x)dx}{s} \qquad (10.37)$$

where s is the thickness of the material and J the current due to the element of charge, W the velocity with which the charge layer moves and $J(x)$ the local current due to the motion of charge carriers.

According to Ohm's law, the current density is

$$J(x) = \rho\mu E(x) \qquad (10.38)$$

where μ is the mobility and E(x) the electric field at a depth x. The electric field is not uniform due to the presence of space charges within the material and the field can be calculated using the Poisson's equation

$$\frac{dE}{dx} = \frac{\rho_1}{\varepsilon_0 \varepsilon_s} \qquad (10.39)$$

where ρ_1 is the total charge density (free plus trapped) and ε_s is the dielectric constant of the material. If we denote the number of free and trapped charge carriers by n_f and n_t respectively, the current is given by

$$J = \frac{\mu e^2 \delta^2}{2\varepsilon_s s} n_t (n_t + n_f) \tag{10.40}$$

in which δ is the depth of charge penetration, $\delta \ll x$ and e the electronic charge.

In general it is reasonable to assume that $n_f \ll n_t$ and equation (10.40) may be approximated to

$$J = \frac{\mu e^2 \delta^2}{2\varepsilon_s s} n_f n_t \tag{10.41}$$

The released charge may be trapped again and for the case of slow retrapping Creswell and Perlman[19] have shown that

$$J = \frac{\mu e^2 \delta\, n_{to}}{2\varepsilon_o \varepsilon_s} \frac{\tau}{\tau_o} \exp\left[\left(-\frac{\varepsilon_a}{kT} - \frac{2}{\beta \tau_o}\right) \times \int_{I_o}^{T} \exp\left(-\frac{\varepsilon_a}{kT}\right) dT\right] \tag{10.42}$$

where n_{to} is the initial density of charges in traps, $1/\tau_o = \nu$ the attempt to escape frequency, ε_a the trap depth below the conduction band.

The relaxation time is related to temperature according to

$$\tau = \tau_0 \exp\left(\frac{\varepsilon_a}{kT}\right) \tag{3.59}$$

Equation (10.42) may be rewritten in the form

$$J = A \exp\left[-p + \beta \int \exp(-p) p^{-2} dp\right] \tag{10.43}$$

$$A = \left(\frac{\mu e^2 \delta^2 n_{to}}{2\varepsilon_o \varepsilon_s}\right) \frac{\tau}{\tau_o} \tag{10.44}$$

$$B = \frac{2\varepsilon_a}{k\beta\tau_o} \tag{10.45}$$

$$p = \frac{\varepsilon_a}{kT} \tag{10.46}$$

For maximum current we differentiate equation (10.43) and equate it to zero to yield

$$B = p_{max}^2 \exp(p_{max}) \tag{10.47}$$

in which p_{max} is related to T_{max} according to equation (10.46).

Figure 10.9 (Creswell aand Perlman 1970) shows the TSD currents in negatively charged Mylar with silver-paste electrodes, with several rates of heating. The spectra are complicated with a downward shift of the peak as the heating rate is lowered. By a partial heating technique the number of peaks and their magnitude were determined and Arrhenius plots were drawn as shown in Fig.10.10. The slope increases with increasing temperature and the activation energy varies in the range of 0.55 eV at 50°C to 2.2 eV at 110°C in four discrete steps. The low energy traps of 0.55 eV and 0.85 eV are electronic and the trap of depth 1.4 eV is ionic. The 2.2 eV trap is either ionic, interfacial or release of electron to conduction band by complex processes. In interpreting these results it is generally true that trap depths less than 1 eV are electronic and if greater than 1 eV they are ionic. Assuming that the traps are monomolecular trap densities of the order of $10^{22}/m^3$ are obtained.

Fig 10.11 shows the TSD currents in corona charged aramid paper (Sussi and Raju, 1994) which shows the complexity of the spectrum and considerable caution is required in interpreting the results.

10.8 COMPENSATION TEMPERATURE

The Arrhenius equation for the relaxation time, which is the inverse of jump frequency between two activated states, is expressed as

$$\tau = \tau_0 \exp\frac{\varepsilon}{kT} \tag{3.59}$$

where τ_0 is the pre-exponential factor. A plot of log τ against at $1/T$ yields a straight line according to equation (3.59) with a slope of the apparent activation energy, ε. For secondary or low temperature relaxation the activation energy is low in the range of 0.5 −1.0 eV and the values of τ_0 are generally on the order of 10^{-12}s. These ranges of values are found from thermally activated molecular motions. However, high values of ε and very low values of τ_0 are not associated with the picture of molecular jump between two sites separated by a energy barrier. These values of both high ε and low τ_0 are explained on the basis of cooperative movement corresponding to long range confirmational changes, characteristic of the α-relaxation[20].

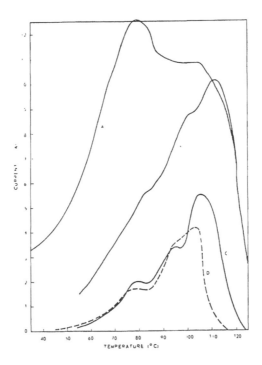

Fig. 10.9 Thermal current spectra for negatively corona-charged Mylar (2m). (a)- 1°/min, after 1.5 hr; (b)-1°/min after 24 hr; (c)- 0.4°/min, after 15 days; (d) at 0.2°/min after 25 days (Creswell and Perlman, 1970, with permission of American Institute of Physics.)

In many materials the plot of log τ against at $1/T$ tends to show lower activation energy, particularly at high temperatures. In this region the relaxation time is often represented by the Vogel-Tammann-Fulcher (VTF) law given by

$$\tau = \tau_0 \exp\frac{A}{(T - T_0)} \qquad (10.48)$$

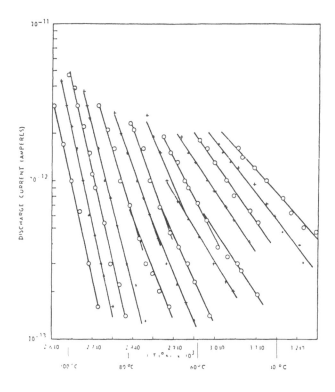

Fig. 10.10 Arrhenius plots of TSD currents in Mylar obtained by partial heating (Creswell and Perlman, 1970, with permission of American Inst. of Physics).

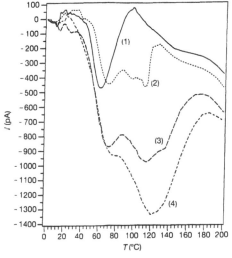

Fig. 10.11 TSD current in corona charged at 16.38 kV in 76mm aramid paper. Influence of electrode material is shown. Charging time and electrode material are: (1) 10 min., Al. (2) 10 min., Ag. (3) 20 min., Al (4) 20 min., Ag [Sussi and Raju 1994]. (with permission of Chapman and Hall)

where A and T_0 are constants. A is related to either activation energy or the thermal expansion co-efficient of the free volume[21]. When $T = T_0$ the relaxation time τ becomes infinite and this is interpreted as the glass transition temperature. In the vicinity of and below the glass transition temperature, τ deviates strongly from the VTF law[22] In some cases a plot of $\log\tau$ versus $1/T$, when extrapolated, converges to a single point (T_c, τ_c) as shown in Fig.10.12[23]. This behavior is expressed by a compensation law according to

$$\tau(T) = \tau_c \exp\frac{\varepsilon}{k}(\frac{1}{T} - \frac{1}{T_c}) \qquad (10.49)$$

where T_c is called the compensation temperature at which all relaxation times have the same value[24]. Compensation is the relationship between the activation energy and the pre-exponential factor in equation (3.59) expressed as

$$\tau_c = \tau_o \exp\left(\frac{\varepsilon}{kT_c}\right) \qquad (10.50)$$

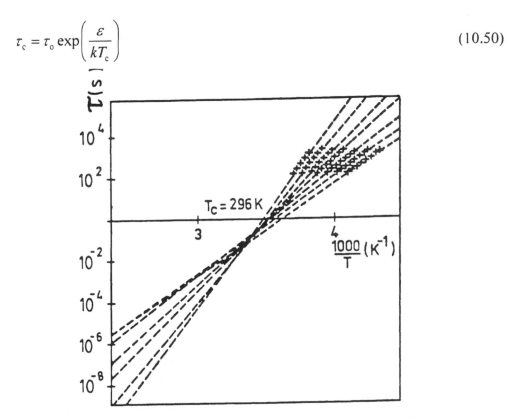

Fig. 10.12 Compensation effect in isotactic polypropylene (Ronarch et. al. 1985, with permission of Am. Inst. Phys.)

According to equation (10.50) a plot of $\ln\tau_0$ versus ε/k gives the reciprocal of the compensation temperature ($-1/T_c$) and the intercept is related to the compensation time τ_c (Fig. 10.13, Colmenero et. al., 1985). The compensation temperature T_c should correspond to the phase transition temperature with good approximation. However a significant departure is observed in semi-crystalline and amorphous polymers as shown in Tables 10.1 and 10.2 (Teysseder, 1997); T_c is observed to be always higher than T_c though the difference is not found to depend systematically on crystallinity. The physical meaning of T_c is not clear though attempts have been made to relate it to changes in the material as it passes from solid to liquid state. Fig. 10.14 shows a collection of the compensation behavior in several polymers; PET data are taken from Teysseder (1997), PPS data are taken from Shimuzu and Nagayama (1993)[25]. Special mention must be made of Polycarbonate which shows Arrhenius behavior over a wide range of temperatures (b) though at temperatures close to T_g a non-Arrhenius behavior is observed.

10.9 METHODS AND ANALYSES

The methods employed to analyze the TSD currents have been improved considerably though the severe restrictions that apply to the method have not been entirely overcome. The TSD currents obtained in an ideal dielectric with a single peak symmetrical about the line passing through the peak and parallel to the ordinate (y-axis) is the simplest situation. The various relationships that apply here are discussed below.

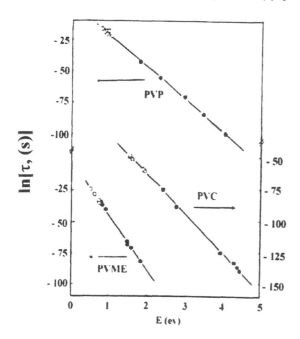

Fig. 10.13 Compensation plots corresponding to poly (N-vinyl-2-pyrrolidone) (PVP), poly(vinylchloride) (PVC), and poly(vinylmethylether) (Colmenero, 1987; with permission of Am. Inst. Phys.)

Table 10.1

Parameters for Various Semicrystalline Polymers (Teyssedre, 1997).
(with permission of Am. Ch. Soc.)

Polymer	$X_c\%$	$T_g°C$	$T_c\text{-}T_g(°C)$	$\tau_c(s)$
Polypropylene[1]	50	-10	33	2.0×10^{-2}
VDF[2]	50	-42	27	5.8×10^{-3}
P(VDF-TrFE) 75/25	55	-36	25	3.0×10^{-3}
65/35[3]	52	-33	22	1.6×10^{-2}
50/50	48	-28	30	5.0×10^{-4}
Polyamides 12	40	40	44	7.8×10^{-3}
6.6[4]	40	57	48	2.5×10^{-2}
PEEK	12	144	17	2.0
	31[5]	157	8	2.0
PPS[6]	10	94	24	2.5
PCITrFE[7]	10	52	74	0.25
PET[8]	45	100	15	10

$X_c\%$ is percentage crystallinity

[1] Reference [Ronarch. 1985]
[2] G. Teyssedre. A. Bernès. C. Lacabanne. J. Poly. Sci: Phys. ed. **31** (1993) 2027
[3] G. Teyssedre. A. Bernès. C. Lacabanne. J. Poly. Sci: Phys. ed. **33** (1993) 2419
[4] F. Sharif. Ph. D. Thesis. University of Toulouse. 1984
[5] M. Mourgues. A. Bernès. C. Lacabanne. Thermochim. acta. 226 (1993) 7
[6] H. Shimizu. K. Nakayama. J. Appl. Phys., **74** (1993) 1597
[7] H. Shimizu. K. Nakayama. J. Appl. Phys.. **28** (1989) L1616
[8] A. Bernès. D. Chatain. C. Lacabanne. Thermochim. Acta. **204** (1992) 69

At low temperatures the term within square brackets in equation (10.21) is small and we can approximate the equation to

$$J = \frac{J_o}{\tau_o}\exp\left(-\frac{\varepsilon}{kT}\right) \qquad (10.51)$$

where

$$J_o = \frac{N\mu^2 E_p}{3kT} \qquad (10.52)$$

Equation (10.51) has the form of the well known Arrhenius equation and the slope of the log J against 1/T gives the energy. This method is known as the initial rise method[26].

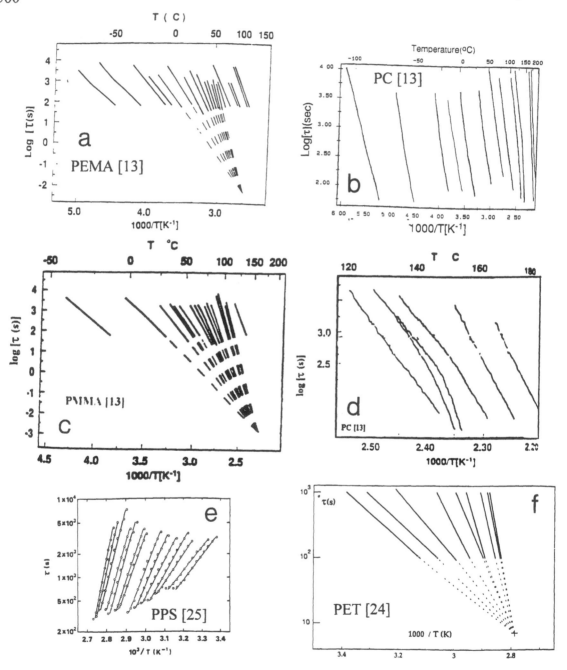

Fig. 10.14 Compensation behavior in several polymers. Polymer and the reference shown on each. (with permission of Polymer: a, b, c, d; Am. Phys. Inst; e; Am. Ch. Inst; f)

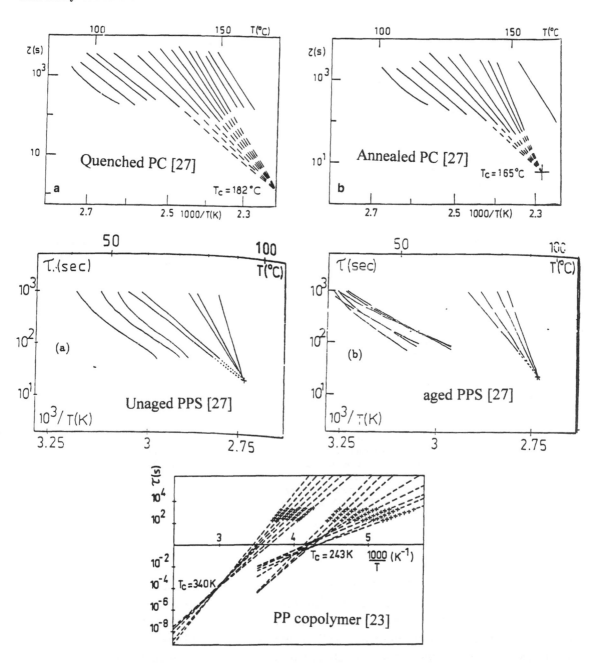

Fig. 10.14 (contd.) Compensation temperature in selected polymers. (g) and (h) in quenched and annealed polycarbonate respectively (Bernès, et al., 1992, with permission of polymer); (j) and (k) in unaged and aged amorphous oriented PPS respectively (Mourgues-Martin, et al., 1992, with permission of IEEE©); (l) Polypropylene block copolymer (Ronarc'h, et al., 1985, with permission of Amer. Inst. Phys.)

Table 10.2
Parameters for various amorphous polymers (Teyssedre, 1997).
(with permission of Am. Ch. Soc.)

Polymer	$T_g °C$	$T_c-T_g(°C)$	$\tau_c(s)$
Poly(methyl methacrylate)[1]	110	18	0.26
Polystyrene[2]	100	2	10
PET[3, a]	82	5	6.2
Polycarbonate[4]	150	32	0.7
PPS[5, b]	91	2	36
Polyurethane[6]	-69	12	0.94
DGEBA-DDS[6, c]	133	18	3.2
PEMA[7]	54	54	6.8×10^{-3}
Poly(vinyl chloride)[8]	74	10	22
Polyacrylate[9]	178	25	16
Poly(Vinyl methyl ether)[7, 9]	248	15	0.6
Polysulfone[9]	189	7	4.7

a = Poly(ethylene terephthalate)
b = Poly(p-phynelene-sulphide)
c = Diglycidyl ether of bisphenol A

[1] J. P. Ibar. Thermochim. Acta. **192** (1991) 91
[2] A. Bernès. R. F. Boyer. C. Lacabanne. J. P. Ibar. "Order in the Amorphous state of the Polymers", Ed: S. E. Keinath, R. L. Miller, J. K. Rieke, Plenum, New York, 1986. p. 305
[3] A. Bernès. D. Chatain. C. Lacabanne. Thermochim. Acta. **204** (1992) 69
[4] A. Bernès. D. Chatain. C. Lacabanne. Polymer. **33** (1992) 4682
[5] M. Mourgues-Martin. A. Bernès. C. Lacabanne. O. Nouvel. G. Seytre, IEEE Trans. Elec. Insul.. **27** (1992) 795
[6] C. Dessaux. J. Dugas. D. Chatain. C. Lacabanne. Journées d' Etude de Polymères XXIII: Toulouse. France, 1995
[7] B. Sauer. P. Avakian. Polymer. **33** (1992) 5128
[8] J. J. Del Val. A. Alegria. J. Colmenero. C. Lacabanne. J. Appl. Phys.. **59** (1986) 3829
[9] J. Colmenero. A. Alegria. J. M. Alberdi. J. J. Del Val. G. Ucar. Phys. Rev. **B, 35** (1987) 3995

Evaluation of the parameters in the case of a single peak is relatively simple since the observed peak is due to a single relaxation time. However the analysis is complex when we know that there are several relaxation times and the TSD current due to each process overlaps, resulting in a broad or not well defined peak. The global TSD currents must be separated into its components. To achieve this a peak cleaning method is employed in which the temperature is suddenly lowered to the initial temperature, T_o, immediately after a peak is observed. A second TSD measurement will then show the subsequent peak which is not a superposition due to the previous peak. Fig. 10.15 shows such a method in which a global curve has been found to be composed of as many as nine separate relaxations[27].

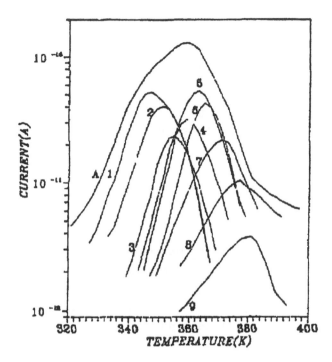

Fig. 10.15 TSD current in amorphous blend of PVC and ABS. Global curve A is composed of several peaks. (Megahed et. al. 1994, with permission of Institute of Physics, England)

A recent technique that has been adopted[28] to separate the global TSD curve into its components makes use of a computational procedure. The TSD current expressed by equation (10.21) may be simplified as

$$\ln(I) = b \ln\left(\frac{n}{n_o}\right) + \ln\left(\frac{n_o}{\tau}\right) - \frac{\varepsilon}{kT} \qquad (10.53)$$

where b is the order of the kinetics, n the number of carrier in the traps and n_o the number of initial carriers in the traps. For the first order kinetics, $b = 1$, and for the second order kinetics, $b = 2$. Taking any two arbitrary points on the TSD curve (I_1, T_1) and (I_2, T_2) and substituting these in equation (10.53), then subtracting one resulting equation from the other yields

$$\ln\left(\frac{I_2}{I_1}\right) = b \ln\left(\frac{n}{n_1}\right) - \frac{\varepsilon_a}{k}\left(\frac{1}{T_2} - \frac{1}{T_1}\right) \qquad (10.54)$$

A similar equation is obtained from a third point (I_3, T_3)

$$\ln\left(\frac{I_3}{I_1}\right) = b\ln\left(\frac{n_3}{n_1}\right) - \frac{\varepsilon_a}{k}\left(\frac{1}{T_3} - \frac{1}{T_1}\right) \tag{10.55}$$

These equations are treated as simultaneous equations with b and ε_a as unknown. The relaxation time is related to the temperature, T_m, at which peak current is observed according to equation (10.22)

$$\tau = \frac{kT_m^2}{\beta\varepsilon_a}\exp\left(-\frac{\varepsilon_a}{kT_m}\right) \tag{10.56}$$

Fan, et al. (1999) adopt the following procedure. Using the high temperature tail of the TSD current-T curve, and equations (10.54), (10.55) and (10.56), the three parameters ε_a, b and τ are determined. Substituting these values in (10.21) an isolated TSD curve is obtained. This curve is subtracted from the global TSD curve and the procedure repeated. The global TSD and separated curves in polyimide are shown in Fig. 10.16.

The relaxation time may be obtained from the TSD curves by employing the relationship

$$\tau(T) = \frac{P(T)}{J(T)} = \int_t^\infty \frac{J(t')\,dt'}{J(t)} \tag{10.57}$$

where $\tau(T)$ is the relaxation time. $J(t)$ is the current density at time t and the numerator is the total charge density remaining in the polarized dielectric[29].

Equation (10.57) has been applied in several polymers to determine the relaxation time and compensation parameters. The upper limit of integral in equation (10.57) is infinity, meaning that the TSD currents should be determined till it is reduced to an immeasurably small quantity. However in some of the high temperature materials, as in aramid paper (Raju 1992), it may not be possible to extend the current measurement to such high temperatures. Fig. 10.17 shows the physical meaning of equation (10.57); in a low temperature dielectric which can be heated up to its glass transition temperature, the TSD currents decrease to zero, whereas it is not easy to reach this point in the case of high temperature polymers.

One of the experimental parameters in the measurement of TSD currents is the heating rate, β. By rearranging equation (10.56) the heating rate is expressed as

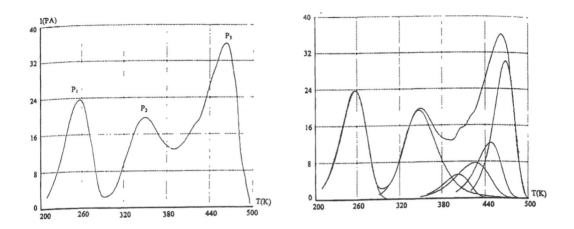

Fig. 10.16 TSD currents in polyimide. (a) global (b) Peak cleaned. (Fan, et al. 1999, with permission of Inst. of Physics. England)

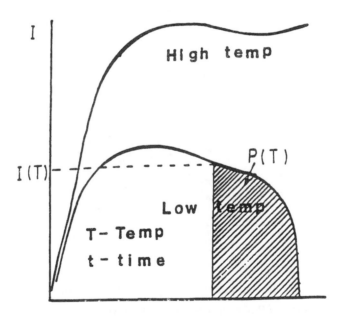

Fig. 10.17 Schematic diagram for calculating the relaxation time from TSD currents. In low temperature polymers the glass transition temperature is reached rendering the TSD current to zero. In high temperature dielectrics there is considerable charge remaining at A.

$$\beta = \left(\frac{k}{\varepsilon_a \tau}\right) T_m^2 \exp\left(-\frac{\varepsilon_a}{kT_m}\right) \qquad (10.56)$$

To determine the remaining charge at the end of the TSD run, as at A in Fig. 10.18 we have to adopt a steady state method to determine the remaining charge. By such a procedure, the released charge and the relaxation time is determined.

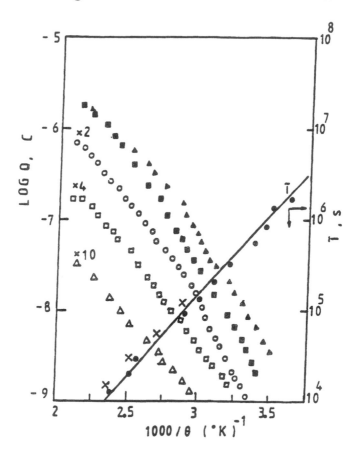

Fig. 10.18 Relaxation time and released charge as a function of 1000/T in aramid paper. Poling field and heating rate: ▲ 7.9 MV/m, 1 K/min; ■ 11.8 MV/m, 2 K/min.; O 7.9 MV/m, 2 K/min.; ☐ 3.9 MV/m, 2K/min.; △ 2 MV/m, 2K/min.; ● 7.9 MV/m, 1 K/min.; ✗1 K/min. (Raju, 1992, with permission of IEEE).

Increasing the heating rate during TSD measurement increases the right side of equation (10.56) by the same amount. This causes T_m to shift upwards. By measuring TSD currents at two different heating rates β_1 and β_2 and determining T_{m1} and T_{m2}, respectively, the parameters ε_a and τ may be determined. Expressing equation (10.56) twice, one for each heating rate, and taking the ratio we get the activation energy as

$$\varepsilon_a = k \left(\frac{T_{m1} T_{m2}}{T_{m1} - T_{m2}} \right) \ln \left[\frac{\beta_1}{\beta_2} \left(\frac{T_{m2}}{T_{m1}} \right)^2 \right] \qquad (10.58)$$

This value of the activation energy may be inserted into equation (10.56) to determine τ. A linear relationship has also been noted, with several linear heating rates as

$$\ln \left(\frac{\tau_m^2}{\beta} \right) = \ln \left(\frac{\varepsilon_a \tau}{k} \right) + \frac{\varepsilon_a}{kT_m} \qquad (10.59)$$

A plot of $\ln (\tau_m^2/\beta)$ versus $(1/\tau_m)$ results in a straight line; the slope gives (ε_a/k) and the intercept gives τ.

Careful considerations should be given to identify dipolar relaxation separated from charge carrier traps or from interfacial polarization. To achieve this purpose, TSD currents are measured at various poling fields; a linear variation of the peak current, I_m, with E_p considered to be evidence of dipolar mechanism for polarization. A linear relationship between charge released to the external circuit, this is the integral of I(T)-t curve, with E_p considered to be evidence of dipolar mechanism. In the case of dipolar relaxation measurements, with reversed polarity of E_p, they generate TSD currents that are exact mirror images, since the influence of electrodes in charge injection is negligible.

Fig. 10.19[30-35] shows a collection of TSD currents in several polymers. TSD spectra are influenced by a number of external factors such as adsorbed gases, moisture, oxidized layers, aged and unaged samples, etc. Notwithstanding the fact that the results are obtained by controlling only a few of these factors, this collection is considered to be appropriate for a quick reference. A comprehensive list of nearly 500 references has been compiled by Lavergne and Lacabanne covering the literature up to the year 1990.

10.10 TSD AND AC DIELECTRIC PROPERTIES

We conclude this chapter by briefly referring to the derivation of the complex dielectric constant $\varepsilon* = \varepsilon' - j\varepsilon''$ from the TSD spectra. The linking parameter is the relaxation time

τ (T) and the appropriate equations are (3.50) and (3.51) in chapter 3. The dielectric decrement for the ith relaxation process is $\Delta\varepsilon_i = (\varepsilon_s - \varepsilon_\infty)_i$ and it can be evaluated from

$$\Delta\varepsilon_i = \frac{1}{\varepsilon_0 E} \int_{t_0}^{\infty} J(t)dt = \frac{1}{\varepsilon_0 E\beta} \int_{T_0}^{T} J(T)dT \qquad (10.60)$$

The usefulness of equation (10.60) lies in the fact that the dielectric spectra are obtained at sub-harmonic frequencies which is a range at which bridge measurements are not easy to carry out. The data obtained from TSD measurements may be used to examine the low frequency phenomena such as sub-α-relaxation (below T_g) or interfacial polarization. Figs. 10-20 and 10-21 (Shimizu, 1993) show the frequency and temperature dependence of calculated values in PPS. Curves A, B and C of Fig. 10-20 show a sub-T_g relaxation at frequencies of 10^{-9}, 10^{-7} and 10^{-5} Hz respectively at low temperatures as indicated in the legend. At higher temperatures, curves D-G, a sharp peak is observed at T_g due to the α-relaxation. In comparing the derived complex dielectric constant $\varepsilon*$ with the measured values, the frequency that is equivalent for comparison is given by equation (10.56), (Van Turnhout, 1975)

$$\tau = \frac{kT_m^2}{\beta\varepsilon_a} = \frac{1}{2\pi f_{eq}} \qquad (10.61)$$

From equation (10.61) we obtain the equivalent frequency as

$$f_{eq} = \frac{\beta\varepsilon_a}{2\pi kT_m^2} \qquad (10.62)$$

Substituting values appropriate to PPS, $T_m = 370$ K, $\varepsilon_a = 2.6$ eV, $\beta = 0.04$ K/s yields $f_{eq} = 10^{-3}$ Hz. Since the low frequency ac data is usually limited to 1 Hz, the advantages of using the TSD spectra, particularly for sub-T_g relaxation become evident.

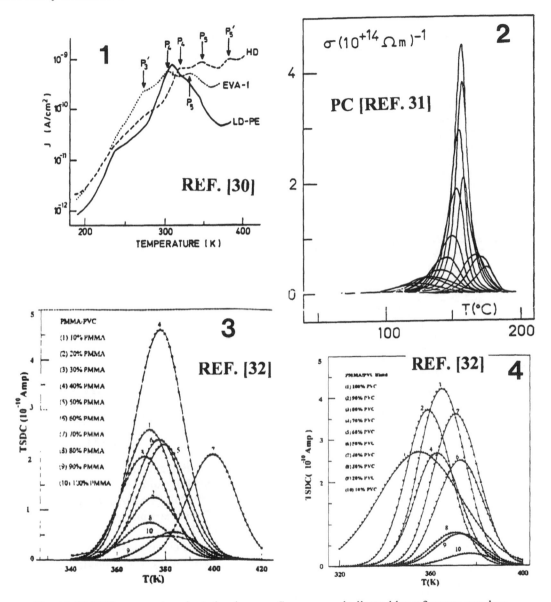

Fig. 10.19 TSD spectra in selected polymers. Sources are indicated by reference number. (with permission: 1-J. Appl. Phys., 2-Polymer, 3 & 4-J. Phys. D: Appl. Phys., England)

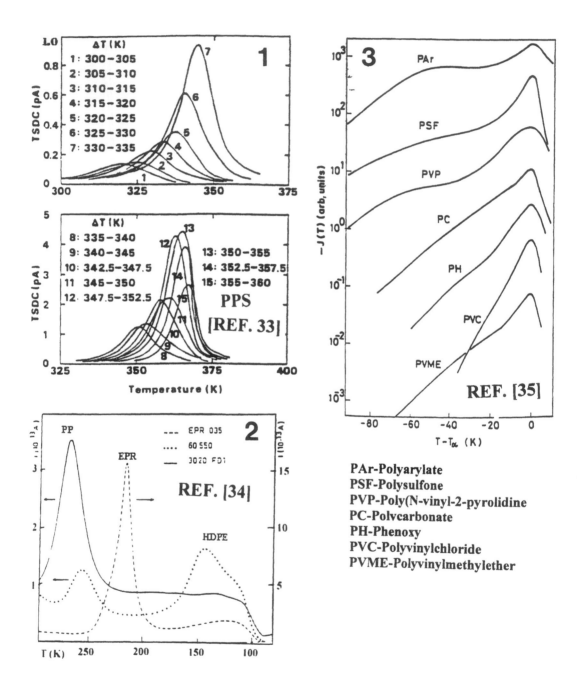

Fig. 10.19 (contd.) TSD spectra in selected polymers. (With permission: 1 &2-J. Appl. Phys., 3-Phys. Rev.)

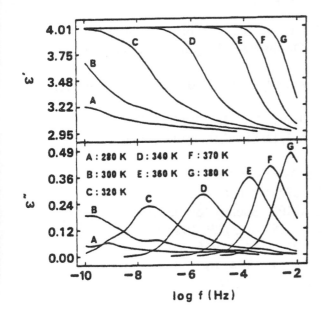

Fig. 10.20 Complex dielectric constant of PPS versus frequency with temperature as the parameter, derived from TSD spectra of Fig. 10.19 (label 1). (Shimizu and Nagayama, 1993; with permission of J. Appl. Phys.)

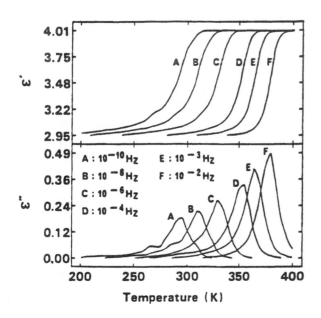

Fig. 10.21 Complex dielectric constant of PPS versus frequency with frequency as the parameter, derived from TSD spectra of Fig. 10.19 (label 1) [33]. (Shimizu and Nagayama, 1993; with permission of J. Appl. Phys.)

10.11 REFERENCES

1 Electrets: Topics in Applied Physics, Ed: G. M. Sessler, Springer-Verlag, Berlin, 1980.
2 J. Van Turnhout, Thermally Stimulated Discharge of Polymer Electrets, Elsevier, Amsterdam, 1975.
3 R. Chen and Y. Kirsch, "Analysis of Thermally Stimulated Processes, Pergamon Press, London, 1981.
4 P. Kelly, M. J. Kaubitz and P. Braülich, Phys. Rev., **B4** (1960) 1971.
5 S. Haridoss, J. Comp. Phys., **26** (1978) 232; D. Shenker and R. Chen, J. Comp. Phys., 10 (1972) 272.
6 M. A. Sussi, Ph. D. Thesis, University of Windsor, 1992.
7 G. R. Govinda Raju, IEEE Trans. on EI, **27** (1992) 162-173.
8 M. A. Sussi and G. R. Govinda Raju, SAMPE (Society for the Advancement of Material and Process Engineering) Journal, **28** (1992) 29; **30** (1994) 1.
9 C. Bucci, R. Fieschi and G. Guidi, Phys. Rev., **148** (1966) 816.
   C. Bucci and R. Fieschi, Phys. Rev. Lett., **12** (1964) 16.
10 A. K. Jonscher, J. Phys. D: Appl. Phys., **24** (1991) 1633.
11 E. Neagu and D. K. Das-Gupta, IEEE Transactions on **EI, 24** (1989) 489.
12 H. Frohlich, "Theory of Dielectrics"' Clarandon Press, Oxford, 1986.
13 B. B. Sauer and P. Avakian, Polymer, **33** (1992) 5128.
14 A. K. Jonscher, "Dielectric Relaxation in Solids", Chelsea Dielectric Press, London, 1983.
15 I. Kunzer and P. Müller, Phys. stat. sol. **(a), 13** (1972) 197.
16 P. Müller, Phys. Stat. Sol., **(a), 23** (1974) 579.
17 R. A. Creswell and M. M. Perlman, J. Appl. Phys., **41** (1970) 2365.
18 M. A. Sussi and G. R. Govinda Raju, J. Matl. Sci., **29** (1994) 73.
19 M. M. Perlman, J. Electrochem. Soc: Solid State Sci. Technol. **119** (1972) 892.
20 J. Colmenero, A. Alegria, J. M. Alberdi, J. J. Del Val, G. Ucar, Phys. Rev. **B, 35** (1987) 3995.
21 M. H. Cohen and D. J. Turnbull, J. Chem. Phys., **31** (1959) 1164.
22 M. D. Migahed, M. T. Ahmed and A. E. Kotp, J. Phys. d: Appl. Phys., **33** (2000) 2108.
23 D. Ronarc'h, P. Audren and J. L. Moura, J. Appl. Phys., **58** (1985) 474.
24 G. Teyssedre, S. Mezghani, A. Bbbernes, and C. Lacabanne in Dielectric Spectroscopy of Polymeric Materials, Ed: J. P. Runt and J. J. Fitzgerald, American Chemical Soc., Washington, D. C., 1997.
25 H. Shimuzu and K. Nagayama, **74** (1993) 1597-1605.
26 G. F. J. Garlick and A. F. Gibson, Proc. Phys. Soc., **60** (1948) 574.
27 Data for PET:M. D. Megahed, M. Ishra and T. Fahmy, J. Phys. D: Appl. Phys., **27** (1994) 2216.

Data for glassy polycarbonate: A. Bernés, D. Chatain and C. Lacabanne, Polymer, **33** (1992) 4682.

Data for amorphous oriented PPS: M. Mourgues-Martin, A. Bernés and C. Lacabanne, IEEE Trans. Elec. Insu., **27** (1992) 795.

28 Y. Fan, X. Wang, W. Zhang and Q. Lei, J. Phys. D: Appl. Phys., **32** (1999) 2809.

29 C. Bucci, R. Fieschi and G. Guidi, Phys. Rev., **148**, 816 (1966). see page 819.

30 T. Mizutani, Y. Suzuoki and M. Ieda, J. Appl. Phys., **48** (1977) 2408.

31. A. Bernes, D. Chatain and C. Lacabanne, Polymer, **33** (1992) 4681.

32. M. D. Migahed, M. T. Ahmed and K. E. Kotp, J. Phys. D: Appl. Phys., **33** (2000) 2108, see Fig. 3.

33. H. Shimizu and K. Nagayama, J. Appl. Phys., **74(3)**, (1993) 1597.

34. D. Ronarc'h and P. Audren, J. Appl. Phys., **58(1)**, (1985) 466.

35. J. Colmenero, A. Alegria, J. M. Alberdi, J. J. del Val, and G. Ucar, Phys. Rev. B, **35 (8)**, (1987) 3995.

36. C. Lavergne and C. Lacabanne, IEEE Electrical Insulation Magazine, **9 (2)**, (1993) 5.

Why, sometimes I've believed as many as six impossible things before breakfast.
-Lewis Carroll (Through the Looking Glass)

11

SPACE CHARGE IN SOLID DIELECTRICS

This chapter is devoted to the study of space charge build up and measurement of charge density within the dielectric in the condensed phase. When an electric field is applied to the dielectric polarization occurs, and so far we have treated the polarization mechanisms as uniform within the volume. However, in the presence of space charge the local internal field is both a function of time and space introducing non-linearities that influence the behavior of the dielectrics. This chapter is devoted to the recent advances in experimental techniques of measuring space charge, methods of calculation and the role of space charge in enhancing breakdown probability. A precise knowledge of the mechanism of space charge formation is invaluable in the analysis of the polarization processes and transport phenomena.

11.1 THE MEANING OF SPACE CHARGE

Space charge occurs whenever the rate of charge accumulation is different from the rate of removal. The charge accumulation may be due to generation, trapping of charges, drift or diffusion into the volume. The space charge may be due to electrons or ions depending upon the mechanism of charge transfer. Space charge arises both due to moving charges and trapped charges.

Fig. 11.1 shows the formation of space charge due to three processes in a dielectric that is subjected to an electric field[1].

(a) The electric field orients the dipoles in the case of a homogenous material and the associated space charge is a sharp step function with two peaks at the electrodes.
(b) Ion migration occurs under the influence of the electric field, with negative charges migrating to the positive electrode and vice-versa. The mobility of the various carriers

are not equal and therefore the accumulation of negative charges in the top half is random. Similarly the accumulation space charge due to positive charges in the bottom portion is also random and the voltage due to this space charge is also arbitrary. The space charge is called "heterocharges".

(c) Charges injected at the electrodes generate a space charge when the mobility is low. The charges have the same polarity as the electrode and are called "homocharges."

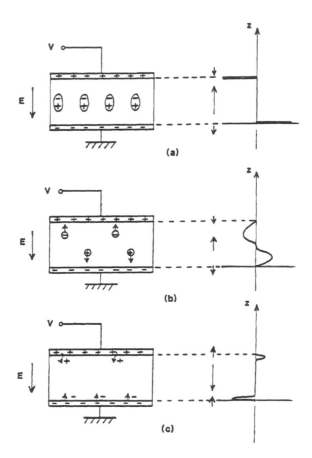

Fig. 11.1 Development of charge distribution ρ (z) in a dielectric material subjected to an electric field. (a) dipole orientation, (b) ion migration, (c) charge transfer at the interfaces (Lewiner, 1986, © IEEE).

A modern treatment of space charge phenomenon has been presented by Blaise and Sarjeant[2] who compare the space charge densities in metal oxide conductors (MOS) and high voltage capacitors (Table 11.1). The effect of moving charges is far less in charging of the dielectric and only the trapped charges influence the internal field.

11.2 POLARONS AND TRAPS

The classical picture of a solid having trapping sites for both polarities of charge carriers is shown earlier in Fig. 1.11. The concept of a polaron is useful in understanding the change in polarization that occurs due to a moving charge.

Table 11.1

Electronic space charge densities in MOS and HV capacitors (Blaise and Sargent, 1998) (with permission of IEEE)

Parameter	unit	MOS		HV capacitors	
		Mobile	Trapped	Mobile charges	Trapped charges
mobility	m^2/Vs	~20×10^{-4}		10^{-7}-10^{-4}	
Current density	A/m^2	10-10^4		10^{-2}-0.1	
Applied field	MV/m	100-1200	100-1200	10-100	10-100
Charge density	C/m^3	20μ-0.02	300-30,000	200μ-0.02	-
Charge conc.	/m^3	10^{-8}-10^{-5}	0.1-0.01	2×10^{-8}-2×10^{-6}	10^{-3}-10

An electron moving through a solid causes the nearby positive charges to shift towards it and the negative charges to shift away. This distortion of the otherwise regular array of atoms causes a region of polarization that moves with the electron. As the electron moves away, polarization vanishes in the previous location, and that region returns to normal. The polarized region acts as a negatively charged particle, called polaron, and its mass is higher than that of the isolated charge. The polarization in the region due to the charge is a function of the distance from the charge. Very close to the charge, ($r < r_e$), where r is the distance from the charge and r_e is the radius of the sphere that separates the polarized region from the unpolarized region. When $r > r_e$ electronic polarization becomes effective and when $r > r_i$ ion polarization occurs.

Let us consider a polaron of radius r_p in a dielectric medium in which a fixed charge q exists. The distance from the charge is designated as r and the dielectric constant of the medium varies radially from ε_∞ at $r_1 < r_p$ to ε_s at $r_2 > r_p$. The binding energy of the polaron is, according to Landau[3]

$$W = -\frac{1}{4\pi}\left[\frac{1}{\varepsilon_1} - \frac{1}{\varepsilon_2}\right]\frac{q^2}{r_p} \tag{11.1}$$

where r_p is the radius of the polaron, ε_∞ and ε_s are the dielectric constants which shows that smaller values of r_p increase the binding energy. This is interpreted as a more localized charge. The localization of the electron may therefore be viewed as a coupling between the charge and the polarization fields. This coupling causes lowering of the potential energy of the electron.

The kinetic energy determines the velocity of the electron which in turn determines the time required to cross the distance of a unit cell. If this time is greater than the characteristic relaxation time of electron in the ultraviolet region, then the polarization induced by the electron will follow the electron almost instantaneously. The oscillation frequencies of electron polarons is in the range of 10^{15}-10^{16} Hz. If we now consider the atomic polarization which has resonance in infrared frequencies, a lower energy electron will couple with the polarization fields and a lattice polaron is formed. The infrared frequency domain is 10^{12}-10^{13} Hz and therefore the energy of the electron for the formation of a lattice polaron is lower, on the order of lattice vibration energy. The lattice polaron has a radius, which, for example in metal oxides, is less than the interatomic distance.

Having considered the formation of polarons we devote some attention to the role of the polarons in the crystal structure. Fig. 11.2(a) shows the band structure in which the band corresponding to the polaron energy level is shown as $2J_p$ [Blaise and Sargent, 1998]. At a specific site i (11.2b) due to the lattice deformation the trap depth is increased and therefore the binding energy is increased. This is equivalent to reducing the radius of the polaron, according to equation (11.1), and therefore a more localization of the electron. This variation of local electronic polarizability is the initiation of the trapping mechanism.

Trapping centers in the condensed phase may be classified into passive and active centers. Passive centers are those associated with anion vacancies, that can be identified optically by absorption and emission lines. Active trapping centers are those associated with substituted cations. These are generally of low energy (~1eV) and are difficult to observe optically. These traps are the focus of our attention.

11.3 A CONCEPTUAL APPROACH

Focusing our attention on solids, a simple experimental setup to study space charge is shown in fig. 11.3[4]. The dielectric has a metallic electrode at one end and is covered by a conducting layer which acts as a shield. The current is measured through the metallic end. The charges may be injected into the solid by irradiation from a beam of photons, X-rays or gamma rays. Photons in the energy range up to about 300 keV interact with a solid, preferentially by the photoelectric effect. Photons above this energy interact by Compton effect; an increase of wavelength of electromagnetic radiation due to scattering by free or loosely bound electrons, resulting in absorption of energy (Gross, 1978). The secondary electrons are scattered mainly in the forward direction. The electrons move a certain distance within the dielectric, building up a space charge density and an internal electric field which may be quite intense to cause breakdown.

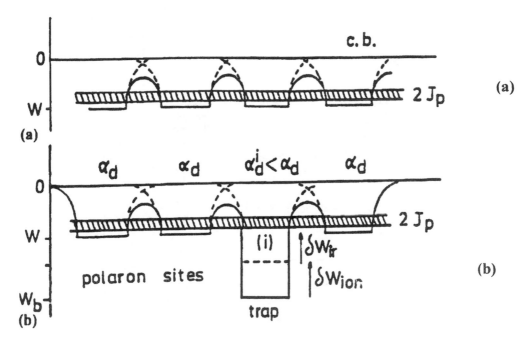

Fig. 11.2 (a) Potential wells associated with polaron sites in a medium of uniform polarizability, forming a polaron band of width 2Jp. (b) Trapping effect due to a slight decrease of electronic polarizability on a specific site i, ($\alpha_{di} < \alpha_d$). The charge is stabilized at the site due to lattice deformation. This leads to the increase of trap depth by an amount dW_{ion}. The total binding energy is $W_b = \delta W_{ir} + \delta W_{ion}$ (Blaise and Sargent, 1998, © IEEE).

The space charge build up due to irradiation with an electron beam is accomplished by a simple technique known as the 'Faraday cup'. This method is described to expose the principle of space charge measurements. Fig. 11.4 shows the experimental arrangement used by Gross, et al[5]. A dielectric is provided with vacuum deposited electrodes and irradiated with an electron beam. The metallic coating on the dielectric should be thin enough to prevent absorption of the incident electrons. The electrode on which the irradiation falls is called the "front" electrode and the other electrode, "back electrode". Both electrodes are insulated from ground and connected to ground through separate current measuring instruments. The measurements are carried out in either current mode or voltage mode and the method of analysis is given by Gross, et al.[6]

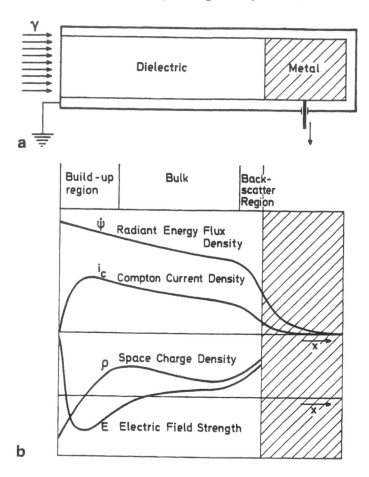

Fig. 11.3 (a) Technique for measurement of current due to charge injection. (b) Schematic for variation of space charge density and electric field strength (Gross, 1978, ©IEEE).

Electrical field, particularly at high temperatures, also augments injection of charges into the bulk creating space charge. The charge responsible for this space charge may be determined by the TSD current measurements described in the previous chapter. In amorphous and semicrystalline polymers space charge has a polarity opposite to that of the electrode polarity; positive polarity charges in the case of negative poling voltage and vice-versa. The space charge of opposite polarity is termed heterocharge whereas space charge of the same polarity is termed homocharge. In the case of the hetero charges the local space charge field will intensify the applied field, whereas in the case of homo charges there will be a reduction of the net field. In the former case of heterocharges, polarization that occurs in crystalline regions will also be intensified.

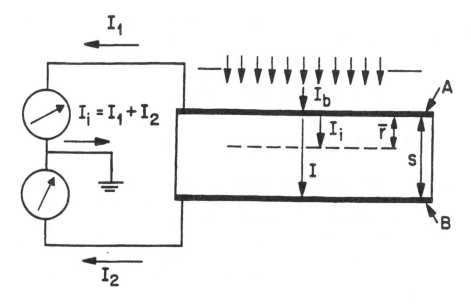

Fig. 11.4 Split Faraday cup arrangement for measurement of charge build up and decay. A-Front electrode, B-back electrode, s-thickness of dielectric, $\bar{r}$ -center of gravity of space charge layer. The currents are: I_i-injection current, I_1-front electrode current, I_2=rear current, I=dielectric current (Gross et. al. 1973, with permission of A. Inst. Phys.).

The increase in internal electric field leads to an increase of the dielectric constant ε' at high temperatures and low frequencies, as has been noted in PVDF and PVF[7]. It is important to note that the space charge build up at the electrode-dielectric interface also leads to an increase of both ε' and ε'' due to interfacial polarization as shown in section 4.4. It is quite difficult to determine the precise mechanism for the increase of dielectric constant; whether the space charge build up occurs at the electrodes or in the bulk. Obviously techniques capable of measuring the depth of the space charge layer shed light into these complexities.

The objectives of space charge measurement may be stated as follows:

(1) To measure the charge intensities and their polarities, with a view to understanding the variation of the electric field within the dielectric due to the applied field.
(2) To determine the depth of the charge layer and the distribution of the charge within that layer.
(3) To determine the mechanism of polarization and its role in charge accumulation.

(4) To interpret the space charge build up in terms of the morphology and chemical structure of the polymer

In the sections that follow, the experimental techniques and the methods employed to analyze the results are dealt with. Ahmed and Srinivas[8] have published a comprehensive review of space charge measurements, and we follow their treatment to describe the experimental techniques and a sample of results obtained using these techniques. Table 11.2 presents an overview of the methods and capabilities.

11.4 THE THERMAL PULSE METHOD OF COLLINS

The thermal pulse method was first proposed by Collins[9] and has been applied, with improvements, by several authors. The principle of the method is that a thermal pulse is applied to one end of the electret by means of a light flash. The flash used by Collins had a duration of 8μs. The thermal pulse travels through the thickness of the polymer, diffusing along its path. The current, measured as a function of time, is analyzed to determine the charge distribution within the volume of the dielectric. The experimental arrangement is shown in Fig. 11.5.

The electret is metallized on both sides (40 nm thick) or on one side only (lower fig. 11.5), with an air gap between the electret, and a measuring electrode on the other. By this method voltage changes across the sample are capacitively coupled to the electrode. The gap between the electrode and the electret should be small to increase the coupling. The heat diffuses through the sample and changes in the voltage across the dielectric, $\Delta V(t)$, due to non-uniform thermal expansion and the local change in the permittivity, are measured as a function of time. The external voltage source required is used to obtain the zero field condition which is required for equations (11.3) and (11.4) (see below).

Immediately after the heat pulse is applied, temperature changes in the electret are confined to a region close to the heated surface. The extent of the heated zone can be made small by applying a shorter duration pulse. The process of metallizing retains heat and the proportion of the retained heat can be made small by reducing the thickness of the metallizing. In the ideal case of a short pulse and thin metallized layer, the voltage change after a heat pulse applied is given by

$$\rho_T = \int_o^d \rho(x)dx \qquad (11.2)$$

where ρ_T is the total charge density (C/m^2). Determination of the total charge in the electret does not require a deconvolution process.

Table 11.2
Overview of space charge measuring techniques and comments (Ahmed and Srinivas, 1997). R is the spatial resolution and t the sample thickness.
(with permission of IEEE)

Method	Disturbance	Scan mechanism	Detection process	r (μm)	t (μm)	Comments
Thermal pulse method	Absorption of short-light pulse in front electrode	Diffusion according to heat-conduction equations	Voltage change across sample	⩾ 2	~ 200	High resolution requires deconvolution.
Laser intensity modulation method	Absorption of modulated light in front electrode	Frequency-dependent steady-state heat profile	Current between sample electrodes	⩾ 2	~ 25	Numerical deconvolution is required
Laser induced pressure pulse method	Absorption of short laser light pulse in front electrode	Propagation with longitudinal sound velocity	Current between sample electrodes	1	100 – 1000	No deconvolution is required
Thermoelastically generated LIPP	Absorption of short laser light pulse in thin buried layer	Propagation with longitudinal sound velocity	Current or voltage between sample electrodes	1	50 – 70	Deconvolution is required
Pressure wave propagation method	Absorption of short laser light pulse in metal target	Propagation with longitudinal sound velocity	Voltage or current between sample electrode	10	5 – 200	Resolution improved with deconvolution. Also used for surface charge measurements
Non-structured acoustic pulse method	HV spark between conductor and metal diaphragm	Propagation with longitudinal sound velocity	Voltage between sample electrode	1000	⩽ 10000	Used for solid and liquid dielectric. Higher resolution with deconvolution
Laser generated acoustic pulse method	Absorption of short laser light pulse in thin paper target	Propagation with longitudinal sound velocity	Voltage between sample electrodes	50	⩽ 3000	Deconvolution is required. Target and sample immersed in dielectric liquid
Acoustic probe method	Absorption of laser light pulse in front electrode	Propagation with longitudinal sound velocity	Voltage between sample electrodes	200	2000 – 6000	
Piezoelectrically-generated pressure step method	Electrical excitation of piezoelectric quartz plate	Propagation with longitudinal sound velocity	Current between sample electrodes	1	25	Deconvolution is required
Thermal step method	Applying two isothermal sources across sample	Thermal expansion of the sample	Current between sample electrodes	150	2000 – 20000	Deconvolution is required
Electro-acoustic stress pulse method	Force of modulated electric field on charges in sample	Propagation with longitudinal sound velocity	Piezoelectric transducer at sample electrode	100	⩽ 10000	Deconvolution is required. Also used for surface charge measurements.
Photoconductivity method	Absorption of narrow light beam in sample	External movement of light beam	Current between sample electrodes	⩾ 1.5	–	Nondestructive for short illumination time
Space charge mapping	Interaction of polarized light with field	parallel illumination of sample volume or movement of light beam or sample	Photographic record	200	–	Mostly used on transparent dielectric liquids
Spectroscopy	Absorption of exciting radiation in sample	External movement of radiation source or sample	Relative change in the observed spectrum	⩾ 50	–	Few applications
Field probe	None	Capacitive coupling to the field	Current	1000	⩽ 20000	Destructive

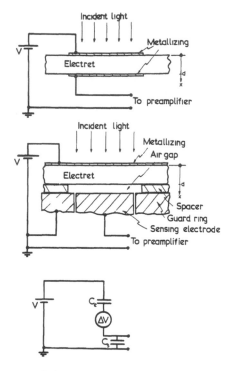

Fig. 11.5 Schematic diagram of the apparatus for the thermal pulsing experiment in the double metallizing and single metallizing configurations. (Collins, 1980, Am. Inst. Phys.)

The observed properties of the electret are in general related to the internal distribution of charge $\rho(x)$ and polarization $P(x)$ through an integral over the thickness of the sample. The potential difference V_0 across the electret under open circuit conditions (zero external field) is given by

$$V_0 = \frac{1}{\varepsilon_s \varepsilon_0} \int_0^d \left[x\rho(x) + P(x) \right] dx \tag{11.3}$$

where $\rho(x)$ is the charge density in C/m^3 and d the thickness of the sample.

Collins (1980) derived the expression

$$\Delta V(t) = \frac{1}{\varepsilon_s \varepsilon_0} \int_0^d \left\{ \left[A\rho(x) - B\frac{dP}{dx} \right] \int_0^x \Delta T(x') dx' \right\} dx \tag{11.4}$$

where $A = \alpha_x - \alpha_e$ and $B = \alpha_p - \alpha_x - \alpha_e$, x is the spatial coordinate with $x = 0$ at the pulsed electrode. $\rho(x)$ and $P(x)$ are the spatial distributions of charge and polarization. The symbols α mean the following:

$\quad \alpha_x$ = Thermal coefficient of expansion

$\quad \alpha_\varepsilon$ = Temperature coefficient of the dielectric constant

$\quad \alpha_p$ = Temperature coefficient of the polarization

There are two integrals, one a function of charge and the other a function of temperature. Two special cases are of interest. For a non-polar dielectric with only induced polarization $P = 0$, equation (11.4) reduces to

$$\Delta V(t) = \frac{(\alpha_x - \alpha_e)}{\varepsilon_s \varepsilon_0} \int_0^d \left\{ \left[\rho(x) \right] \int_0^x \Delta T(x') dx' \right\} dx \qquad (11.5)$$

For an electret with zero internal field

$$\rho(x) = +\frac{dP}{dx} \qquad (11.6)$$

$$\Delta V(t) = \frac{\alpha_p}{\varepsilon_s \varepsilon_0} \int_0^d P(x) \Delta T(x,t) dx \qquad (11.7)$$

Collins used a summation procedure to evaluate the integral in equation (11.5). The continuous charge distribution, $\rho(x)$ is replaced by a set of N discrete charge layers ρ_n with center of gravity of each layer at mid point of the layer and having coordinate $x_j = (j - \frac{1}{2})d/N$ with $j = 1, 2, ...N$. The integral with the upper limit x in equation (11.5) is replaced with the summation up to the corresponding layer x_j. Equation (11.5) then simplifies to

$$\Delta V(t) = \frac{(\alpha_x - \alpha_e)}{\varepsilon_s \varepsilon_0} \sum_{j=1}^N \rho_j \left(\sum_{i=1}^j \Delta T_i \right) \qquad (11.8)$$

Assuming a discrete charge distribution the shape of the voltage pulse is calculated using equation (11.8) and compared with the measured pulse shape. The procedure is repeated till satisfactory agreement is obtained. Collins' procedure does not yield a unique distribution of charge as a deconvolution process is involved.

The technique was applied to fluoroethylenepropylene (FEP, Teflon™) electrets and the depth of charge layer obtained was found to be satisfactory. Polyvinylidene fluoride (PVDF) shows piezo/pyroelectric effects, which are dependent on the poling conditions. A copolymer of vinylidene fluoride and tetrafluoroethylene (VF$_2$-TFE) also has very large piezoelectric and pyroelectric coefficients. The thermal poling method has revealed the poling conditions that determine these properties of the polymers. For example, in PVDF, a sample poled at lower temperatures has a large spatial non-uniformity in the polarization across its thickness. Even at the highest poling temperature some non-uniformity exists in the spatial distribution of polarization. Significant differences are observed in the polarization distribution, even though the samples were prepared from the same sheet.

Seggern[10] has examined the thermal pulse technique and discussed the accuracy of the method. It is claimed that the computer simulations show that the only accurate information available from this method is the charge distribution and the first few Fourier coefficients.

11.5 DEREGGI'S ANALYSIS

DeReggi et al.[11] improved the analysis of Collins (1980) by demonstrating that the voltage response could be expressed as a Fourier series. Expressions for the open circuit conditions and short circuit conditions are slightly different, and in what follows, we consider the former[12].

The initial temperature at (x,0) after application of thermal pulse at x=0, t=0 may be expressed as

$$T(x,0) = T_1 + \Delta T(x,0) \tag{11.9}$$

where T$_1$ is the uniform temperature of the sample before the thermal pulse is applied, and $\Delta T(x,0)$ is the change due to the pulse. $\Delta T(x,0)$ is a sharp pulse extending from $x = 0$ with a width s<<d. From equation (11.9) it follows that the temperature at x after the application of the pulse is

$$T(x,t) = T_1 + \Delta T(x,t) \tag{11.10}$$

where

$$\Delta T(x,t) = a_0 + \sum_{n=1}^{\infty} a_n \cos\left[\frac{n\pi x}{d}\right] \exp\left[\frac{-n^2 t}{T_1}\right] \tag{11.11}$$

$$a_0 = d^{-1} \int_0^d \Delta T(x,0) dx = \lim_{\infty} \Delta T(x,t) \tag{11.12}$$

$$a_n = \frac{2}{d} \int \Delta T(x,0) \cos\left[\frac{n\pi x}{d}\right] dx; \quad n = 1,2,.. \tag{11.13}$$

The temperature at the surface is given by

$$\Delta T(0,t) = a_0 + \sum_{n=1}^{\infty} a_n \exp\left[\frac{-n^2 t}{\tau}\right] \tag{11.14}$$

$$\Delta T(x,t) = a_0 + \sum_{n=1}^{\infty} (-1)^n a_n \exp\left[\frac{-n^2 t}{\tau}\right] \tag{11.15}$$

where $\tau = d^2/\pi^2 K$ and k is called thermal diffusivity. The dimensionless quantities $\Delta T(0,t)/a_0$ and $\Delta T(d, t)/a_0$ can be obtained by measuring the transient resistance of one or both the electrodes. Then the ratios a_n/a_0 and τ_1 can be determined without knowing the detailed shape of the light pulse.

Substituting equation (11.15) into (11.5) the voltage at time t is given as

$$\Delta V(t) = \frac{\alpha_x - \alpha_\varepsilon}{\varepsilon_s \varepsilon_0}\left[\alpha_0 A_0 + \frac{d}{n}\sum_{n=1}^{\infty} \frac{a_n A_n}{n} \exp\left(\frac{-n^2 t}{T_1}\right)\right] \tag{11.16}$$

where the following relationships hold.

$$A_0 = \int_0^d x\rho(x) dx \tag{11.17}$$

$$A_n = \int_0^d \rho(x)\sin(\frac{n\pi x}{d}) dx \tag{11.18}$$

The terms a_n and A_n are the coefficients of Fourier series expansions for $\Delta T(x,0)$ and $\rho(x)$ respectively, if these are expanded as cosine and sine terms, respectively.

For the short circuit conditions, equations for the charge distribution and the polarization distribution are given by Mopsik and DeReggi (1982, 1984). About 10-15 coefficients could be obtained for real samples, based on the width of the light pulse. The polarization distribution determined will be unique as a deconvolution procedure is not resorted to. Fig. 11.6 shows the results for a nearly uniformly poled polyvinylidene fluoride (PVF_2) which was pulsed alternately on both sides. An interesting observation in this study is that there is a small peak just before the polarization falls off.

A further improvement of the thermal pulse technique is due to Suzuoki et. al.[13] who treat the heat flow in a slab in the same way as electrical current in an R-C circuit with distributed capacitance. The electrical resistance, capacitance, current and voltage are replaced by the thermal resistance R_t, thermal capacitance C_i, heat flow q (x, t) and temperature T(x, t), respectively.

The basic equations are:

$$-\frac{\partial q(x,t)}{\partial x} = C_1 \frac{\partial T(x,t)}{\partial t} \tag{11.19}$$

$$-\frac{\partial T(x,t)}{\partial x} = R_1 \frac{\partial q(x,t)}{\partial t} \tag{11.20}$$

The heat flux is given by

$$\frac{q(x,t)}{q_0} = 1 - \frac{x}{l} - \frac{2}{\pi} \sum_{k=1}^{k=\infty} \frac{1}{k} \exp\left(-\frac{k^2 \pi^2 t}{\tau}\right)\left(\frac{knx}{l}\right) \tag{11.21}$$

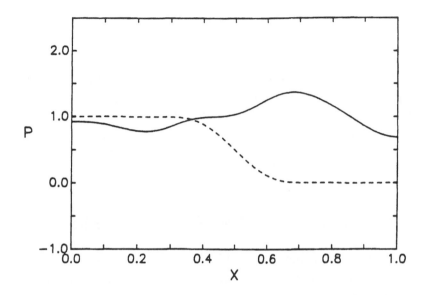

Fig.11.6 Polarization in PVF2 sample. The solid line is experimental distribution. The dashed line is the resolution expected for a step function, at x = 0.5 (DeReggi, et al., 1982, with permission of J. Appl. Phys.).

The total current in the external circuit at t = 0, when the specimen is illuminated at $x = 0$ is given by

$$j_1(0) = -\frac{\alpha q_0}{C_i l^2} \int_0^d (d - x)\rho(x)dx \qquad (11.22)$$

Similarly, the current, when the specimen is illuminated at x = d, is

$$j_2(0) = -\frac{\alpha q_0}{C_i l^2} \int_0^d x\rho(x)dx \qquad (11.23)$$

The total amount of space charge is

$$Q_t = -\frac{C_i l}{\alpha q_0}[j_1(0) + j_2(0)] \qquad (11.24)$$

The mean position of the space charge is

$$\frac{\overline{x}}{l} = \frac{j_2(0)}{j_1(0) + j_2(0)} \tag{11.25}$$

The thermal pulse was applied using a xenon lamp and the pulse had a rise time of 100 μs, width 500μs. Since the calculated thermal time constant was about 5ms, the light pulse is an approximation for a rectangular pulse. The materials investigated were HDPE and HDPE doped with an antistatic agent. Fig. 11.7 shows the measured currents in doped HDPE. Homocharges were identified at the anode and in doped HDPE a strong heterocharge, not seen in undoped HDPE, was formed near the cathode.

11.6 LASER INTENSITY MODULATION METHOD (LIMM)

Lang and Das Gupta[14] have developed this method which is robust in terms of data accuracy and requires only conventional equipment, as opposed to a high speed transient recorder, which is essential for the thermal pulse method. A thin polymer film coated with evaporated opaque electrodes at both surfaces is freely suspended in an evacuated chamber containing a window through which radiant energy is admitted. Each surface of the sample, in turn, is exposed to a periodically modulated radiant energy source such as a laser. The absorbed energy produces temperature waves which are attenuated and retarded in phase as they propagate through the thickness of the specimen. Because of the attenuation, the dipoles or space charges are subjected to a non uniform thermal force to generate a pyro-electric current which is a unique function of the modulation frequency and the polarization distribution.

Let ω rad s^{-1} be the frequency of the sinusoidally modulated laser beam and the specimen illuminated at $x = d$. The surface at $x = 0$ is thermally insulated. The heat flux absorbed by the electrode is q (d. t) which is a function of the temperature gradient along the thickness. The one dimensional heat flow equations are solved to obtain the current as

$$I(\omega) = \frac{C\omega(j+1)}{D\sinh D(j+1)d} \int_0^d P(x)\cosh D(j+1)x\,dx \tag{11.26}$$

where $D = (\omega/2K)^{1/2}$, j is the complex number operator and C contains all the position and frequency-dependent parameters. The current generated lags the heat flux because of the phase retardation of the thermal wave as it progresses through the film. The current therefore has a component in phase and in quadrature to the heat flux.

The mathematical treatment of measured currents at a number of frequencies for determining P(x) involves the following steps: The integral sign in equation (11.26) may be replaced by a summation by dividing the film into n incremental thickness, each layer having its polarization, P_j, where j=1,2,....n. The matrix equation [I] = [G] [P] where

$$[I] = I(\omega_k);$$
$$[G] = \frac{[C\omega(i+1)\cosh D(i+1)x]}{[D\sinh D(i+1)d]} \qquad (11.27)$$
$$[p] = P_k$$

is solved. The in-phase component of measured current is used with the real part of G and the quadrature component is used with the imaginary part. It is advantageous to measure $I(\omega)$ at more than n frequencies and apply the least square method to solve for P.

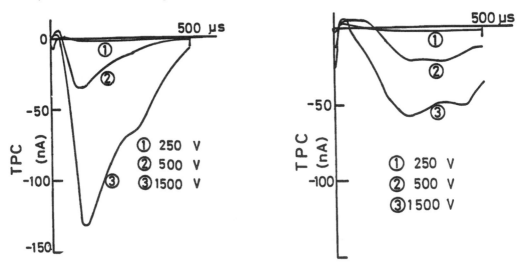

Fig. 11.7 Experimental thermal pulse currents in doped HDPE for (a) cathode illumination (b) anode illumination. Negative currents show the existence of a positive space charge in the sample (Suzuoki et al, 1985; with permission of Jap. J. Appl. Phys.)

Fig. 11.8 shows the polarization distributions and pyroelectric currents versus frequency. Because of the impossibility of producing an experimentally precise type of polarization, a triangular distribution was assumed and the currents were synthesized. Using the in-phase and quadrature components of these currents, the polarization distribution was calculated as shown by points. The parameters used for these calculations are d = 25.4 µm, K = 0.1×10^{-7} m^{2}s^{-1}, $10^2 <$ f $< 10^5$ Hz (101 values), obtaining 51 values of P_k. Lang and Das Gupta (1981) have used the LIMM technique to study spatial distribution of polarization in PVDF and thermally poled polyethylene.

11. 7 THE PRESSURE PULSE METHOD

The principle of the pressure pulse method was originally proposed by Laurenceau, et al.[15] and will be described first. There have been several improvements in techniques that will be dealt with later. The pressure probe within a dielectric causes a measurable electrical signal, due to the fact that the capacitance of a layer is altered in the presence of a stress wave. The pressure pulse contributes in two ways towards the increase of capacitance of a dielectric layer. First, the layer is thinner than the unperturbed thickness due to the mechanical displacement carried by the wave. Second, the dielectric constant of the compressed layer is increased due to electrostriction caused by the pulse[16].

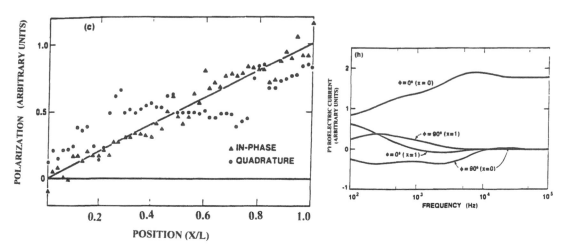

Fig. 11.8 (a) Pyroelectric current versus frequency (x = 0 and x = d refers to heating from x = 0 and x = d side of the film. ϕ = 0 and ϕ = π/2 refers to in phase or in quadrature with heat flux respectively. (b) Polarization distributions (solid line) and calculated distributions (points). Selected data from (Lang and Das Gupta, 1981, with permission of Ferroelectrics).

A dielectric slab of thickness d, area A, and infinite-frequency dielectric constant ε_∞ with electrodes a and b in contact with the sample, is considered. The sample has acquired, due to charging, a charge density $\rho(x)$ and the potential distribution within the dielectric is V (x). All variables are considered to be constant at constant x; the electrode *a* is grounded, electrode *b* is at potential V. The charge densities σ_a and σ_b are given by

$$\sigma_a = -\frac{d - \langle x \rangle}{d} \frac{Q}{A} - \varepsilon_\infty \frac{V}{d} \qquad (11.28)$$

$$\sigma_b = \frac{\langle x \rangle}{d} \frac{Q}{A} + \varepsilon \frac{V}{d} \qquad (11.29)$$

where

$$\langle x \rangle = \frac{\int_0^d x \rho(x) dx}{\int_0^d \rho(x) dx} \qquad \text{and} \qquad \frac{Q}{A} = \int_0^d \rho(x) dx \qquad (11.30)$$

Expressions (11.28) and (11.29) show that if $V = 0$ and if the sample is not piezoelectric, a uniform deformation along the x axis does not alter the charges on the electrodes since $(d - \langle x \rangle)/d$ remains constant. This implies that in order to obtain the potential or charge profiles, a non-homogeneous deformation must be used. A step function compressional wave propagating through the sample with a velocity v, from electrode **a** towards **b**, provides such a deformation. As long as the wave front has not reached the opposite electrode, the right side of the sample is compressed while the left part remains unaffected (Fig. 11.9). The charge induced on electrode **b** is a function of the charge profile, of the position of the wave front in the sample, but also of the boundary conditions at the electrodes: Open circuit or short circuit conditions. In the first case the observable parameter is the voltage, in the second case, the external current.

Let the unperturbed thickness of the sample be d_0 and Δp the magnitude of pressure excess in the compressed region, β the compressibility of the dielectric defined as the fractional change in volume per unit excess pressure, $\beta = -\Delta V/(V\Delta p)$. The compressed part of the dielectric has a permittivity of ε' and x_f is the position of the wave front at time t, which can be expressed as $x_f = d - v_0 t$. In the compressed region charges, which are supposed to be bound to the lattice, are shifted towards the left by a quantity $u(x,t) = -\beta \Delta p (x - x_f)$. In the uncompressed part the charges remain in the original position.

The electric field in the uncompressed part is $E(x,t)$ and $E'(x,t)$ in the compressed part. At the interface between these regions the boundary condition that applies is

$$\varepsilon E(x_f, t) = \varepsilon' E'(x_f, t) \qquad (11.31)$$

The boundary condition for the voltage is

$$V(d,t) - V(0,t) = -\int_0^d E(x,t)\,dx \qquad (11.32)$$

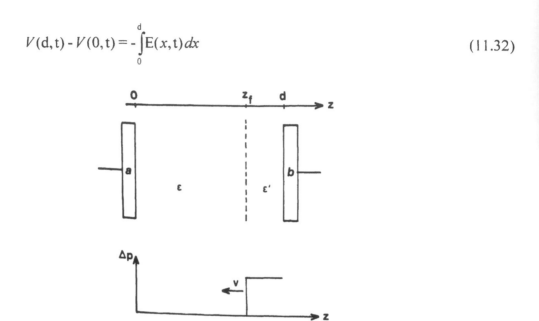

Fig. 11.9 Charge in a dielectric between two electrodes, divided into a compressed region of permittivity ε' and an uncompressed region of permittivity ε; the step function compression travels from right to left at the velocity of sound. The position of the wave front is x_f. The undisturbed part has a thickness d_0 (Laurenceau, 1977, with permission of A. Inst. of Phy.).

Laurenceau, et al. (1977) provide the solution for the voltage under open circuit conditions as

$$V(d,t) = \left[1 + \left(\frac{\varepsilon}{\varepsilon'}\right)(\beta\Delta p - 1)\right] V(x_f, 0) \qquad (11.33)$$

The current under short circuit conditions is

$$\frac{1}{A}\int_0^t J(\tau)\,d\tau = \varepsilon \frac{1 + (\varepsilon/\varepsilon')(\beta\Delta p - 1)}{x_f + (\varepsilon/\varepsilon')(d - x_f)} V(x_f, 0) \qquad (11.34)$$

Equations (11.33) and (11.34) show that the time variation of both the voltage and current is an image of the spatial distributions of voltage and current inside the sample prior to perturbation. The front of the pressure wave acts as a virtual moving probe sweeping across the thickness at the velocity of sound.

Laurenceau, et al. (1977) proved that the pressure pulse method gives satisfactory results: a compressional step wave was generated by shock waves and a previously charged polyethylene plate of 1 mm thickness was exposed to the wave. The short circuit current measured had the shape expected for a corona injected charge, reversed polarity, when charges of opposite sign were injected. Further, the charges were released thermally and the current was reduced considerably, as expected.

Lewiner (1986) has extended the pressure pulse method to include charges due to polarization P resulting in a total charge density

$$\rho(x) = \rho_s(x) - \frac{dP}{dx} \qquad (11.35)$$

where ρ_s is the charge density due to the space charge. The open circuit voltage between the two electrodes is given by

$$V(t) = \beta G(\varepsilon_r) \int_0^{x_f} E(x,0)p(x,t)dx \qquad (11.36)$$

where $x_f = vt$ is the wave front which is moving towards the opposite with a velocity v, $G(\varepsilon_r)$ is a function of the relative permittivity which in turn is a function of pressure. In short circuit conditions the current I(t) in the external circuit is related to the electric field distribution by

$$I(t) = \beta C_o G(\varepsilon_r) \int_0^{x_f} E(x,0)\frac{\partial}{\partial t}p(x,t)dx \qquad (11.37)$$

where C_0 is the uncompressed geometric capacitance, $C_o = \varepsilon_o \varepsilon_r A/d$. Equations (11.36) and (11.37) show that if p(x, t) is known, the electric field distribution may be obtained from the measurement of V(t) or I(t). If the pressure wave is a step like function of amplitude Δp (fig. 11-9) then V(t) will be a mirror image of the spatial distribution of the potential in the sample as discovered by Laurenceau, et al. (1977), whereas I(t) is directly related to the electric field. If the pressure wave is a short duration pulse, then V(t) and I(t) give directly the spatial distributions of the electric field and charge density. If the pressure wave profiles change during its propagation through the sample, this effect can be taken into account by a proper description of p(x, t). The techniques used to generate a short rise time pressure waves are shock wave tubes, discharge of capacitors in fluids, piezoelectric transducers and short rise time laser pulses. Fig. 11.10 shows a typical experimental set up for the laser pulse pressure pulse method.

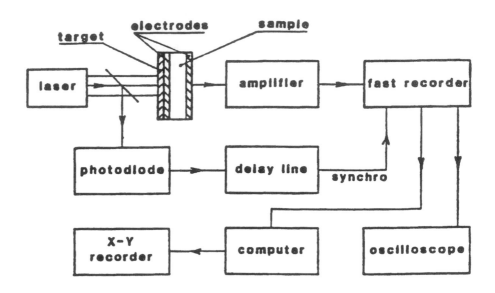

Fig. 11.10 Experimental set up for the measurement of space charge (Lewiner, 1986, © IEEE)

The choice of the laser is governed by two conditions. First, the homogeneity of the beam must be as good as possible to give a uniform pressure pulse over the entire irradiated area. Second, the duration of the laser pulse is determined by the thickness of the sample to be studied. For thin samples, < 100 μm, short duration pulses of 0.1-10 ns duration are appropriate. For thicker samples broader pulses are preferred since there is less deformation of the associated pressure pulse as it propagates through the thickness. The power density of the laser beam between 10^6-10^8 W/cm^2 yields good results.

The measured voltage and current in 50 μm thick Teflon (FEP) film charged with negative corona up to a surface potential is of 1250 V is shown in Fig. 11.11. The charge decay as the temperature of the charged sample is raised. is shown in Fig.11.12. The charged surface retains the charge longer than the opposite surface, and higher temperature is required to remove the charge entirely. The LIPP technique is applied with several variations depending upon the method of generating the pressure pulse. The methods are briefly described below.

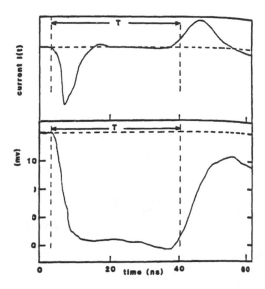

Fig. 11.11 Current and voltage wave forms measured during the propagation of a pressure pulse through a negative corona charged FEP film of 50 μm thickness. T is the time for the pulse to reach the charged surface (Lewiner, 1986, © IEEE).

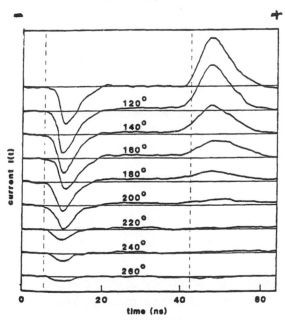

Fig. 11.12 Charge decay with temperature in negative corona charged FEP film. Charged side retains charges longer (LEWINER, 1986, © IEEE)

11.7.1 LASER INDUCED PRESSURE PULSE METHOD (LIPP)

A metal layer on one side of a dielectric absorbs energy when laser light falls upon it. This causes stress effects and a pressure pulse, < 500 ps duration, is launched, which propagates through the sample with the velocity of sound. Fig. 11.13 shows the experimental arrangement used by Sessler, et al.[17]. The method uses one sided metallized samples and it is charged at the unmetallized end by a corona discharge. The laser light pulses, focused on the metallized surface, having a duration of 30-70 ps and 1-10 mJ energy, are generated by a Nd:YAG laser.

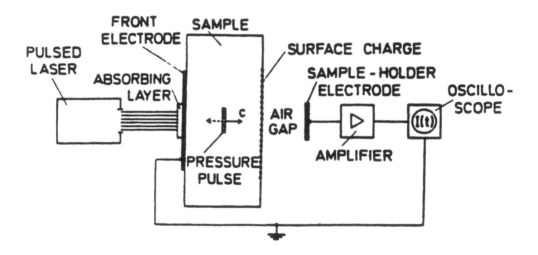

Fig. 11.13 Experimental setup for the laser-induced pressure-pulse (LIPP) method for one sided metallized samples (Sessler, et al., 1986, © IEEE).

The pressure pulse generates, under short circuit conditions, the current signal

$$I(t) = \frac{Ap\tau}{\rho_0(s + \varepsilon_s g)}\left[\left(1 + \frac{(\varepsilon_\infty + 2)(\varepsilon_\infty - 1)}{3\varepsilon_s}\right)\rho(x) - \frac{d\,e(x)}{dx}\right]_{x=vt} \qquad (11.38)$$

where A is the sample area, p the amplitude of the pressure, τ the duration of the pressure pulse, ρ_0 the density of the material, s the sample thickness, g the air gap thickness, e (x) the piezoelectric constant of the material, ε_s the static (dc) dielectric constant and ε_∞ the infinity frequency dielectric constant.

11.7.2 THERMOELASTIC STRESS WAVES

This method has been adopted by Anderson and Kurtz (1984). When some portion of an elastic medium is suddenly heated thermoelastic stress waves are generated. A laser pulse of negligible duration enters a transparent solid and encounters a buried, optically absorbing layer, causing a sudden appearance of a spatially dependent temperature rise which is proportional to the absorbed energy.

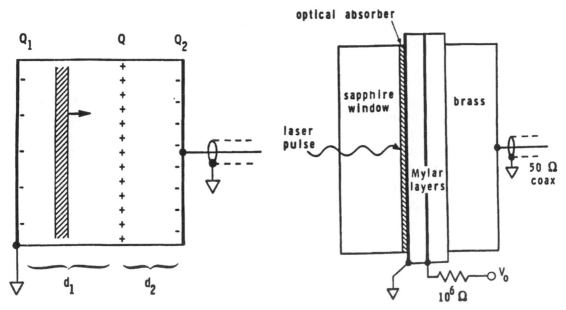

Fig. 11-14 (a) Pressure pulse in a slab of dielectric containing a plane of charge Q. The pulse travels to the right. Electrode 2 is connected to ground through a co-axial cable and measuring instrument. (b) Experimental arrangement for measuring injected space charge. The Mylar film adjacent to the sapphire window acquires internal charge as a result of being subjected to high-field stress prior to installation in the measurement cell. Thicknesses shown are not to scale. (Anderson and Kurtz, 1984 © Am. Inst. Phys.)

The thickness of the sample in the x direction is assumed to be small compared to the dimensions along the y and z directions so that we have a one dimensional situation. At the instant of energy absorption the solid has inertia for thermal expansion and hence compressive stress appears in the solid. The stress is then relaxed by propagation, in the opposite direction, of a pair of planar, longitudinal acoustic pulses which replicate the initial stress distribution. Each of these pressure pulses carries away half of the mechanical displacement needed to relax the heated region. The measured signal is the voltage as a function of time and a deconvolution procedure is required to determine the charge density (Anderson and Kurtz, 1984).

11.7.3 PRESSURE WAVE PROPAGATION (PWP) METHOD

In this method[18] the pressure wave is generated by focusing a laser beam to a metal target bonded to the dielectric sheet under investigation. Earlier, Laurenceau, et al. (1977) had used a step pressure wave to create non-homogeniety as described in section 12.7 above. Though the step function has the advantage that the observed signal is the replica of the potential distribution within the volume before the pressure wave arrived, the difficulty of producing an exact step function limited the usefulness of the method. When a high spatial resolution is required the pulse shape should be small, which is difficult to achieve with a shock tube for two reasons.

First, the shock wave travels in the shock tube with a velocity of a few hundred meters per second; a small angle between the wave front and the sample results in a strong decrease of the spatial resolution. Second, the reproducibility of the wave shape is poor. In the PWP method Alquie, et al. (1981) generalize the calculation to an arbitrary shape of the pressure pulse. The time dependence of the voltage across the sample has a unique solution for the electric field distribution within the sample. Fig. 11.15 shows the sample which has a floating electrode in the middle and two identical samples of FEP on either side. The solution for the electric field shows a sharp discontinuity at the point where there is a charge reversal as expected.

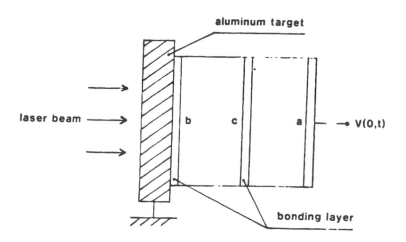

Fig. 11.15 Sample preparation for field distribution study in FEP. Two discs metallized on both sides are joined in the middle creating a floating electrode through which the sample was charged (Alquie, 1981, © IEEE).

11.7.4 NONSTRUCTURED ACOUSTIC PROBE METHOD

The measurement of electric field within a charged foil of relatively small thickness, ~25 μm-1mm, offers more flexibility in the choice of methods because the pulses are not attenuated as much as in thicker samples. For larger systems or thicker samples a different approach, in which the electric field is measured by using a nonstructured acoustic pulse to compress locally the dielectric of interest, has been developed by Migliori and Thompson[19]. In this technique the pulse shape is unimportant, and attenuation effects are easily accounted for, thereby increasing the effective range of the probe to several tens of centimeters in polymers. Because the probe is sensitive to electric fields, small variations in electric fields and space charge are detectable. Fig. 11.16 shows the experimental arrangement. An acoustic pulse is generated using a spark gap which is located in a tube and situated at about 0.1 mm from a replaceable metal diaphragm. An energy storage capacitor, also located in the same tube (fig. 11.16a) provides the energy for a spark.

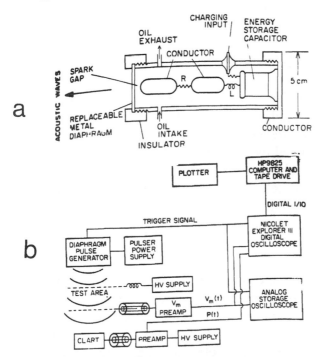

Fig. 11.16. (a) Diaphragm-type acoustic pulse generator used to generate non-structured acoustic pulses for electric field measurements in oil and polymers. (b) Block diagram of the instrumentation used to acquire and process the acoustic and electrical signals required for an acoustic electric field measurement. The box labeled CLART represents the capacitance- like acoustic receiving transducer, and the two pre-amplifiers are described in the text (Migliori and Thompson, 1980, with permission of J. Appl. Phys.).

The other end of the tube is closed with a thick brass plate and the tube is filled with transformer oil. The capacitor is charged until breakdown occurs between the conductor and the metal foil. The spark current (~100A) has a rise time of a fraction of a μs and launches an acoustic wave having a peak amplitude of ~10^5 Pa.

The acoustic signal is measured by a capacitance-like acoustic transducer, preamplifier and bias supply immersed in oil in its own tank, which is shielded. This tank is shown as CLART in fig. (11-16b). The measured voltage due to the acoustic pulse is given by

$$V_m(t) = -\left(2 - \frac{1}{\varepsilon}\right)\beta p v \tau E(vt) \tag{11.39}$$

where τ is the time required to pass a particular point. The technique was employed to measure space charge in an oil filled parallel plate capacitor, and 16mm thick poly(methyl methacrylate) (PMMA).

11.7.5 LASER GENERATED ACOUSTIC PULSE METHOD

The non-structured acoustic pulse method described in the previous section has the disadvantage that the generated pulse is not reproducible. It also suffers from insufficient bandwidth and the ratio of low frequency energy to high frequency energy is too high. An improved technique is adopted by Migliori and Hofler[20] using a laser to generate the acoustic pulses (Fig. 11.17). Light from a ruby laser which generates pulses of 1.5 J and 15 ns half width is directed through a plate glass window into an aluminum tank filled with liquid freon. The tank contains a laser beam absorber to convert radiation to pressure, a pressure transducer and pre-amplifier, the sample under test and its preamplifier.

The beam strikes an absorber which consists of a carbon paper stretched flat and with the carbon side towards the film. Alignment of the laser perpendicular to the beam is not critical. The paper absorbs radiation and the freon entrained between the paper fibres is heated. The pressure wave has a rise time of about 15 ns and a peak of 0.5 Mpa. The method is sensitive enough to determine the electric fields in the range of 50 kV/m, with a spatial resolution of 50 μm in samples 3 mm thick.

11.7.6 ACOUSTIC PULSE GENERATED BY MECHANICAL EXCITATION

Mechanical excitation may also be used to generate acoustic pulses for probing the electric field within a charged dielectric and this method has been adopted by Roznov and Gromov[21]. A Q-switched ruby laser is used to illuminate a graphite disk with 30 ns pulses and an energy density of 0.5 J/cm^2.

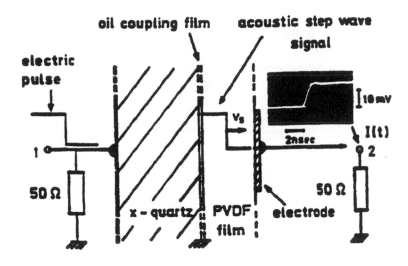

Fig. 11.17 Laser generated acoustic pulse method to measure the electric field inside a solid dielectric (Migliori and Hofler, 1982, with permission of Rev. Sci. Instr.).

11. 7. 7 PIEZO-ELECTRIC PRESSURE STEP METHOD (PPS)

The pressure pulse step method used by Gerhardt-Multhaupt, et al.[22] is shown in fig. 11.18. Charging a cable and discharging through a relay-trigger, generates a square pulse of 400-600 V and 100 ns duration. The sequence of the positive and negative pulse is applied to a 3 mm thick piezoelectric quartz pulse. A thin (~100-200 nm) layer of silicone oil couples the generated pressure pulse with the metallized surface of the sample. This sandwich of quartz-oil-sample ensures good acoustic contact. The unmetallized surface is in contact with a conducting rubber disc which is connected to a preamplifier and an oscilloscope. The sample is charged by electron beam of varying energy. This method of charging has the advantage that increasing the energy of the beam can vary the depth of penetration of charges.

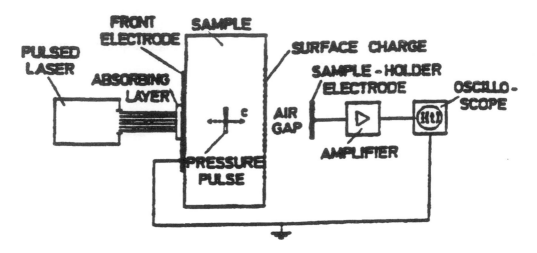

Fig. 11.18 Principle of Piezoelectric pressure step method (PPS) (Sessler [23], 1997, © IEEE).

11.7.8 PULSED ELECTRO-ACOUSTIC STRESS METHOD

The method adopted by Takada and Sakai[24] is based on the principle that an electrical pulse applied to a dielectric with a stored charge launches an acoustic pulse that originates from the bulk charge. A dc electric field is applied to deposit charges in the bulk and an ac field of a much smaller magnitude, having 1 MHz frequency, is applied to launch the acoustic wave. The dc supply and ac oscillators are isolated appropriately by a combination of resistance and capacitance.

A piezoelectric transducer is attached to each electrode and the acoustic field excited by the alternating field propagates through the electrodes to the transducer. The electrical signal detected by the transducer is amplified and rectified. The method measures the electric fields in the interface of a dielectric, and utilizes a transducer and associated equipment at each electrode. Further, the signal from the grounded electrode may be measured by directly connecting to a recorder, but the signal from the high voltage electrode needs to be isolated by an optical guide and appropriate data transfer modules.

The technique was improved subsequently by Takada, et al[25] whose experimental arrangement is shown in fig. 11.19. The oscillator of the previous method is replaced with a high voltage pulser and the signal from the piezoelectric transducer is recorded on a dual beam oscilloscope for further analysis.

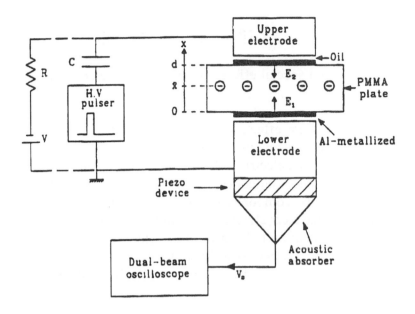

Fig. 11.19 Pulsed Electro-Acoustic Method [Takada et. al., 1987, © IEEE)

11.7.9 ELECTRON BEAM METHOD

This method was used by Sessler, et al. prior to the methods described above[26]. A mono-energetic electron beam is incident through the front electrode under short circuit conditions (both electrodes connected to ground through small impedance). A brief comment about the method of charging is appropriate here. Arkhipov, et al.[27] have shown, by theoretical analysis, that if the rear electrode is grounded and the front (radiated) electrode is open circuited, the space charge build up takes place mostly in the un-irradiated region. For a grounded front electrode, with the rear open circuited, there is only weak charging in the irradiated region. Consequently, a dielectric is much less charged for the second mode than the first.

The beam generates a radiation induced conductivity to a depth determined by its energy. At the plane of the maximum penetration maximum conductivity occurs and the plane is treated as a virtual electrode (thickness ~5μm). Since the depth of the plane depends on the energy, the virtual electrode may be swept through the dielectric by increasing the energy of the electrons. The currents are measured at each electrode from which the spatial distribution of charges are determined. The method is destructive in the sense that the charge is removed after experimentation.

11.7.10 SPECIAL TECHNIQUES

Techniques dependent on special effects or restricted to selected dielectric materials are described below.

1. Kerr Effect: The Kerr effect, generally applied to birefringent liquids such as nitrobenzene, consists of passing plane polarized light (called a polarizer) through a cell containing the dielectric. The electric field in the dielectric splits the plane polarized light into two components, one component traveling faster than the other. The phase difference between the two components makes the emerging light circularly polarized. This effect is known as the Kerr effect and the intensity of the output light measured by a photomultiplier is dependent on the square of the electric field. The Kerr effect, when it occurs in solids is usually referred to as Pockels effect. This method has been employed by Zahn, et al.[28] to study electron beam charged PMMA, having a self field of 1-2.5 MV/m.

2. Spectroscopic Method: In an electric field the spectral lines shift or split and this phenomenon is known as the Stark effect. By observing the spectrum in a spectroscope the field strength may be determined[29].

A variation of the Stark effect is to make use of the principle that very weakly absorbed visible monochromatic light liberates carriers that move under the field of trapped charges. By detection of the photocurrent produced one could infer the charge distribution[30].

3. Field Probe and Scanning Microscopic Methods: Measurements of an electric field by probes have been used to compute the charge distribution (Tavares, 1973). A scanning microscope has also been used to determine the charging behavior of PMMA under electron beam irradiation[31].

4. Vapor Induced Depolarization Currents (VIDC): This method is based on the principle that, when a charged dielectric is exposed on one of its faces to a saturated vapor of a solvent, diffusion occurs and the trapped charges are released. A current is detected in the external circuit from which the magnitude of the trapped charge may be evaluated[32].

11.8 EXPERIMENTAL RESULTS

The experimental results in fluoropolymers have been succinctly summarized by Sessler (1997), considering different methods of charging and different techniques of measurement. Table 11.3 provides references to selected data. In the following we briefly discuss some polymers of general interest.

Poly(ethylene) (PE): (a) unirradiated

The PEA method has been applied by Mizutani, et al.[33] to study space charge in three different grades of PE at electric field strengths up to 90 MV/m. Homocharges were observed and the space charge was dependent on the electrode material suggesting carrier injection from both electrodes. These results confirm the results of the earlier field probe method adopted by Khalil and Hansen[34]. To separate injected charges from the ionized impurities, Tanaka, et al. have adopted a technique of charge injection suppression layers[35]. The pulsed acoustic technique was employed to measure spa charge density. In LDPE with suppression layers on both electrodes heterocharges are observed. However, by removing charge suppression homo charges were observed. This technique yields the true charge injected by the electrodes into PE.

In HDPE homocharges are identified at the anode, and in doped HDPE a strong heterocharge, not seen in undoped HDPE, is formed near the cathode (Suzuoki et. al. (1995).

Poly(ethylene) (PE): (b) irradiated

Chen, et al.[36] have observed, by LIPP technique, that the energetic ionizing radiations such as a γ-source can alter the chemical structure and may also give rise to the presence of trapped charges. Low radiation doses (≤ 10 kGy) were observed to affect the low density polyethylene differently when compared with high doses (≥ 50 kGy).

A. CROSS LINKED POLYETHYLENE (XLPE)

XLPE is being used increasingly in the power cable industry and two main chemical methods are employed to produce it from low density polyethylene. One method is to use a peroxide (dicumyl peroxide, DCP) which decomposes to form free radicals or a silane based grafting process. The space charge accumulation has been measured by the LIPP technique[37] and the observations are:

1. The cross linking method appears to determine the polarity of the space charge adjacent to the electrodes. The DCP cross linking favors heterocharge, and silane cross linking favors homocharge. The charge densities do not vary appreciably.

2. Reversing the applied voltage polarity results in a near-perfect inversion of the space charge across insulator/polymer interface. This is interpreted as evidence for electron transfer by tunneling from donor to acceptor states centered on the Fermi level.

B. FLUOROETHYLENE PROPYLENE (FEP)

The LIPP technique of Sessler et. al. (1977) is applied to FEP, which is charged with electron beams, energy ranging from 10 to 50 keV, as permanent polarization does not occur. Surface charges are observed and very little trapping occurs within the sample volume, in agreement with the results of laser induced acoustic pulse method. Das Gupta, et al[38] have measured the spatial distribution of charges injected by monoenergetic electrons by both LIPP and LIMM techniques and good agreement is obtained between the two methods (Fig. 11.20).

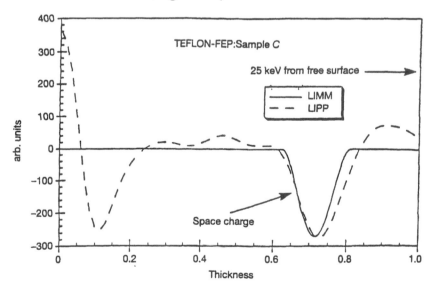

Fig. 11.20 LIPP and LIMM techniques applied to 25 μm FEP with 40 keV electrons from non-metallized side (Das Gupta, et al. 1996, with permission of Inst. Phys., England).

Fig. 11.21 shows the evolution of charge characteristic in FEP charged by electron beam at 120° C. As the annealing duration is increased the charge peak broadens with the charge depth increasing from about 10 μm with no annealing, to about 22 μm with annealing at 120°C. This broadening is caused by charge release at the higher temperature, charge drift in the self-field extending towards the rear electrode and fast retrapping. The calculated available trap density is 2.5×10^{22} m^{-3} but the highest filled-trap density in electron beam experiments is ~ 6×10^{21} m^{-3} (Sessler, 1977).

In a recent study Bloss, et al.[39] have studied electron beam irradiated FEP by using both the thermal pulse (TP) and LIPP method. The TP method has high near-surface resolution, better than 100 nm. The Lipp method has nearly constant resolution with depth. The electron beam deposits a negative charge and in addition, a negative charge layer was formed close to the electrode-polymer interface if the electron beam entered the dielectric through the interface. This effect is attributed to the fact that the metal electrode scatters the electron beam, producing secondary electrons. The secondary electrons have much less energy than the primary ones and do not penetrate deep in to the volume. Since there is no scattering at the unmetallized surface, there are no secondary electrons and only the primary electrons enter the sample.

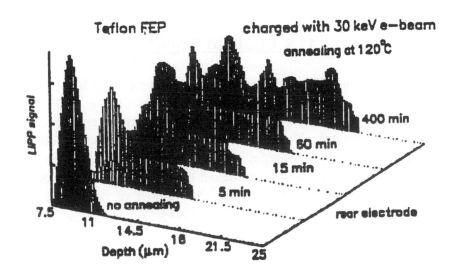

Fig. 11.21 Evolution of charge distribution in 30 keV electron-beam charged Teflon FEP with annealing time at 120°C, as obtained by LIPP method, (Sessler1997, © IEEE).

C. POLY(VINYLIDENEFLUORIDE) PVDF

The LIPP technique of Sessler, et al. (1986) has shown that the piezoelectric material does not form permanent polarization till a threshold electrical field of 100 MV/m is attained. The permanent component of the polarization increases more than proportionally to the voltage, which is characteristic of ferroelectric material. The permanent polarization is reasonably uniform in the volume, which is typical for high field room temperature poling of PVDF. The same result has also been obtained by the thermal pulse method of De Reggi (1984). The polarization is due to combined effects of instantaneous polarization, electrode charges and space charges accumulated near the electrodes.

D. POLY(TETRAFLUOROETHYLENE)

Corona charged PTFE shows that the trapped charge does not spread as measured by the thermal pulse method. However, PPS measurements of positively and negatively corona charged laminates of the dielectric show that there is some minor spreading towards the electrodes. At high humidity the centroid of the charge has even been observed beneath the sample surface. Positive and negative corona charging at temperatures of 260°C shows a larger and definite bulk charge which is non-uniformly distributed.

11.9 CLOSING REMARKS

As discussed above, a number of methods have been developed for the measurement of space charge in both thin and relatively thick samples (Table 11.3). The method of analyses of the signals have also increased the accuracy of determination of the charge distribution and charge depth. The thermal methods require a deconvolution technique to derive the spatial distribution of charges, $\rho(x)$, from the measured current as a function of time, $I(t)$. The deconvolution may be achieved by the Fourier coefficients of $\rho(x)$. The accuracy of the current measurement limits the number of coefficients to ~10, if the sample is pulsed from both sides (Sessler, et al., 1986).

Table 11.3
Selected Space Charge Measurement in Polymers

The number within square brackets refers to references at the end of the section.

Method	Polymer	Reference
Thermal Pulse Method	Polyethylene	Suzuoki et. al., (1985), [40]
	Fluoroethylenepropylene (FEP)	Collins (1980), Bloss et. al. (2000)
DeReggi's Method	PVDF	DeReggi et. al. (1982, 1984) [41]
LIMM	Polyethylene	
	Fluoroethylenepropylene (FEP)	Das Gupta et. al. (1996)
	PVDF	Lang and Das Gupta (1981)
Pressure Pulse Method	Poly(ethylene)	Laurenceau (1977)
	PET	Anderson and Kurtz (1984).
Laser Induced Pressure Pulse Method (LIPP)	Fluoroethylenepropylene	Alquie et. al. (1981), Lewiner (1986), Das Gupta et. al. (1996)
	PVDF	Sessler (1997)
	Polyimide (Kapton)	Sessler (1997)
	PET (Mylar)	Sessler (1997)
	Cross linked polyethylene (XLPE)	Bambery and Fleming (1998)
	Polyethylene	Lewiner (1986),[42]
Thermoelastic Stress Waves	PET	Sessler (1997)
Pressure Wave Propagation (PWP) Method	Fluoroethylenepropylene	
Nonstructured Acoustic Probe method	PMMA	Migliori and Thompson (1980)
Mechanically generated acoustic Probe Method	PMMA	Rozno and Gromov (1986)
	Polyethylene	Rozno and Gromov (1986)

Table 11.3 Continued

Method	Polymer	Reference
Piezoelectric Pulse Step (PPS)	Fluoroethylenepropylene	Gerhard-Multhaupt et. al. (1983)
	Mylar	Gerhard-Multhaupt et. al. (1983)
	P(VDF-TFE)	[43]
Pulsed Electro-acoustic Stress Method	PE, PET, PS, PVDF, XLPE,	(Takada et. al. 1983), [44], [45]
	LDPE	Tanaka et. al. (1995)
	PMMA	[46]
Electron beam sweeping	FEP	Sessler et. al. (1977)
Kerr Effect	PMMA	Zahn et. al. (1987)
Field Probe	LDPE	Khalil et. al (1988)

The thermal methods have the advantage of a high depth resolution for thickness less than ~ 100μm. The acoustic methods have the advantages of a good resolution throughout the bulk (~1-2 μm). The Thermal Pulse method, LIMM, LIPP and PEA are being increasingly used in preference to other methods. It is now possible to measure the space charge by more than one technique and these complementary measurements have been carried out in FEP, using the LIMM and LIPP techniques (Das Gupta et. al., 1996), and LIPP and TP methods (Bloss, et al. 2000). Such techniques help to isolate experimental artifacts from the characteristics of the dielectric under study.

Damamme, et al.[47] have discussed the points to which attention should be paid in calibrating the experimental setup. Calibration generally falls into two broad categories: one, the parameters that are associated with the charging method, charge, potential, temperature, energy of injected electrons and their total charge, etc. The latter may be measured to an accuracy of ± 1 pc. The charge response resolution is ~ 2 f C but it may be particularly difficult to achieve this accuracy in case of short laser pulses. The second is connected with the measurement technique used. The analyses of measurements require several physical properties of the polymer, such as the speed of sound, the thermal diffusivity, the permittivity, the high frequency dielectric constant, etc. These properties should be measured in situ, preferably, to remove ambiguity in their values. Since a host of experimental techniques are required it may not always be possible to accomplish this. In such situations the properties obtained for the same sample with particular reference to humidity, previous history, etc. should be used.

The role of the space charge in the performance characteristics of dielectrics has been discussed by Bartnikas[48] in both communication and power application areas. Future work should be concentrated on establishing a clear connection between charging characteristic and operational performance in the technological applications.

11.10 REFERENCES

[1] J. Lewiner, IEEE Trans. on EI, **12** (1986) 351.

[2] G. Blaise and W. J. Sarjeant, "IEEE Trans. on Diel. EI., **5** (1998) 779.

[3] H. Frölich, "Introduction to the theory of the polaron", in "Polarons and Excitons", Eds : C. G. Kwyer and G. D.Whitefield,: Oliver & Boyd, London, 1963.

[4] B. Gross, IEEE Trans. on Nucl. Sci., **NS-25** (1978) 1048.

[5] B. Gross, J. Dow, S. V. Nablo, J. Appl. Phys., **44** (1973) 2459.

[6] B. Gross, G. M. Sessler, J. E. West, Appl. Phys. Lett., **22** (1973) 315.

[7] S. Osaki, S. Uemura, Y. Ishida, J. Poly. Sci. A-2, 9 (1971) 585.

[8] N. H. Ahmed and N. N. Srinivas , IEEE Trans. Diel. EI., 4 (1997) 644.

[9] R. E. Collins, J. Appl. Phys., **51** (1980) 2973.

[10] H. Von Seggern, Appl. Phys. Lett., **33** (1978) 134.

[11] A. S. DeReggi, C. M. Guttmann, F. I. Mopsik, G. T. Davis and M. G. Broadhurst, Phys. Rev. Lett., **40** (1978) 413.

[12] F. I. Mopsik, and A. S. DeReggi, J. Appl. Phys., **53** (1982) 4333; Appl. Phys. Lett., **44** (1984) 65.

[13] Y. Suzuoki, H. Muto, T. Mizutani, J. Appl. Phys., **24** (1985) 604.

[14] S. B. Lang and D. K. Das Gupta, Ferroelectrics, **39** (1981) 1249.

[15] P. Laurenceau, G. Dreyfus, and J. Lewiner, Phys. Rev. Lett., **38** (1977) 46.

[16] R. A. Anderson and S. R. Kurtz, J. Appl. Phys., **56** (1984) 2856.

[17] G. M. Sessler, R. Gerhardt-Multhaupt, von Seggern and J. W. West, IEEE Trans. Elec. Insu., **EI-21**, (1986) 411.

[18] C. Alquie, G. Dreyfus and J. Lewiner, Phys. Rev. Lett., **47** (1981) 1483.

[19] A. Migliori and J. D. Thompson, J. Appl. Phys., **51** (1980) 479.

[20] A. Migliori and T. Hofler, Rev. Sci. Instrum., **53** (1982) 662.

[21] A. G. Rozno and V. V. Gromov, IEEE Trans. Elec. Insul., **EI-21**, (1986) 417-423

[22] R. Gerhardt-Multhaupt, M. Haardt, W. Eisinger and E. M. Sessler, J. Phys. D: Appl. Phys., **16** (1983) 2247.

[23] G. M. Sessler, IEEE Trans. Diel. Elec. Insu. 4 (1997) 614.

[24] (a) T. Takada and T. Sakai, IEEE Trans. Elec. Insu., **EI-18**, (1983) 619.
    (b) T. Takada, IEEE Trans. Diel. Elec. Insul., **6** (1999) 519.

[25] T. Takada, T. Maeno and Kushibe, IEEE Trans. Elec. Insul., **22** (1987) 497.

[26] G. M. Sessler, J. E. West, D. A. Berkeley and G. Morgenstern, Phys. Rev. Lett., **38** (1977) 368.

[27] V. I. Arkhipov, A. I. Rudenko and G. M. Sessler, J. Phys. D: Appl. Phys., **24** (1991) 731.

[28] M. Zahn, M. Hikita, K. A. Wright, C. M. Cooke and J. Brennan, IEEE Trans. Elec. Insu., **22** (1987) 181.

[29] S. J. Sheng and D. M. Hanson, J. Appl. Phys., 45 (1974) 4954.

[30] A. D. Tavares, J. Chem. Phys., **59** (1973) 2154.

[31] Z. G. Song, H. Gong and C. K. Ong, J. Appl. Phys., **30** (1997) 1561.

[32] M. Falck, G. Dreyfus, and J. Lewiner, Phys. Rev., **25** (1982) 550.

[33] T. Mizutani, H. Semi and K. Kaneko, IEEE Trans. Diel. Elec. Insul., **7** (2000) 503.

[34] M. S. Khalil, B. S. Hansen, IEEE Trans. Elec. Insu., **23** (1988) 441.

[35] Y. Tanaka, Y. Li, T. Takada and M. Ikeda, J. Phys. D: Appl. Phys., **28** (1995).

[36] G. Chen, H. M. Banford, A. E. Davies, IEEE Trans. Diel. Elec. Insul., **5** (1998) 51.

[37] K. R. Bambery and R. J. Fleming, IEEE Trans. Diel. Elec. Insul., **5** (1998) 103.

[38] D. K. Das Gupta, J. S. Hornsby, G. M. Yang and G. M. Sessler, J. Phys. D: Appl. Phys., **29** (1996) 3113.

[39] P. Bloss, A. S. DeReggi, G. M. Yand, G. M. Sessler and H. Schäfer, J. Phys. D: Appl. Phys., **33** (2000) 430.

[40] A. Cherfi, M. Abou Dakka, A. Toureille, IEEE Trans. Elec. Insu. 27 (1992) 1152.

[41] M. Wübbenhorst, J. Hornsby, M. Stachen, D. K. Das Gupta, A. Bulinski, S. Bamji, IEEE Trans. Diel. Elec. Insul., **5** (1998) 9.

[42] T. Mizutani, IEEE Trans. Diel. Elec. Insul., **1** (1994) 923.

[43] S. N. Fedonov, A. E. Sergeeva, G. Eberle and W. Eisenmenger, J. Phys. D: Appl. Phys., **29** (1996) 3122.

[44] Y. Li, M. Yasuda, T. Takada, IEEE Trans. Diel. Elec. Insul., **1** (1994) 188.

[45] X. Wang, N. Yoshimura, Y. Tanaka, K. Murata and T. Takada, J. Phys D: Appl. Phys., **31** (1998) 2057.

[46] T. Maeno, T. Futami, H. Kushibe, T. Takada and C. M. Cooke and J. Brennan, IEEE Trans. Elec. Insul., **23** (1988) 433.

[47] G. Damamme, C. Le Gressus and A. S. De Reggi, IEEE Trans. Diel. Elec. Insu., **4** (1997) 558.

[48] R. Bartnikas, Trans. Diel. Elec. Insul. **4**, (1997) 544.

APPENDIX 1

TRADE NAMES OF POLYMERS

Chemical name	Trade name	Formula	Remarks
Poly(tetrafluoroethylene)	Teflon PTFE Halon fluon	$(CF_2\text{-}CF_2)_x$	rods, sheets, Film, extrusion
Poly(fluoroethylene-propylene)	Teflon FEP	$(CF_2\text{-}CF_2)_x\,(C_3F_6)_y$	Film, molding and extrusion
Poly(3-perfluoropropoxy-perfluorohexamethylene)	Teflon PFA	$(C_2F_4)_x\,(C_2F_3O)_y\,C_3F_4$	Film, extrusion
Poly(ethylene terephthalate) (PET)	Mylar	$(CO_2\,C_6H_6\,CO_2\,C_2H_4)_x$	Film, tubing
Aromatic polyamide	Nomex, Kevlar	$(NH\,C_6H_6\,NH\,CO\,C_6H_6\,CO)_x$	Paper, laminates
Polyimide (PI)	Kapton Upilex Pyre-ML	$[(C_6H_6)_3\,O\,(C_2NO_2)_2]_x$	Film, enamel, resin
Polyetherimide (PEI)	Ultem	$[(C_2\,NO_2)_2\,(C_6H_6)_5\,O\,C_3H_6]_x$	Film, molding resin
Poly(methyl methacrylate) (PMMA)	Perspex Acrylic Lucite Acrylite Plexiglas	$(C_5\,H_8\,O_2)_x$	molding resin, sheets

Chemical name	Trade name	Formula	Remarks
ethylene tetrafluoroethylene (ETF)	Tefzel Hostaflon	$(C_4H_4F_4)_x$	Film, extrusion
Polyetherketone	PEEK	$[(C_6H_6)_3\ C_2O_3]_x$	molding resin
Polyvinylidene fluoride (PVDF)	Kynar	$(C_2\ H_2\ N)_n$	Film,
Polychlorotrifluoro-ethylene (PCTFE)	Kel-F	$(C_2\ F_3\ Cl)_n$	Film, extrusion
Polystyrene-butadiene	K-resin		molding resin
Polycarbonate (PC)	Lexan, Calibre, Makrolon	$H\ [(C_6H_6)_2C_4H_6O_3]_n\ OH$	Molding compound
Polybutylene Terephthalate (PBT) or Polytetramethylene terephthalate (PTMT)	Ultradur Valox Celanex	$[(C_6H_6)\ C_6\ O_4\ H_8)]_n$	Molding Resin
Nylon	Capron, Rilsan, Ultramid, Celanese, Zytel and Vydyne	$[C_{12}H_{22}O_2N_2]_n$	Molding Resin, extrusion, enamel
Polysulfone	Udel, Ultrason S	$[(C_6H_6)_4\ C_3H_6\ O\ SO_2]_n$	Molding resin
Phenol formaldehyde	Bakelite	$[(C_6H_6)_4\ C_3\ O_3\ H_8]_n$	Resin, Laminate
Polyester-imide (PEI)	Teramid Isomid	$[(C_6H_6)_4\ C_{18}\ H_8\ O_{18}\ N_8]$	enamel

APPENDIX 2
General Classification of Polymer Dielectrics

Thermoplastic		Elastomers	Thermosetting	Ceramic/glass
RESINS	Films			
Acetal	Cellulose Acetate (CA)	Natural rubber (polyisoprene)	Alkyds or thermosetting polyester	Alumina
Acrylic		Butyl rubber	Allyls	Aluminum nitride
Amide (Nylon)	Cellulose triacetate	Styrene Butadiene Rubber	Aminos	Aluminum silicate
Polyetherimide	PTFE	Neoprene	Epoxy	Beryllia
Polyarylate	PFA	EPDM	Phenolics	Boron nitride
Polybutylene Terephthalate	FEP	EPR		Cordierite
Polycarbonate	ETFE	chlorosulfonated polyethylene (CSM)		Diamond
Polyetheretherketone	ECTFE	Chlorinated polyethylene (CP)		Magnesia
Polyethylene ether	PVDF			Porcelain
Polyphenylene	PCTFE			Sapphire
Polypropylene	PC			Silica
Polystyrene	PET			Steatite
Styrene-Butadiene	PE			Zircon

Thermoplastic	Elastomers	Thermosetting	Ceramic/glass	Thermoplastic
Styrene acrylonitrile	PI			
Acrylonitride butadiene styrene	PPA			
Polysulfone	PP			
Polyether sulfone	PS			
	PVC			
	PS			
	PES			
	PEI			

APPENDIX 3
Selected Properties of Insulating Materials

E = Dielectric Strength (MV/m), * indicates short term

ρ = Volume Resistance, Ω m

tan δ values should be multiplied by 10^{-4}, i.e. 40 means 0.004

Material	E	ρ	ε (50/60 Hz)	ε (1 kHz)	ε (1 M Hz)	tan δ (50/60 Hz)	tan δ (1 kHz)	tan δ (1 M Hz)
Acetol (homopolymer)	15	1×10^{13}	3.7	-	3.7	-	-	48
Acetol, (co-polymer)	15	1×10^{12}	3.7	-	3.7	10	-	10
Acrylonitrile butadiene styrene (ABS)	17[1]*	2×10^{14}	2.6	-	2.6	40	11	60
Cellulose Acetate	160*	10^{13}		3.6	3.2	-	130	380
Polyethylene (PE)	40-60	10^{14}	2.3	2.3	2.3	2-10	2-10	2-10
Polyethylene (LD)	200*	10^{14}	2.3	2.2	2.2	2-10	3	3
Polyethylene (HD)	200	10^{14}	2.35		2.34	2.4	-	2-7
Polyethylene (XL)	220	10^{14}			2.3	-	-	3
Polypropylene (PP), (Unoriented)	-	3×10^{13}	2.27	2.2	2.2	4	3	3
Polypropylene (PP), (Biaxially oriented)	200*	3×10^{12}	2.27	2.2	2.2	-	3	3
Polypropylene/polyethylene copolymer	24	4×10^{13}			2.24			3
Polystyrene (PS)	200*	10^{14}	2.6	2.4-2.7	2.4-2.7	3	5	5
Polyvinyl chloride (PVC), (plasticized)	72	10^{12}	-	4-8	3.3-4.5	-	700-1600	400-1400

[1] *Handbook of Electrical and Electronic Insulating Materials*: W. T. Shugg, IEEE Press, 1995.

Material	E	ρ	ε (50/60 Hz)	ε (1 kHz)	ε (1 M Hz)	tan δ (50/60 Hz)	tan δ (1 kHz)	tan δ (1 M Hz)
Polyvinyl chloride (PVC)	40	10^{13}	3.4	3.4	3.1	200	180	150
Polymethyl methacrylate (PMMA)	16-20	1×10^{16}	3.7-4.0	3.5	2.2	500	500	300
Polytetra fluoroethylene (PTFE)	88-176	10^{16}	2.1	2.0	2.0	2	1	1
Teflon-PFA	176-200	10^{16}	-	2.0	2.0	-	2	2
Teflon-FEP	200	10^{16}	-	2.0	2.0	-	2	3
Teflon-ETFE	200			2.6	2.6	-	80	5
PVDF	10.2	2×10^{12}	-	8.4	6.6	-	180	1700
ECTFE	200-225	10^{14}	-	2.6	2.6	-	20	130
PCTFE	120-156	10^{16}	-	2.6	2.3	210 (100 Hz)	230	120
Polycarbonate (PC)	252	10^{14}	-	3.0	2.9	-	15	100
Polyester (PET)	275-300	10^{16}	-	3.2	3.0	-	50	160
Polyimide (PI)	280	10^{16}	-	3.5	3.4	-	25	100
Polyphenylene sulphide (PPS)	15.2	10^{14}	3.1	-	3.2	3	-	7
Polyphenylene ether (PPE)	22	10^{14}	2.6	-	2.6	4	-	9
Nylon 6/6	24	10^{11}	8.0		4.6	2000	2000	1000
Nylon 11	30	10^{12}	3.9		3.1	400	500	500
Polyetherimide(PEI)	19	7×10^{15}	3.2	3.2	3.1	40		60

APPENDIX 3 (Contd.)

Material	E	ρ	ε (50/60 Hz)	ε (1 kHz)	ε (1 M Hz)	tan δ (50/60 Hz)	tan δ (1 kHz)	tan δ (1 M Hz)
Polybutylene Terephthalate (PBT)	17	4×10^{14}	3.3	3.2-3.3 (100 Hz)	3.1	20	-	200
Polysulfone (PS)	17	1×10^{15}	3.1		3.0	8		34
Poly(ether sulfone) (PES)	16	$>1\times10^{15}$	3.5	3.5	3.5	10	35	40
Epoxy resin	25-45	$10^{13}-10^{15}$	3.5-3.9	-	3.6	35-90	-	150-200
Glass Filled Alkyd/Polyester resin	15	10^{11}	5.6	-	4.6	1000	-	200
Glass filled Melamine Molding resin	12	2×10^{9}	8.0	-	6.2	-	-	200
Glass Filled Epoxy Molding Resin	15	1×10^{11}	5.0	-	4.6	100	-	100
Natural Rubber (hard)	15-40	1×10^{14}	3.2-4	-	2.5-3.5	30-150	50-150	-
Natural Rubber (soft)	15-50	-	3.5-5	-	-	20-800	-	-
Butyl Rubber	24	10^{15}	2.1-2.4	-	-	-	30	-
Neoperene	16-28	1×10^{9}	6-8	-	-	-	30	-
Styrene Butadiene Rubber (SBR)		1×10^{13}	3.0-3.5				30	
Ethylene propylene diene rubber (EPDM)	20-	1×10^{14}	2.5-3.5	-	-	-	70	-
Silicone Rubber	20-30		2.5-3.2	2.5-3.2	3.0-3.6	4-25	3-10	20-50
Polyurethane	20	4×10^{9}		10	8		800	1000
Silicone resin	22	2×10^{12}	2.7		2.7	10		10

APPENDIX 3 (Contd.)

Material	E	ρ	ε (50/60 Hz)	ε (1 kHz)	ε (1 M Hz)	tan δ (50/60 Hz)	tan δ (1 kHz)	tan δ (1 M Hz)
Aramid Paper (Calendered)			2.4			60		
Aramid Paper (uncalendered)			1.3				30	
Aramid and mica paper			3.6			1200		
Alumina ceramic	12.5	1×10^{12}			8.8-10			20
Beryllia ceramics	12	1×10^{12}			7.0			50
Glass bonded mica	16	1×10^{10}			6.7			18

CERAMICS

(Reproduced : Handbook of Electrical and Electronic Insulating Materials, II ed. W. T. Shugg, IEEE)

Material	Dielectric strength (MV/m)	Dielectric constant (1 MHz)	tan δ (1 MHz) × 10^4
Alumina (99.9% Al_2O_3)	13.5	10.1	2
Aluminum silicate	6	4.1	27
Berylia (99% BeO)	14	6.4	1
Boron nitride	38	4.2	3.4
Cordierite	200	4.8	50
Magnesia	-	5.4	<3
Porcelain	-	8.5	50
Quartz	-	3.8	38
Sapphire	-	9.3-11.5	0.3-8.6
Silica (fused)	-	3.2	45 @ 10GHz
Steatite	9-16	5.0	23

APPENDIX 3 (Contd.)

Material	Dielectric strength (MV/m)	Dielectric constant (1 MHz)	tan δ (1 MHz) × 10⁴
Zircon	-	5.0	23

MICA and MICA PRODUCTS

(Reproduced : Handbook of Electrical and Electronic Insulating Materials, II ed. W. T. Shugg, IEEE)

Material	Dielectric strength (MV/m)	Dielectric constant (1 MHz)	tan δ (1 MHz) × 10⁴
Muscovite Ruby (natural)	120	6.5-8.7	-
Phlogopite amber (natural)	120	5-6	-
Fluoro-phlogopite synthetic	120	6.5	2
Glass bonded mica	14-16	6.7-12.5	12-60

REFERENCES

1. W. T. Shugg, "Handbook of Electrical and Electronic Insulating Materials, IEEE Press, 1995.
2. C. C. Ku and R. Liepins, "Electrical Properties of Polymers", Hanser Publishers, Munich, 1987.
3. A. Von Hippel, "Dielectric Materials and Applications", The Technology Press of M. I. T. and John Wiley and Sons, New York, 1958.
4. F. E. Karasz, Ed.: "Dielectric Properties of Polymers", Plenum Press, New York, 1972.
5. Vishn Shah, "Handbook of Plastics Testing Technology", Wiley-Interscience, New York, 1984.
6. J. Arganoff, Ed.: Modern Plastics Encyclopedia, Vol. 60, McGraw Hill, New York, (1983-84).

APPENDIX 4
Relative Ranking of Thermoplastic Polymers
1= Most desirable, 18= Least desirable

POLYMER	Dielectric strength	Dielectric constant	Dissipation factor	Volume resistivity	Arc resistivity
Acetal homo-polymer	15	14	7	15	4
Acetal copolymer	15	14	8	16	2
Acrylic	5	13	17	1	1
Nylon 6/6	1	18	18	18	7
Nylon 11	8	16	16	16	9
Polyamide-imide	3	17	15	5	3
Polyetherimide	7	10	11	2	8
Polybutylene terephthalate	11	11	10	7	5
Polycarbonate	15	9	14	4	11
Modified polyphenylene oxide	15	8	2	12	18
Polyphenylene sulfide	15	8	2	12	18
Polypropylene	1	1	4	9	6
Polystyrene	5	2	1	9	16
Polystyrene butadiene	12	2	9	12	17
Styrene acrylonitrile	8	4	13	9	12
Acrylonitrile butadiene styrene	8	4	12	8	13
Polysulfone	8	7	5	6	10
Polyethersulfone	12	12	6	3	15

Handbook of Electrical and Electronic Insulating Materials, W. T. Shugg, IEEE©, 1995

APPENDIX 5
SELECTED PROPERTIES OF POLYMER INSULATING MATERIALS

POLYMER	Recommended Temp (°C)	Volume resistivity (Ωm)	Arc resistance (s)
Acetal homo-polymer[1]	90	1×10^{13}	220
Acetal copolymer[1]	105	1×10^{14}	240 (burns)
Acrylic[1]	95	1×10^{16}	no tracking
Acrylonitrile butadiene – styrene[1]	95	2×10^{14}	100
Epoxy, Bisphenol A			100^3
Modified polyphenylene – oxide[1]	105	1×10^{14}	75
Nylon 6/6[1]	130	1×10^{13}	130
Nylon 11[1]	65	1×10^{14}	123
Polyamide-imide[1]	260	8×10^{14}	230
Polybutylene terephthalate[1]	150	4×10^{14}	184
Polycarbonate[1]	115	8×10^{14}	120
Polyetherimide[1]	170	7×10^{15}	128
Polyethersulfone[1]	180	$>1\times10^{15}$	70
Polyethylene, LD	80	1×10^{14}	160^2

Polyethylene, HD	90	1×10^{14}	235
Polyimide, unreinforced	240	1×10^{16}	$152\text{-}230^3$
Polyphenylene sulfide[1]	205	1×10^{14}	34
Polypropylene[1]	105	$>1\times10^{14}$	150
Polystyrene[1]	70	1×10^{14}	65
Polysulfone[1]	150	5×10^{14}	122
Poly(tetrafluoroethylene)	260	$>1\times10^{16}$	$>200^3$
Poly(trifluorochloroethylene)	200	$>1\times10^{16}$	$>360^3$
PVC	80	1×10^{12}	$60\text{-}80^2$
Poly(vinylidene fluoride)	135	2×10^{12}	$50\text{-}60^3$
Styrene acrylonitrile[1]	80	$>1\times10^{14}$	115
Styrene butadiene[1]	65	1×10^{14}	50

1. W. T. Shugg, Handbook of Electrical and Electronic Insulating Materials, IEEE, 1995
2. J. Brandrup and E. H. Immergut, Eds., Polymer Handbook, Wiley-Interscience, New York (1977).
3. Vishn Shah, Handbook of Plastics Testing Technology, Wiley-Interscience, New York (1984).

INDEX